W9-DJD-414

PSY 302
EXHE
$37.20

SECOND
EDITION

Adolescence

Eastwood Atwater

PRENTICE HALL
Englewood Cliffs, New Jersey 07632

Library of Congress Cataloging-in-Publication Data

Atwater, Eastwood, (date)
  Adolescence.

  Bibliography: p. 403
  Includes index.
  1. Adolescence. 2. Youth—United States. I. Title.
HQ796.A76 1988   305.2'35   87-25895
ISBN 0-13-008699-1

Editorial/production supervision and interior design: Virginia Rubens
Cover design: Suzanne Bennett & Assoc.
Cover photo: Lou Jones, The Image Bank
Photo Research: Kay Dellosa
Photo Editor: Lorinda Morris-Nantz

© 1988 by Prentice-Hall, Inc.
A Division of Simon & Schuster
Englewood Cliffs, New Jersey 07632

*All rights reserved. No part of this book may be
reproduced, in any form or by any means,
without permission in writing from the publisher.*

Printed in the United States of America

10 9 8 7 6 5 4 3 2

ISBN 0-13-008699-1 01

PRENTICE-HALL INTERNATIONAL (UK) LIMITED, *London*
PRENTICE-HALL OF AUSTRALIA PTY. LIMITED, *Sydney*
PRENTICE-HALL CANADA INC., *Toronto*
PRENTICE-HALL HISPANOAMERICANA, S.A., *Mexico*
PRENTICE-HALL OF INDIA PRIVATE LIMITED, *New Delhi*
PRENTICE-HALL OF JAPAN, INC., *Tokyo*
PRENTICE-HALL OF SOUTHEAST ASIA PTE. LTD., *Singapore*
EDITORA PRENTICE-HALL DO BRASIL, LTDA., *Rio de Janeiro*

To Susan and Gail, whose passage to adulthood has been an inspiration and an education in the transitional years.

Contents

Preface   **xv**

1 Introduction to Adolescence   **1**

ADOLESCENCE IN PERSPECTIVE   2

The Common Experience of Adolescence   2
Rites of Passage   4
Abbreviated Adolescence   5

THE SOCIAL INVENTION OF ADOLESCENCE   6

Approaching the Age of Modern Adolescence   7
Major Social Changes   7
Prolonged Adolescence   9

ADOLESCENCE TODAY   10

The Boundaries of Adolescence   11
Growing Up Faster   12
Contemporary Youth   14

SUMMARY   17
REVIEW QUESTIONS   18

2 Perspectives on Adolescence   **20**

THE BIOLOGICAL PERSPECTIVE   21

Hall's Critical-Stage Theory   21
Gesell's Maturational Theory   22
Puberty and Adolescent Development   24

THE PSYCHOANALYTIC PERSPECTIVE   25

The Freudian View of Development   25
Adolescent Sexual Conflicts   27
Coping with Adolescent Sexuality   28
Newer Developments   29

THE PSYCHOSOCIAL-CULTURAL PERSPECTIVE 30
Havighurst's Developmental Theory 31
Bandura's Social-Learning Theory 32
Mead's Cultural Theory 33
Adolescence as a Cultural Phenomenon 35

ADOLESCENT STORM AND STRESS IN PERSPECTIVE 38
Reevaluating Adolescent Storm and Stress 38
Patterns of Adolescent Growth 39
The Impact of Social Change 39

SUMMARY 40
REVIEW QUESTIONS 42

3 **Physical Development**   **43**
PUBERTY 44
The Process of Puberty 44
Hormonal Changes 45
Trend toward Earlier Maturation 46

PHYSICAL CHANGES 48
Height and Weight 48
Skeletal and Muscular Changes 49
Physical Changes and Body Image 51

SEXUAL MATURATION 53
Girls 53
Menarche 54
Boys 58
First Seminal Emission 60

EARLY VERSUS LATE MATURERS 60
Early Versus Late-Maturing Girls 61
Early versus Late-Maturing Boys 63
Individual Differences 63

SUMMARY 64
REVIEW QUESTIONS 66

4 **Cognitive Development**   **67**
PIAGET'S VIEW OF COGNITIVE DEVELOPMENT 68
Piaget's Concept of Intelligence 68
Childhood Thought 70
Adolescent Thought 70

ADOLESCENT THOUGHT AND LEARNING 71

Academic Studies 73
The Developmental Gap 74
Facilitating Formal Thought 76

ADOLESCENT THOUGHT AND SOCIALIZATION 77

Socialization 77
Egocentrism 78
Cognitive Maturity 80

CREATIVITY 81

Types of Creativity 81
Characteristics of Creative Adolescents 82
Creativity, Intelligence, and Achievement 84

SUMMARY 86
REVIEW QUESTIONS 87

5 The Family

88

CHANGES IN THE AMERICAN FAMILY 89

Smaller, More Mobile Families 89
Changing Functions of the Family 89
More Flexible Roles 91

DUAL-CAREER FAMILIES 93

Working Mothers 93
Household Chores 95
Latchkey Teens 97

DIVORCE 98

Adolescents in Strife-ridden Homes 98
How Adolescents Are Affected by Divorce 99
Long-Term Effects of Divorce 100

SINGLE-PARENT AND REMARRIED FAMILIES 101

Single-Mother Families 101
Single-Father Families 105
Remarried Families 106

SUMMARY 107
REVIEW QUESTIONS 109

6 Family Relationships and Adolescent Autonomy   110

PARENT—ADOLESCENT RELATIONSHIPS   111
How Adolescents Feel about Their Parents   111
Parental Control   112
Adolescents' Perception of Parental Control   114

PARENT—ADOLESCENT COMMUNICATION   116
Attitudes and Values   116
Parent—Adolescent Conflicts   118
Improving Parent—Adolescent Communication   120

ADOLESCENT AUTONOMY   122
The Process of Achieving Autonomy   122
Types of Autonomy   124
Autonomous Decision Making   125

SUMMARY   127
REVIEW QUESTIONS   128

7 The Self-Concept and Identity   129

THE SELF-CONCEPT   130
Changes during Adolescence   130
Self-Esteem   131
Change and Stability in Self-Esteem   134

THE SEARCH FOR IDENTITY   138
Erikson's Psychosocial Theory   138
The Adolescent Identity Crisis   138
Problems in Identity Formation   141

VARIATIONS IN IDENTITY STATUS   142
Types of Identity Status   143
Changes during Adolescence   145
Factors Facilitating Identity Achievement   147
Sex Differences in Identity   148

SUMMARY   150
REVIEW QUESTIONS   151

8 Peers   153

PEER RELATIONS   154
Parent and Peer Influence   154
Changing Patterns of Parent/Peer Influence   155

Function of Peers 157
Social Acceptance 158

PEER GROUPS 161
Cliques and Crowds 161
Changes in Peer Groups 162
Peer Conformity 164

FRIENDSHIP 165
Intimacy with Parents and Peers 166
Changing Patterns of Friendship 167
Same-Sex and Opposite-Sex Friendships 169

DATING AND EARLY MARRIAGE 170
Dating 171
Dating Patterns 171
Adolescent Marriage 173

SUMMARY 176
REVIEW QUESTIONS 177

9 Adolescents at School   178
SECONDARY SCHOOLS 179
Compulsory Education 179
The Curriculum 181
Schools for Early Adolescents 182

THE SCHOOL ENVIRONMENT 184
School and Class Size 185
Teachers 186
School and Classroom Climate 189

SELECTED FACTORS AFFECTING SCHOOL ACHIEVEMENT 190
Learning Ability 191
Exceptional Students 193
Family and Socioeconomic Background 196
Dropping Out 197

IMPROVING THE SCHOOLS 199
What People Think of Public Schools 200
New Directions 202
The School Adolescents Would Like 203

SUMMARY 204
REVIEW QUESTIONS 206

10 Sexuality   207

SEXUAL ATTITUDES 208
Changing Attitudes toward Sex 208
How Consistent Are Attitudes and Behavior? 210
Sex and Love 212

SEXUAL BEHAVIOR 214
Masturbation 215
Physical Intimacy and Sexual Intercourse 216
Homosexuality 217

PROBLEMS IN SEXUAL BEHAVIOR 219
Use of Contraceptives 219
Premarital Pregnancy 222
Sexually Transmitted Diseases 224

SEX EDUCATION 226
Sources of Sex Information 226
Sex Education in the Home 228
Sex Education in the School 228

SUMMARY 230
REVIEW QUESTIONS 232

11 Work and Career Choice   233

ADOLESCENTS IN THE WORKPLACE 234
Teenage Employment 234
Combining School and Work 235
The Value of Work Experience 237

THE PROCESS OF CAREER CHOICE 239
Stages of Career Choice 239
Career Maturity 241
Identifying a Compatible Career 241

INFLUENCES ON CAREER CHOICE 244
The Family 244
Socioeconomic Influences 246
Sex Differences 247

PREPARING FOR A CAREER 250
Career Education 250
Going to College 253
Career Outlook 255

SUMMARY 257
REVIEW QUESTIONS 258

12 Moral Development and Religion 260

THE DEVELOPMENT OF MORAL REASONING 261
Kohlberg's Stages of Moral Reasoning 263
The Attainment of Moral Reasoning 264
A Critique of Kohlberg's Theory 266

MORAL BEHAVIOR 267
Resistance to Wrongdoing 268
Helping Behavior 271
Personal and Moral Values 274

RELIGION 276
The Developmental Sequence 276
Cults 279
Religion and Moral Behavior 282

SUMMARY 284
REVIEW QUESTIONS 285

13 Delinquency 287

THE EXTENT OF DELINQUENCY 288
Age Trends 288
Types of Offenses 290
Sex Differences 292

CONTRIBUTING FACTORS 294
The Family 295
Socioeconomic Factors 296
Personality 299
Psychopathology 300
Drugs 302

TREATMENT AND PREVENTION OF DELINQUENCY 304
Police and the Juvenile Court 305
Current Approaches to Treatment 306
Preventing Delinquency 308

SUMMARY 309
REVIEW QUESTIONS 311

14 Youth and Drugs 312

USE OF DRUGS 313
The Extent of Drug Use 313

The Development of Drug Use 316
Why Youths Use Drugs 318

ALCOHOL AND TOBACCO 320
The Effects of Alcohol 320
Use of Alcohol 320
Psychological and Social Factors 323
Tobacco and Smoking 324

SELECTED ILLEGAL DRUGS 326
Marijuana 327
Stimulants 331
Depressants 334
Other Drugs 335

PUTTING DRUGS IN PERSPECTIVE 337
Relationships with Drugs 337
Treatment of Drug Abuse 340
Alternatives to Drugs 341

SUMMARY 345
REVIEW QUESTIONS 346

15 Psychological Disorders   **348**
PSYCHOLOGICAL DISTURBANCE IN ADOLESCENCE 349
ANXIETY AND EATING DISORDERS 351
Anxiety Reactions 352
Phobias 352
Obsessive-Compulsive Disorders 353
Anorexia Nervosa 354
Bulimia 356

DEPRESSION AND SUICIDE 357
Manifestations of Depression 358
The Range of Depression 359
Adolescent Suicide 360
Contributing Factors 362
The Prevention of Suicide 363

SCHIZOPHRENIA 364
Adolescent Schizophrenia 365
Contributing Factors 365
Outlook 367

THERAPY 368
SUMMARY 371
REVIEW QUESTIONS 373

16 Transition to Adulthood 374

LEAVING HOME 375
Peaceful and Stormy Departures 375
A Family Matter 376
Leaving for College 377
Returning to the Nest 379

TAKING HOLD IN THE ADULT WORLD 380
Autonomous Decision Making 381
Preparing for Economic Independence 383
Establishing Close Relationships 385

FROM YOUTH TO ADULTHOOD 388
Continuity 388
Change 389
Individual Differences 390

SUMMARY 392
REVIEW QUESTIONS 393

Appendix: Methods of Studying Adolescents 395

SCIENTIFIC INQUIRY 395
METHODS OF STUDY 396
Observation 397
The Case-Study Method 397
Survey Research 398
Correlational Studies 399
Experimental Studies 399
Ex Post Facto Designs 400

CROSS-SECTIONAL AND LONGITUDINAL STUDIES 400
Cross-Sectional Method 400
Longitudinal Method 401
Combined Cross-Sectional/Longitudinal Method 401

References 403

Subject Index 417

Name Index 425

Preface

This revised edition is appropriate for an introductory course in adolescence. It seeks to present a well-balanced account of adolescence, one based on established principles of development and recent research findings. The sequence of chapters reflects a developmental approach, but I have tried to relate every aspect of adolescent development to its larger social and cultural context, including the impact of social change. My experiences in teaching a course on adolescence, conducting therapy with older youths, and being a parent of two adolescents now reaching adulthood have greatly enriched my understanding. I hope this shows through in the text, along with the personal examples I have given.

In this revision I have made a special effort to update the basic information by using recent research studies throughout the book and incorporating new developments, such as the changing patterns of drug use. I have completely rewritten the chapters on the family, family relationships and adolescent autonomy, self-concept and identity, school, sexuality, work and career choice, moral development, and drugs. I've also rewritten the appendix on methods of studying adolescents, illustrating the various methods of study cited.

The end-of-chapter summaries include the main points of each chapter, numbered and arranged according to the major headings, thereby helping the reader to grasp the chapter as a whole in relation to its parts. I have also added a list of review questions, useful as an aid in reviewing the material for tests, especially for essay and short-answer questions.

Glossary definitions of new or technical terms, normally found at the back of a book, have been placed in the margins near appropriate passages where the terms are initially used, making them more convenient to use.

To highlight material of special interest and to increase the reader's personal involvement with the material, I have provided boxed items throughout the book. These include personal examples and exercises as well as pertinent information on adolescence. The boxed items, together with the figures and tables, help to round out the total presentation.

I have benefited greatly from criticisms of the first edition of *Adolescence,* and would especially like to thank the following reviewers: Thomas Moeschl of Broward Community College, and Stephen B. Wilson of Triton College. I would also like to thank the people in my college library as well as those in neighboring libraries who have been so cooperative in my continuing search for appropriate material for this book. Thanks also to Randy Clouser for his help with the chapter on drugs.

But the main thanks go to you, my readers, and I trust that the time spent with this book will be rewarded by a new understanding of the intriguing subject of adolescence.

EASTWOOD ATWATER

1 Introduction to Adolescence

- **ADOLESCENCE IN PERSPECTIVE**
  The Common Experience of Adolescence
  Rites of Passage
  Abbreviated Adolescence

- **THE SOCIAL INVENTION OF ADOLESCENCE**
  Approaching the Age of Modern Adolescence
  Major Social Changes
  Prolonged Adolescence

- **ADOLESCENCE TODAY**
  The Boundaries of Adolescence
  Growing Up Faster
  Contemporary Youth

- **SUMMARY**

- **REVIEW QUESTIONS**

Motion-picture cameras are out of focus when they zero in on teenagers, claims Justine Bateman of the *Family Ties* series on television. Adolescents are misrepresented as sex-crazed, mindless hedonists who engage in outrageous activities. A major barrier to teen credibility on the screen is age. For a variety of reasons, including child-labor laws, teenagers are commonly played by actors and actresses in their 20s. Then too, the producers will take one small aspect of teen life and exaggerate it all out of proportion to reality. Movies keep showing kids at parties trashing someone's home—but you never see the parents reacting to it. In real life that just doesn't happen. Also, who is going to let someone ride a motorcycle through their house? Sometimes the dialogue reflects the way teenagers really talk. But more often young people speak in clichés, mostly because the writers are adults who can't keep the expressions and slang up to date. Movies tend to be more realistic about teenagers' dress. Probably the best thing movies reflect is the music, usually because the producers simply use music from the Top 40 hits. What Bateman most resents is the tendency of movie and television producers to lump all teenagers into a few categories like jocks and cheerleaders. (Scott, 1985)

ADOLESCENCE IN PERSPECTIVE

Motion-picture and television cameras are not the only ones out of focus when it comes to teenagers. Parents, teenagers, and the media sometimes unwittingly misrepresent young people. This is especially apparent when putting youth in historical perspective. For instance, a history teacher friend of mine—who subscribes to the ancient Greek view that there's nothing new under the sun—claims that kids today are no different from those of the past. And there's probably a grain of truth in this view, in that young people in every era are motivated by certain common concerns, such as becoming independent of their parents. But other people—who are more inclined to view what is new as necessarily different from the past—regard today's youth as unique, as a generation that thinks and behaves in ways unlike those in any past era. One need only look at the widespread impact of computers and the fascination with life in outer space to realize the truth of this view.

adolescence: The period of rapid growth between childhood and adulthood; includes psychosocial as well as physical growth.

Actually, adolescents today are both like and unlike those of the past. Contemporary adolescents are like their counterparts in earlier eras in that they undergo similar biological changes at puberty as well as the psychological and social adjustments in transition to adult status that accompany such changes. But today's adolescents also differ in many ways from those of the past, primarily because of the social and cultural context in which young people grow up today. We may gain a better understanding of how today's young people are like and unlike those of the past by taking a look at adolescents in historical perspective, including the emergence of our contemporary view of adolescence.

The Common Experience of Adolescence

development: The relatively enduring changes in people's capacities and behavior as they grow older, because of the biological growth process and people's interaction with the environment.

To some extent, adolescence is a universal experience because of the rapid increase in the growth process that characteristically occurs at puberty and the impact this has on the individual's development. For instance, consider the description of youth given by Aristotle, the ancient Greek philosopher, in the boxed item "Aristotle on Youth." It is impressive when we remember that Aristotle wrote these words more than 2300 years ago! Although much more can certainly be said about the positive side of adolescents, the description is still relevant today mostly because of the uneven growth and the associated problem of self-control that characteristically occur at puberty.

puberty: Technically the attainment of sexual maturity or reproductive powers; more generally, the entire process of glandular and bodily changes accompanying sexual maturation.

Joan Newman (1985) points out that much of the obnoxiousness commonly seen among youth stems from the uneven growth process that occurs at puberty. The adolescent's awkward movements and postures stemming from uneven acquisition of physical competencies at this age are familiar enough. Even more important is the uneven development of other competencies, such as the ability to think in an adult, abstract manner; the ability to see things from other people's perspective; the ability to control and

2

ARISTOTLE ON YOUTH

Young men have strong passions, and tend to gratify them indiscriminately. Of the bodily desires, it is the sexual by which they are most swayed and in which they show absence of self-control. They are changeable and fickle in their desires, which are violent while they last, but quickly over. . . . They are hot-tempered and quick-tempered, and apt to give way to their anger; bad temper often gets the better of them, for owing to their love of honor they cannot bear being slighted, and are indignant if they imagine themselves unfairly treated. . . . They have exalted notions, because they have not yet been humbled by life or learnt its necessary limitations; moreover, their hopeful disposition makes them think themselves equal to great things—and that means exalted notions. . . . They are fonder of their friends, intimates, and companions than older men are, because they like spending their days in the company of others, and have not yet come to value their friends or anything else by their usefulness to themselves. All their mistakes are in the direction of doing things excessively and vehemently. . . . They love too much and hate too much, and the same with everything else. They think they know everything; and are always quite sure about it; this, in fact, is why they overdo everything.

Rhetoric, Book II, chapter 12

---

express one's emotions appropriately; the ability to understand right and wrong on the basis of ethical principles; and the practical knowledge of the world. For instance, the adolescent who has grown taller and heavier but who remains emotionally and socially immature may behave in a more extreme and offensive way than a younger child. Instead of making noise with his toys or throwing them, he may play his stereo loudly, become violent, and throw bricks. Furthermore, his misjudgments can lead to permanent harm. He may drop out of school, jeopardize his health and safety through the abuse of drugs, make someone pregnant, and take someone's life through careless or drunken driving. Adolescents who are *always* obnoxious need help. But even normal, healthy adolescents will act this way sometimes, largely because of the uneven development of competencies at this age.

Even people who eventually become famous do not escape the trials and tribulations of adolescence. Norman Kiell (1967) has collected dozens of illustrations from the adolescent periods of men and women who later achieved fame to show that their adolescences were not that different, if at all, from those of ordinary young people. In most instances, the genius or the special gifts one might expect to see were apparent neither to the individuals nor to the people closest to them. Instead, their autobiographies reflect many of the characteristics commonly associated with adolescence.

Charles Darwin once admitted that at the age of 16 school was simply a "blank" for him, and that he was considered an ordinary student by his teachers and a "very ordinary boy" by his father. Darwin reports being deeply

Rite of passage for a young man in Savu Island, Indonesia.

hurt when told by his father, "You care for nothing but shooting, dogs, and rat-catching, and you will be a disgrace to yourself and all your family." Yet by the time Darwin reached adulthood, he found a constructive outlet for his interest in nature and eventually achieved worldwide fame for his scientific support of evolution.

Rites of Passage

rites of passage: Ceremonial acts marking the passage from one developmental or cultural stage to another, with the initiation into adulthood sometimes known as "puberty rites."

Partly because of the uncertainties surrounding the transition to adulthood, the attainment of adult status is marked by formalized *rites of passage* in many preindustrial societies. These ceremonies and rites differ somewhat among boys and girls. For instance, in many primitive societies the initial rites for adolescent males involve circumcision, either in an individual or group setting. After the boy recovers, he may undergo tests of manliness, such as surviving alone in the forest or kneeling down on hot coals. Rites for girls are often occasioned by the experience of their first menstrual period. At this time the girl may be confined to a separate part of the house for about a week. At the end of this period, she takes part in a series of ceremonies, such as spinning cotton to be used in a hammock by someone in her family. She is then dressed, adorned with jewelry, and allowed to take part in festivities of drinking, dancing, and singing. Afterwards she is treated like an adult and assumes the privileges of adult women.

A major purpose of these rites of passage is to confer adult status on the young in a way that is clear both to the individual and the rest of the community. Thus, the boy who has withstood the pain of circumcision bravely may know that he is ready to join his fathers and brothers in hunting. Similarly, the newly initiated girl may continue living in her parents' home while being assured that she will be treated as an adult. Rites of passage also help pass on the appropriate male and female sex roles of a given society. However, because youth do not learn all the necessary skills for adulthood in such ceremonial activities, the rites of passage are often an integral part of a long series of practices designed to prepare the young for adulthood.

By contrast, our society provides no clear rites of passage to adulthood. Instead, adult status is achieved unevenly in one realm of life after another over a long period of time. Thus, religious ceremonies, such as confirmation and bar mitzvah, denote adult status in the religious community. But there is considerable variation in the age at which such ceremonies occur, not to mention the fact that almost half the population does not belong to any religious organization. A more common means of recognizing adult status can be seen in the variety of laws marking legal adulthood. Yet, these laws are often inconsistent from one aspect of society to another. Thus, young people may begin to drive a car at one age, marry without parental permission at another age, but not own property until still another age. Then too, young people may be eligible for military service before being allowed to buy a drink in a bar. In sum, the various rites of passage are so inconsistent and so stretched out over a long period of time it's no wonder that young people remain unsure when they have come of age.

Lacking clear-cut rites of passage to adulthood, American youth often engage in informal rites of their own. The hazing that sometimes accompanies membership in clubs and peer groups may resemble the tests of manliness inflicted on tribal youth. The teenager's preoccupation with bodily adornment, whether hair styles, pierced ears, or dress may serve a similar purpose. Also, the achievement of adult status is frequently marked by funtional rites of passage, such as getting a driver's license, dating, and being allowed to stay out late at night. Graduation from high school, getting a job, and becoming married also serve as important rites of passage to adulthood. Youth who are denied the adult responsibilities of work and marriage are likely to take on some of the superficial signs of adulthood, such as smoking and drinking.

sex role: The complex set of expectations and behaviors associated with males and females in a given society.

confirmation: A Christian ceremony in which a person is admitted to full membership in a church, having affirmed vows made at baptism.

Bar mitzvah: Literally "son of the commandment"; a Jewish boy who has arrived at the age of religious responsibility—13 years; the ceremony celebrating this event.

Bat mitzvah: A Jewish girl who undergoes a ceremony analogous to that of a bar mitzvah; the ceremony celebrating this event.

Abbreviated Adolescence

Until the mid- to late nineteenth century, the transition to adulthood in the United States was relatively fast, making for a brief adolescence. There were a

apprenticeship: A period of training that prepares individuals for a skilled trade such as carpentry.

number of reasons for this. For one thing, the average life span extended only into the late forties or early fifties, with marriage and childbearing taking up a greater proportion of a lifetime. Then too, a long period of schooling was not as widely available or required as it is today. A substantial proportion of youth moved away from home in their midteens to take part in apprenticeship training programs or to board near their places of work. Probably the single most important influence at this time was the shortage of labor, encouraging an early entry into the labor market.

The transition to adulthood among the working class was brief and harsh, with exploitation of child labor a common practice. At one point in the nineteenth century, children made up 40 percent of the factory workers in New England. The decline in farming and the rise of manufacturing resulted in many farmers' daughters finding work in factories. Young girls would begin work at 4:30 in the morning. Each girl would tend three or more looms in a dust-filled room until 7 o'clock in the evening, taking only a short break for breakfast and lunch. At the end of the day the girls would return to their boarding houses. They were usually so tired after the evening meal that they went right to bed, often sleeping six to a room, with little or no privacy.

Among the privileged classes the passage to adulthood was equally brief but not nearly as harsh. A British visitor to nineteenth-century Boston observed that the young people were born "middle-aged." They were sent to college at 14 or 15 years of age and earned their degrees at 17. They were immediately launched into their careers as merchants, teachers, or physicians. Consequently they had no time for the athletic games so characteristic of the British gentleman. Instead they had only the pastimes of "chewing, smoking, drinking, driving hired horses in wretched gigs with cruel velocity." All their energies were spent on working and making money. The visitor concluded that young people in America were "a melancholy picture of prematurity" and that the main business of life seemed to be "to grow old as fast as possible" (Grattan, 1859/1987).

THE SOCIAL INVENTION OF ADOLESCENCE

Likewise during the nineteenth century, a number of important changes were taking place in American society that would eventually prolong the transition to adulthood. In explaining the impact of these changes, Joseph Kett (1977) suggests that adolescence became more of a "social invention" than a discovery. That is, the prolonged period of adolescence so familiar to us is essentially a conception of behavior and development that was imposed on youth in response to the needs of an emerging urban-industrial society. It was not something that was based primarily on an empirical evaluation of the way young people actually behaved. Many of the influences that led to the social invention of adolescence were already apparent by the latter part of the nineteenth century.

A DELAYED ADOLESCENCE

Throughout the 1970s the comedian George Carlin delighted audiences with his perceptive wit and outrageous appearance, especially his hair held tight in a pony tail. Now, with shorter hair and a trimmer beard, Carlin looks back at the '70s—and his 30s—as a vital stage of his personal and professional development.

"I was still working out and acting out my adolescence in my 30s," he said, "and I happened to have been able to do it publicly and to be paid for it. As an adolescent, I never acted like one; I was really doing adult things then, so it was a deferred adolescence, and when I finally got through it all, I was able to feel more mellow, as anyone does when you go through some turbulence."

Gary Ronberg, *The Philadelphia Inquirer*, April 25, 1980, p. 8B.

Approaching the Age of Modern Adolescence

During the last half of the nineteenth century there was an increasing tendency for young people to remain at home well beyond puberty, reversing the long-standing parental practice of sending children out to live, study, or board in someone else's household between puberty and marriage. A major reason for this change was the end of the labor shortage, making young people more of an economic liability than the asset they had been in earlier eras. Other factors also played a part, including the decline of apprenticeship training, the extension of mass schooling, and a change in popular attitudes toward the young. For instance, in one community, nearly half the 11 to 15 years olds and one-fourth of the 16 to 20 year olds were neither enrolled in school nor working. More adolescents were roaming the streets with little or nothing to do, increasing adult concerns about the youth in their community (Elder, 1980).

A major step was provided by Stanley Hall, who—in his monumental two-volume work on adolescence published in 1904—first proposed the idea that adolescence is an important separate stage of development. Hall's belief that adolescence is a critical stage that requires an enriched environment for the normal, healthy development of adolescents became a major force in the formation of new institutions for youth, such as the YMCA, YWCA, Boy and Girl Scouts, and various other youth clubs.

Major Social Changes

Several social changes were especially important in prolonging the period of adolescence, namely the introduction of compulsory education, child-labor laws, and the concept of juvenile justice. These changes were motivated, at least on the conscious level, by humanitarian principles such as protecting the young from exploitation in the workplace and giving them a better education. Yet they also had the unintended effect of prolonging young people's dependence on adults.

compulsory education: Required attendance at school because of compulsory attendance laws by the state.

Compulsory education for children and adolescents between the ages of 6 and 16 was introduced widely in the United States in the late nineteenth century. At about the same time, the period of schooling was extended to include high school. In both instances, the main reason was to give young people the necessary skills for exercising the rights and privileges of democracy. Access to public education also provided an arena for individuals to improve themselves, thereby helping to diminish the growing gap between youths in the different social classes. However, compulsory education extended the legal power of school authorities over students, thereby prolonging their dependence on adults.

The reversal of the labor shortage in the last half of the nineteenth century led to extensive child-labor legislation. The newly created laws were aimed at protecting the young from exploitation and providing them with opportunities to acquire the skills needed for living in an industrial society. At the same time, such laws have had the effect of isolating young people from the workplace, making it difficult for them to find their role in it and excluding them from meaningful employment in the labor market. While a delayed adolescence tends to benefit youth from affluent homes who plan to continue their education in college, it often has the opposite effect on those who seek an early entry into the workplace.

juvenile justice: The concept that youthful offenders who have not reached legal age should be treated by different rules and principles than the legal system applied to adults.

Humane considerations also led to the idea of juvenile justice. Up until the turn of the century, teenagers who broke the law were often punished as adults. Then in 1899 the Illinois legislature passed the first Juvenile Court Act, which provided for informal hearings, confidential records, and separate detention of young people. On the positive side, the juvenile justice movement has helped to protect youthful offenders from the harsher penalties of the adult law. But, as we'll see in a later chapter, the juvenile court has also resulted in inconsistent and often unfair treatment of youth, thereby prolonging the period of life before young people can assume the rights and responsibilities of adults.

Other influences also contributed to the present notion of adolescence. The continuing move of people from the farm to the city brought about more interaction at school and in the neighborhood among boys and girls of the same age, giving rise to the increased importance of peer groups in adolescence. Also, the increase in human longevity promoted an extended adolescence. Because the average person only lived until about 50 years of age at the turn of the century, a relatively rapid transition to adulthood made sense. But with the continuing increase in the average life expectancy, individuals could afford to spend a longer time in adolescence. Finally, the emerging scientific understanding of the growth process has contributed to the refinement of the developmental stages.

Prolonged Adolescence

All these changes have had the effect of prolonging adolescence, essentially to fulfill the aims of the new urban-industrial society which developed after the Civil War. Such a society needs individuals who are educated, vocationally skilled, and capable of exercising the rights and responsibilities of citizens in a democratic society. The new concept of adolescence provided the prolonged period of growth necessary to achieve these aims.

psychosocial moratorium: Erikson's concept of modern adolescence as a socially approved delay in development granted to those who are not ready to assume adult commitments.

Adolescence in this sense involves a period of psychosocial growth which extends beyond the attainment of physical and sexual maturity. Erik Erikson (1980) has characterized contemporary adolescence as a "psychosocial moratorium" granted to those who are not ready to assume adult commitments like work and marriage. Thus, adolecence has become a time when youth are given greater freedom for experimenting with their life roles. Such an extended period of development provides young people with certain advantages. First, they are able to spend longer years in school, thereby gaining a better education. Also, a larger proportion of youth than in the past may prepare for skilled careers, enabling them to rise to a higher socioeconomic level than their parents. Then too, the postponement of adult commitments gives young people greater freedom for exploring their life roles and values, thereby promoting a more individuated personal identity in relation to their family of origin than in the past. Erikson points out that youth from affluent homes who are ambitious, interested in learning, and concerned with self-fulfillment tend to make better use of these opportunities than less privileged youth who have grown up accustomed to following in their parents' footsteps.

The extension of adolescence also has its disadvantages. First, the extended period of formal schooling isolates young people from the workplace, making it difficult for them to know what people do at work, as well as complicating the task of making a wise career choice. And too, youths who already know what they want to do in life, especially those who prefer to work in the manual trades, often resent having to spend long years in the classroom studying topics that don't interest them. Furthermore, a delayed adolescence prolongs young peoples' dependence on adults, thereby promoting needless discontent and rebelliousness. Finally, inherent in the protracted period of youth is the hazard of getting locked into adolescent roles and lifestyles, with its pleasure-seeking activities and irresponsibility.

In earlier eras, young people could hardly wait to grow up because of the privileges that went with adulthood. But today individuals are often reluctant to give up the greater freedom that goes with youth. Adulthood is seen as a time for commitments—a nine-to-five job, marriage, children, and paying taxes. Youth means freedom from these commitments and the opportunity to explore and aspire to a better life. Thus, society has passed from a period which ignored adolescence to a period in which youth has become its favorite age.

ANOTHER CHANCE

Have you ever wished, as most of us have, that you could relive your high-school days, knowing what you do now? It is just a fantasy for most of us. But Patrick Lajko actually did it.

It all began when Lajko (pronounced Like-oh), a graduate of Iowa State University and a member of the University's national championship gymnast team, began practicing with local high-school gymnasts at the YMCA in Wichita Kansas, where he had taken a job. Soon he was working out with the high-school gymnast team so regularly that someone jokingly suggested that he should enroll in school. The idea appealed to him, partly because, like Peter Pan, he feared growing old. But like many of us, he also wanted a fresh start, another chance at high school knowing what he now knew. Lajko assumed the name of Scott Johnson, and applied to the local high school. With the help of friends who knew his true identity, he intercepted a letter requesting his high-school transcript in order to falsify it.

Lajko, then in his twenties rather than his teens, began school in the spring semester, studied biology and English, and returned in the fall, taking courses in speech and government. He made an "A" average. He also played

on swimming and gymnast teams, winning eleven medals and a first in vaulting, his old Iowa State speciality. As Scott Johnson, he got caught up in the spirit of things, dating high-school girls, and sharing the usual worries of his teenage friends. He also became a high-school hero. But one evening a former colleague at work saw him perform on TV and blew the whistle. A few days later, just before the swimming team was to leave for a meet, Lajko was taken off the team bus and confronted by the assistant principal.

Lajko has mixed feelings about his attempts to recapture youth. He feels badly that he let down a lot of people who believed in him during his second high-school career. But he also feels that he benefited in many ways. He felt he was smarter, better, and happier in high school the second time around. He also felt it was a rejuvenating experience. He learned that you feel as old as the people you are with. Even though going back to school has not been a Fountain of Youth for him, it has slowed down the aging process, making him feel younger than others his age.

Bill Nack, Newsday Service, *The Philadelphia Inquirer*, Dec. 15, 1976, pp. 1D, 4D.

ADOLESCENCE TODAY

At the same time, our view of adolescence continues to change. Empirical studies in a variety of scientific fields are giving us greater knowledge of human development, including a more refined understanding of adolescence. Then too, our view of adolescence is changing because of our era's rapid social changes and the far-reaching effects this has on society. Changes in the patterns of work, education, and sexual mores, just to name a few key areas of behavior, not only modify the way young people experience the transition to adulthood but also affect our attitudes and expectations toward youth.

In this section, we'll focus on the current view of adolescence, including the difficulties of defining the boundaries of adolescence as a psychosocial developmental process. We'll also examine the tendency of adolescents to grow up faster in recent years, along with some of the implications this has for adolescent development. Then we will take a brief look at contemporary youth.

The Boundaries of Adolescence

The social invention of adolescence as a prolonged psychosocial process has posed new problems in defining its boundaries. That is, when does adolescence begin and when does it end? The traditional definition of adolescence—the period of rapid growth between childhood and adulthood—was based largely on physical change, especially the more obvious manifestations such as increased height and weight. As a result, adolescence became synonymous with the teenage years, roughly from 13 to 18 years of age. However, the realization that adolescence pertains to a process of psychosocial development as well as physical growth implies that adolescence may begin earlier and last longer than the physical changes of puberty.

For all practical purposes, the beginning of adolescence remains closely associated with the onset of puberty, even though it is no longer synonymous with it. But puberty is not a single, sudden event so much as a slow process of biological change that begins at conception and accelerates at adolescence (Petersen & Taylor, 1980). As such, puberty depends more on various developmental changes—such as hormonal changes, growth rate, and one's reaction to such changes—than on chronological age. Also, there are wide individual and sexual variations in the onset of puberty, making it difficult to generalize about the age of onset.

Early adolescents. (Laimute Druskis)

There is even less agreement about when adolescence ends, with much of the debate depending on which criteria of development are used. Those who think in terms of physical development may see the end of adolescence as synonymous with the attainment of puberty in the late teens. Yet, as we shall see in a later chapter, a small proportion of youths continue growing physically into their twenties. Those who adopt legal criteria for the termination of adolescence are inclined to view the entry into early adulthood in terms of legal privileges, such as being able to vote, obtain a marriage license without parental permission, take a drink, and work a 40-hour week. Still others may choose to define adulthood in more functional terms pertaining to the individual's ability to assume responsibility in the adult community, such as graduation from high school or college, holding a full-time job, marrying, having a family, or becoming economically self-supporting. Those who stipulate the attainment of emotional or social maturity have even greater difficulty specifying specific ages for the end of adolescence. Some people never attain maturity in these areas. Actually, it is difficult to specify a precise end to adolescence because this period of life does not end as much as it merges into a subsequent transitional period known variously as late adolescence, youth, or early adulthood, as we'll discuss in the final chapter. For instance, one study of college students found that a majority of them were simultaneously working on at least one developmental task of middle childhood, adolescence, and early adulthood respectively, suggesting that adolescence is a transitional stage with uncertain boundaries (Roscoe & Peterson, 1984).

All things considered, adolescence may be defined as the developmental period of rapid growth between childhood and adulthood, including psychosocial as well as physical growth. Also, it has become common to distinguish different substages, such as early, middle, and late adolescence. Although adolescence does not have sharp boundaries, it generally begins at puberty—about age 10 or 11 in girls and age 12 or 13 in boys—and lasts until the attainment of early adulthood, which may occur anytime between the late teens and the late twenties or even later.

youth: Transitional period between adolescence and adulthood.

early adolescence: The period from the onset of puberty through the middle school or junior high school years, about 10 to 14 years.

middle adolescence: The middle period corresponding to the high school years, about 15 to 17 years.

late adolescence: The period corresponding to the post-high school and college years, about 18 to the early 20s.

Growing Up Faster

Ironically, there is a tendency for young people to grow up faster in recent years than in the past, thereby further modifying our view of adolescence. For instance, in one newspaper ad, teens 11 to 18 were invited to a workshop on the art of makeup, skin and hair care, and fashion. When the sponsors of the ad were questioned about the wisdom of inviting 11 year olds to such a workshop, they hastened to say that they weren't "pushing" makeup for young teens; but they added that since many 11 year olds are already using makeup, the workshop might help them to do it correctly. The idea of holding beauty workshops for 11 year olds appears symptomatic of a larger, deeper problem in present-day society—children and adolescents growing up too fast (Gilliam, 1983).

In *The Disappearance of Childhood,* Neil Postman (1982) describes the increasing pressure on children and adolescents to look and act like adults. Traditionally, one of the differences between children and their parents was that children did not know certain things that adults knew. However, with childrens' increasing exposure to television, this difference is rapidly disappearing. Children and adolescents now see ads featuring skin-tight jeans and toothpaste with "sex appeal." They're also exposed to adult sexuality and violence on soap operas and other television shows. As a result, children and adolescents are adopting the dress, behavior, and mannerisms of adults sooner than in the past.

David Elkind (1981) has observed the same tendency in today's society and has characterized it as the "hurried child syndrome." Elkind points out that parents and the media, reflecting the larger society, are putting more pressure on their children and adolescents to grow up faster. A major reason is that parents are under so much pressure themselves and are so concerned with realizing their own lives that they unwittingly rush their kids to grow up sooner. Consequently, children and adolescents are expected to achieve,

Adolescents are being pushed to perform without regard for their own developmental needs. (Ken Karp)

succeed, and handle responsibility at an earlier age than in the past. Unfortunately, this encourages the wrong kind of growth—young people's performance being judged by adult standards without regard for their own developmental needs. Young people are viewed largely in their roles as students, soccer players, and musicians rather than as whole persons. By the time these individuals find out they are not fully mature, it may be too late. Ironically, because society is changing so rapidly, young people need more developmental time, not less, in order to grow up emotionally. Being forced to behave in an adult manner before they have reached adulthood has become a source of stress for young people. Elkind points out that there is probably a strong relationship between the hurried-child syndrome and the troubles adolescents are experiencing in the 1980s, such as low performance at school, delinquency, drugs, teen pregnancy, and adolescent suicide.

hurried-child syndrome: The tendency for children and adolescents to grow up faster today because of the increased social pressure to do so.

Contemporary Youth

Today's youth are not only under greater pressure to grow up faster, but they are also growing up amid different social conditions than youth in the past. For one thing, youths aged 14 to 24 are less in the public limelight than were their counterparts in the 1960s and 1970s when adolescents were increasing rapidly in numbers and visibility. Because of the aging of the population, youth 14 to 24 years of age are now decreasing in numbers and will constitute a smaller proportion of the general population. See Table 1–1. As a result of such changes, high school enrollment has already begun to decrease in some areas of the country, and the less select colleges are having to work harder to attract students. The crime rate, which is normally highest among those in the 14- to 24-year-old age group, is also decreasing somewhat. Employment patterns are also affected, with less competition for entry-level jobs and heightened competition for jobs among those in the 25- to 54-year-old group (U.S. Department of Labor, 1986).

Most American teenagers (58 percent) believe they will enjoy a higher

Table 1–1 POPULATION AGED 14 TO 24 YEARS OLD, 1960 TO 1990

Year	Size (millions)	Percent Change	Percent of Total Population
1960	27.3	—	15.1
1970	40.6	48.5	19.8
1980	46.4	14.4	20.4
1985	43.6	− 6.0	18.2
1990*	41.0	− 5.9	16.4

*Projection

Source: U.S. Bureau of the Census, *Statistical Abstract of the United States: 1987*, 107th ed. (Washington, DC: Government Printing Office, 1986), pp. 14, 16.

standard of living than their parents. But the majority of them (82 percent) also agree that they will face more environmental problems than did their parents' generations. These are just two of the responses to a cross-cultural survey of 2300 persons between ages 13 and 20 in 59 nations (*U.S. News and World Report,* 1986). When asked to choose the top problems facing youth, practically all the Americans (99 percent) cited premarital sex. American teenagers ranked drug abuse second (85 percent), followed by alcoholism (71 percent), suicide (67 percent), teen pregnancy (44 percent), and teen pornography and prositution (15 percent). Around the world, however, drug abuse was named by 83 percent as the leading problem facing youth—a finding, incidentally, that is supported by the Gallup Youth Survey (1983) of American youth. Just 29 percent of the youth in other countries consider premarital sex a major problem—the lowest ranking of any of the six topics listed in the questionnaire. At the same time, two out of five teenagers in other countries believe there will be a nuclear war in their lifetime, compared to only 28 percent of American teenagers. Yet, more than two-thirds of American teenagers feel that both the United States and the Soviet Union are doing a poor job in handling the nuclear arms race.

Many young peoples' concerns reflect the changed conditions of society and the world, especially the increased importance of equality, as well as the heightened competitiveness and financial pressures of the workplace. This can be seen in a comparison of high school seniors in the 1970s and 1980s (Fetters, Brown & Owings, 1984). In school, almost three-fourths (71 percent) of the seniors in the class of 1980 felt that their academic accomplishments had been hindered by poor study habits, reflecting the increasing concern among educators and parents for improving the quality of secondary education. As a result, there was a sevenfold increase in students taking remedial math courses as well as a large increase in remedial English. At the same time, students were spending fewer hours on homework and extracurricular activities at school—especially academically related extracurricular activities—mostly because of greater time spent on outside jobs. Almost half of high school seniors worked 20 hours or more at an outside job. Among those going on to college, increasing percentages of students planned to enter the fields of business, engineering, and computer sciences. Fewer students planned to pursue teaching, mathematics, and the physical sciences. Although the proportion of males planning on attending college remained the same, about 4 out of 10, the number of college-oriented females increased to 5 out of 10. Furthermore, the preferences of young women for "male dominant" jobs such as manager, administrator, and technician doubled from 10 to 20 percent. There has also been an increase in work-related values, such as finding steady work, being successful at what one does, and feeling it is important to have "lots of money." At the same time, youth in the 1980s continue to express strong concern for personal well-being, especially self-fulfillment, and tend to look for happiness in close relationships and marriage.

extracurricular activities: Activities outside the regular academic program of studies but under the supervision of the school, such as sports or dramatics.

Love and close relationships are especially important to young people. (Laimute Druskis)

For instance, when teenagers in the worldwide survey were asked what would bring them the most happiness, the majority (69 percent in the United States and 60 percent overseas) selected "love" far above "good career," "family," "helping others," "freedom," and "fame" (*U.S. News and World Report,* 1986).

Surveys among college students reflect similar changes among older youths. One of the most reliable indicators of attitudes among college students is the annual study conducted by Alexander Astin and Kenneth Green (1987) involving more than 6 million students and 100,000 faculty at more than 1250 colleges around the country. The study supports the widespread belief that college students are more inclined toward career-oriented programs and making money than were their 1960s counterparts. But the popular notion that today's college students are generally more conservative than those in the past was not borne out by the study. For instance, in the political arena in 1985 about one-fifth (19.5 percent) of the students identified themselves as conservative and another one-fifth (22.4 percent) identified themselves as liberal. But more than half of them (56.7 percent) said they were "middle-of-the-road," up from 45.4 percent in 1970. Interestingly, when asked their views on specific social issues, a substantial majority of students continued to support traditional liberal values including more government intervention in such areas as control of environmental pollution, consumer protection, and energy conservation.

One of the most dramatic changes has been the impact of the women's movement on higher education, penetrating every aspect of college life. Once in the minority in American higher education, women now make up 51.8 percent of the entering classes in 1985, compared with only 43.3 percent in 1969. While there has been a declining interest among male students in professions such as medicine and law, there has been a marked increase among women in these traditionally male-dominated fields. Since 1969, the proportion of women headed for medical careers increased by more than 270 percent, compared to a 10-percent decline in the proportion of males. Interest among women in law careers has increased nearly fivefold in the same period.

A final note concerns the diversity among American youth. Although many of the studies and surveys cited in this book reflect common patterns of behavior among youth, it is well to keep in mind that young people in the 1980s are at least as diverse as their counterparts in the past if not more so. Not all youth in the 1950s were squeaky clean. Nor did all young people in the 1960s and 1970s participate in the social and political activism of the day. The majority of them supported the Vietnam War and the draft. Similarly, not all of today's youth resemble the images of them depicted in the media. As we'll see throughout this book, there is at least as much diversity among youth as there is among adults.

SUMMARY

Adolescence in Perspective

1. We began the chapter by noting that adolescents today are both like and unlike those of earlier eras. Although adolescents today undergo many of the same physical and cognitive changes accompanying puberty as did their counterparts in earlier eras, they experience the transition to adulthood somewhat differently because of the social and cultural context in which they grow up.

2. Much of the sameness of adolescence stems from the biological growth process that occurs during puberty and from the awkwardness brought on by the uneven development of various competencies.

3. In many preindustrial societies, the transition to adulthood was marked by formalized rites of passage. The lack of clear-cut rites of passage in contemporary Western society often results in an uncertainty as to when one has reached adulthood.

4. Prior to the present century, the transition to adulthood was relatively fast, making for a brief, sometimes harsh adolescence, especially among working-class youth.

The Social Invention of Adolescence

5. The present-day view of adolescence as a prolonged period of psycho-social development was essentially a social invention resulting from various social and economic changes that occurred in American society during the late nineteenth and early twentieth centuries.

6. The increasing tendency for young people to remain home well beyond puberty in the last half of the nineteenth century, largely because of fewer available jobs, resulted in a more aimless and indecisive period of youth.

7. The subsequent introduction of compulsory education, child-labor legis-lation, and juvenile justice had the effect of prolonging the period of adoles-cence largely in order to accomplish the aims of the new urban-industrial society.

8. Although the extension of adolescence has afforded youth greater educational and career opportunities, it also has had the adverse effect of prolonging their dependence on adults.

Adolescence Today

9. Adolescence has become a prolonged period of development between childhood and adulthood, including psychosocial as well as physical growth and generally lasting beyond the attainment of physical maturation.

10. However, in recent years there has been increased social pressure for children and adolescents to grow up faster, often at the expense of their developmental needs.

11. Contemporary adolescents are also different from their counterparts in the past in that they are spending a greater amount of time at outside jobs during high school and are more oriented to work-related values, such as success and material affluence.

12. At the same time, it is well to keep in mind that there is at least as much diversity among young people today as there is among adults.

REVIEW QUESTIONS

1. How are today's adolescents like and unlike those of earlier eras?

2. In what sense is the transition to adulthood still marked by rites of passage?

3. Describe the transition to adulthood in late nineteenth-century America.

4. Explain adolescence as a social invention.

5. How was adolescence affected by the introduction of compulsory educa-tion, child-labor legislation, and juvenile justice?

6. What is meant by the concept of adolescence as a psychosocial moratorium?

7. What are some of the advantages and disadvantages of adolescence as a prolonged period of development?

8. How would you define adolescence?

9. In what ways are children and adolescents growing up faster today?

10. What are some of the advantages and disadvantages of high school students' spending 20 or more hours at a part-time job?

2 Perspectives on Adolescence

- **THE BIOLOGICAL PERSPECTIVE**
  Hall's Critical-Stage Theory
  Gesell's Maturational Theory
  Puberty and Adolescent Development

- **THE PSYCHOANALYTICAL PERSPECTIVE**
  The Freudian View of Development
  Adolescent Sexual Conflicts
  Coping with Adolescent Sexuality
  Newer Developments

- **THE PSYCHOSOCIAL-CULTURAL PERSPECTIVE**
  Havighurst's Developmental Theory
  Bandura's Social-Learning Theory
  Mead's Cultural Theory
  Adolescence as a Cultural Phenomenon

- **ADOLESCENT STORM AND STRESS IN PERSPECTIVE**
  Reevaluating Adolescent Storm and Stress
  Patterns of Adolescent Growth
  The Impact of Social Change

- **SUMMARY**

- **REVIEW QUESTIONS**

You may recall James Saxe's (1963) poem about the six learned blind men of India who went to see the elephant. The first happened to fall against the elephant's side and exclaimed, "It's like a wall." Another one, feeling the elephant's trunk, concluded, "It's very much like a snake." Still another touched the elephant's ear and marveled, "It's like a fan." And so these learned men argued loud and long. Although each was partly in the right, all were in the wrong.

Like the six learned men of India, today's scientists and clinicians also approach the understanding of adolescence from different perspectives or angles of vision. And on the basis of their limited evidence they form hypotheses and theories—general explanatory statements to account for the observed phenomena. Yet, they too realize that no one theory captures all aspects of adolescence. Instead, each theory has an optimal range of explanation, explaining some aspects of adolescence better than others. Furthermore, few theories of adolescence are fully proven or disproven. In most instances, they are continually modified with the discovery of new information, as well as with the ongoing dialogue between different viewpoints.

We have arranged the representative theories of adolescence according to the various theoretical perspectives—the biological, psychoanalytical, psychosocial, and cultural viewpoints respectively. We have also added a section examining the recurrent notion of adolescent storm and stress. But keep in mind that there is a great deal of overlapping between these perspectives.

One final note. Many other theories of adolescence have been incorporated into the subsequent chapters of this book, especially those of Erik Erikson, Jean Piaget, and Lawrence Kohlberg.

THE BIOLOGICAL PERSPECTIVE

perspective: A specific point of view in understanding or judging ideas, things, or events.

It has long been assumed that adolescence begins with the biological changes accompanying puberty. It should not be surprising, then, that earlier views of adolescence often assumed a direct, causal link between biological factors and the adolescent's psychological development. Some went a step further and largely equated adolescence with puberty. The theories of Stanley Hall and Arnold Gesell are typical of this general approach.

Hall's Critical-Stage Theory

critical-stage theory: The view that biological factors make the organism more susceptible to certain kinds of experience and learning at one stage of development than another.

The modern scientific study of adolescence began with the work of G. Stanley Hall. After receiving America's first Ph.D. in psychology, Hall pioneered the questionnaire method of study, publishing the results in his two-volume work *Adolescence* in 1904.

Hall, like many other thinkers in his day, was heavily influenced by the evolutionary thought of Charles Darwin, and he believed that all human development is controlled by genetically determined factors. As a result, there is a strong emphasis on the biological basis of adolescence, as seen in Hall's chapters on instincts, evolution, and physical growth. A special feature of Hall's thought is his recapitulation theory, that is, that the individual repeats or relives the major stages of evolution in the course of his or her own development. Accordingly, each individual progresses from the early animal-like past (in childhood) through periods of savagery (from later childhood and adolescence) before entering the more recent civilized ways of life characteristic of adulthood.

At the same time Hall held that adolescence is a critical stage of development, a period when biological factors are especially affected by one's environ-

environment: All the conditions and influences surrounding and affecting the development of individuals.

ment. At no other age were evolutionary instincts so likely to give way to cultural influence. As a result, Hall taught that teenagers should be provided with an enriched environment that would aid their development. Adolescence was seen as a critical time for a "new birth," when such distinctively human traits as love and altrusim could be developed for the first time. But because these traits are the least assured by heredity and are the most dependent on individual effort, Hall believed it was important for parents and teachers to help with their adolescents' development. In the process, adults had to draw a fine line between allowing adolescents to express their own instincts and helping to socialize them. The characteristic result of this struggle was a stressful adolescence. Much of the storm and stress was caused by the abrupt and rapid rate of physical growth; but part of it was also caused by the conflict between instinctual drives and the demands for the adolescent's intellecutal, emotional, and social growth. The turbulence at this stage of life was reflected in the adolescent's emotional development—including mood swings between energy and lethargy, enthusiasm and depressive gloom, selfishness and altruism, tenderness and cruelty, and curiosity and apathy. It was also evident at this age in a marked increase in crimes or delinquency as well as religious conversion.

Although Hall's views had a marked influence on the study of adolescence in his day, many of his ideas have not stood the test of time because of his reliance on now-discarded notions about evolution. Hall underestimated the extent to which adolescence is affected by child development. He also exaggerated the emotional turbulence of adolescence. Nevertheless, few would disagree with his view that adolescence is a critical stage of development.

Gesell's Maturational Theory

maturation: Unfolding of the biological growth process, resulting from genetic predispositions but also affected by the environment.

Arnold Gesell, like Hall, believed that maturation, or the unfolding of heredity, is the main cause of human development. Although he acknowledged the importance of the environment and individual differences, he felt these were overshadowed by biological forces. Unlike Hall, though, Gesell based his understanding on detailed empirical observations of both children and adolescents.

Gesell found that human development proceeds in an uneven manner, with an alternating sequence of stable and unstable periods. He reasoned that each new advance in development disturbed the existing equilibrium and required an awkward period of integration or disequilibrium before reaching a new state of equilibrium. As a result, development consists of a spiral pattern of stable and unstable periods of growth in which similar patterns of behavior are repeated at different levels of maturity.

Adolescence is not necessarily a stormy period in Gesell's view, but consists of alternate periods of tranquility and turbulence. Generally, the even-

STANLEY HALL'S ADOLESCENCE

As a child Hall had always heard the genitals referred to as "the dirty place." Only at school did he learn otherwise. But there he was shocked at how much the boys talked about their genitals, compared them, and played with themselves.

Hall was greatly affected by a story told him by his father, about a youth who masturbated and had sexual intercourse with lewd women; as a result, the youth caught a disease that ate his nose away until there were only two flat holes in his face and he also became an idiot. For a long time after that, Hall admitted that whenever he felt sexually aroused, he carefully examined his nose to see if it was getting flat.

Hall's fears were so intense that he tried to control his sexual excitement by applying bandages to prevent an erection while he slept. Anytime he masturbated he felt intense remorse and prayed for help. At one point he felt he was so abnormal he consulted a physician. After examining him, the doctor laughed at his fears, but also warned him of the consequences of having sexual inter-course. Hall later acknowledged that he could have been spared much needless worry had "some one told me that certain experiences while I slept were as normal for boys in their teens as are the monthly phenomena for girls."

Even in his college years Hall was plagued by a sense of inferiority, especially around girls. This was a big factor in his religious conversion during his sophomore year, which in turn intensified his struggles for sexual purity. Hall also believed that his striving for academic achievement was in part a compensation for his sense of inferiority.

Hall was relieved to discover that his adolescence had been "in no sense abnormal or even exceptional." We can well understand a statement by him in 1904 that the experiences of adolescence are "extremely transitory" and are "often totally lost to the adult consciousness."

Adapted from *Life and Confessions of a Psychologist* by G. Stanley Hall. Copyright 1923 by D. Appleton & Co., renewed 1951 by Robert G. Hall. A Hawthorn book. Reprinted by permission of E. P. Dutton a division of NAL Penguin Inc.

numbered ages are stable periods of growth, whereas the odd-numbered ages are unstable. According to Gesell, Ilg, and Ames (1956), the general characteristics of each age are:

10 year olds: remarkably stable, obedient, fond of home and same-sex friends

11 year olds: moody, negative, and quarrelsome, mostly because of beginning biological changes.

12 year olds: cooperative, positive, sociable, and more interested in peers and the opposite sex.

13 year olds: highly self-conscious, withdrawn, moody, tense, and critical, partly because of changes at puberty.

14 year olds: more self-assured, energetic, and sociable, especially with peers.

15 year olds: difficult to describe because of increasing individual differences, but tend to be independent, rebellious, and unpredictable.

16 year olds: more self-confident, friendly, well-adjusted, and oriented towards the future.

One of the dangers in using Gesell's approach is failing to distinguish between the group average (as described by Gesell) and the individual adolescent. Even though Gesell readily acknowledged that adolescents may deviate widely from the average without being abnormal, it is tempting for parents and teachers to think "something is wrong" with an adolescent because he or she is different. A major criticism of Gesell's approach is that chronological age is not necessarily a sound index to one's actual development, for example, emotional and social growth. Nor does Gesell allow for the fact that girls are generally about two years ahead of boys in terms of their growth. Gesell also fails to take into consideration the effects of early and late maturation, as well as the effects of home, school, peers, and previous experience. In short, he does not adequately allow for the importance of individual differences and environment at adolescence. Yet Gesell did pioneer in the scientific study of adolescence through the use of detailed empirical observations of behavior.

Puberty and Adolescent Development

Hall's and Gesell's views reflect a persistent bias in the understanding of adolescence, namely that adolescent development is largely determined by puberty—the process of glandular and bodily changes culminating in physical and sexual maturity. By assuming that the adolescent's psychological development results from the biological changes of puberty, much of adolescent behavior is assumed to be inevitable and resistant to intervention. Fortunately, this assumption is being increasingly challenged. Anne Petersen (1980) has suggested that propositions about the relationship between puberty and psychological development can be grouped into two broad categories, consisting of (1) the direct-effect models and (2) the mediated-effect models. Theorists such as Hall and Gesell who use the direct-effect approach assume a direct causal link between the biological changes of puberty and the adolescent's psychological development. In contrast, more recent theorists who adopt the mediated-effect approach recognize that the impact of puberty on the adolescent's psychological and social development is mediated by other variables. These influences consist of a variety of internalized psychological factors (such as body image) and a wide range of external factors (such as adolescent socialization practices).

It may be helpful to keep this distinction in mind as we consider the various theories of adolescence throughout this chapter and the rest of the book. For instance, in the next section you'll note that much of the early psychoanalytical thought on adolescence relied heavily on the physical changes of puberty, especially the adolescent's adaptation to the intensified sex drive and the sexual conflicts which resulted. However, more recent psychoanalytical views have reflected a greater role of the psychological and social influences in adolescence, thus moving more toward the mediated-effect model. At the same time, the social-learning theory of Albert Bandura and the cultural theory of Margaret Mead—explained in a subsequent section of this chapter—are clearly based on a mediated-effect model, in which the experience of adolescence is heavily influenced by one's social and cultural environment.

One further note concerns the increasing recognition that puberty itself is a complex process. In the past, theorists have sometimes emphasized one aspect of adolescence, such as sexual development, while neglecting others. However, it is increasingly apparent that puberty has many aspects, including complex neurological and chemical changes in the brain, hormonal changes, changes in size, height, body mass, and shape, as well as changes in the reproductive organs. Furthermore, each of these changes interacts with the others. Such distinctions are essential in that they provide functionally different forms of stimuli to the adolescent's psychological and social development (Petersen, 1980).

THE PSYCHOANALYTICAL PERSPECTIVE

psychoanalysis: The theory of psychosexual development and therapy developed by Sigmund Freud in which human behavior is based on the dynamic interaction of unconscious desires, conflicts, anxiety, and defenses.

Sigmund Freud, the founder of psychoanalysis, was influenced by many of the biological and evolutionary ideas popular in his day. But as a clinician, Freud came to regard childhood as the most formative period of human development and as the source of adult neuroses. His daughter, Anna Freud, applied the psychoanalytical concepts to adolescence in greater detail. Although she held that adolescence is secondary to childhood, she felt that the upheavals of adolescence do not simply mirror the first few years of life but may place a decisive stamp of their own upon adult personality. The following section summarizes ideas from both Sigmund and Anna Freud.

The Freudian View of Development

According to Sigmund Freud (1964), the driving forces of personality come from the *id,* the unconscious reservoir of psychic energy derived from biological instincts; these instincts are primarily sexual and aggressive in nature. Early in the child's life, part of the id is modified through parental control to help inhibit the unruly forces of the unconscious. This part is known as the *superego,* or conscience. It serves to socialize the child before the age of rational, conscious control. Another part of personality that develops more gradually out of the id through the effects of life-long socialization is the *ego.* The ego is

Sigmund Freud, the founder of psychoanalysis.
(UPI/Bettman Newsphotos)

the center of self-consciousness and rational control which serves as the manager of the personality. As such, the ego strives to achieve psychic harmony within the personality as a whole by integrating the conflicting demands of the id, superego, and society. Freud once compared the ego to a rider on horseback, capable of guiding but not entirely controlling the underlying psychic energy from the unconscious.

Freud held that the dynamics of personality depend largely on how the sexual instinct and the ego and superego have been shaped during the formative years of childhood. Expression and control of the sexual instinct were especially crucial. Too little or too much control of the sex drive led to neurotic conflicts, if not in childhood then in adulthood. Personality development consisted of the oral, anal, and phallic stages in the first 6 years of life, followed by a quieter latency period (6 to 11), and then a genital stage (12 to 18).

phallic stage: The third stage of psychosexual development in which the child's genital area becomes the primary source of physical pleasure.

Of special importance is the phallic stage (3 to 6 years) in which children tend to experience the *Oedipus complex*. This is named for the Greek king who unwittingly murdered his father and married his mother without realizing who she was. Similarly, preschool-aged boys and girls entertain unconscious wishes and fantasies about marrying their opposite-sex parents. The boy becomes jealous of the attention his father receives from his mother, and soon fears his father. He also becomes fearful of his own impulses toward his

Oedipus complex: The unconscious tendency of a young boy to be sexually attracted to his mother and hostile toward his father.

Electra complex: The unconscious tendency of a daughter to be sexually attracted to her father and hostile toward her mother.

mother and father. Because this conflict is intolerable to the small child, who usually loves and desires both parents, it tends to be resolved spontaneously and unconsciously by the boy's renouncing his mother as an object of special attention and identifying more strongly with his father. The girl experiences a similar sequence of events in the *Electra complex,* eventually identifying more strongly with her mother. However, when an opposite-sex parent is unduly seductive or a same-sex parent is overly critical, a boy or girl may become fixated at this stage. Some results are the Don Juan-type male who "loves 'em and leaves 'em" because of his fear of true intimacy, or the seductive female who enjoys attracting men but cannot be sexually fulfilled with them.

Adolescent Sexual Conflicts

fixation: Continuing a kind of gratification after one has passed through the psychosexual stage at which this was appropriate.

Anna Freud (1958, 1969) held that adolescence is a turbulent time mostly because of the sexual conflicts brought on by sexual maturation, resulting in both quantitative and qualitative changes in the sex drive.

In the preadolescent years, the intensified sex drive brings about a quantitative increase in all impulse activity that has characterized childhood development. Accordingly, adolescents tend to become hungrier, more aggressive and cruel, more inquisitive, and more egocentric. These increased impulses, in turn, weaken the childhood defenses erected against them, thereby reactivating the psychosexual conflicts of childhood. Thus, adolescents may alternate between clinging dependence and exaggerated independence, taking pleasure in being messy while being preoccupied with cleanliness and orderliness, or exhibiting extreme curiosity about sex along with prudish condemnation of it. The reactivation of these conflicts helps to explain the ambivalent behavior so characteristic of adolescents.

The general increase in impulse activity is followed shortly by a qualitative change in the sex drive—namely, the emergence of genital sexual impulses that accompany the maturation of reproductive organs and secondary sex characteristics. But here the Freuds differ in their interpretations of adolescence. Sigmund Freud held that adolescents' handling of their sexual conflicts is largely determined by the way they have resolved their Oedipal conflicts earlier in life. For example, an adolescent with an unresolved Oedipal fixation might be attracted to an older, domineering, or seductive person reminiscent of the relationship with his or her opposite-sex parent in childhood. However, Anna Freud contends that the adolescent's sexual struggles and the strategies for coping with them are also somewhat different from the child's. For one thing, the adolescent has a more developed personality and ego than the child. In addition, the qualities of adolescent thought and emotion give them more sophisticated defenses for coping with their sexual conflicts. Furthermore, adolescents' sexual struggles must be resolved primarily in relation to their peers rather than their parents, though the latter's emotional involvement is still an important factor.

ego defense mecha- nisms (defenses): Automatic, unconscious behavior that protects us from the awareness of anxiety.

One of the biggest differences between the sexual struggles of the child and those of the adolescent is the heightened inner turmoil of the adolescent. Whereas the sexual conflicts of early childhood had been resolved largely because of such external influences as the disapproving parent, the adolescent experiences an additional inner dimension—namely, the conflict between the ego and superego in control of sexual impulses. Although adolescents may be well aware of parental disapproval in yielding to temptation, much of their conflict comes from their own feelings of guilt or concern about

repression: Unconscious process of excluding unacceptable ideas or feelings from consciousness.

the loss of self-esteem. They must constantly choose a course of action between the two extremes of utter self-denial and complete self-indulgence of their sexual needs. Too much control will result in a rigid and restricted adolescent who remains cut off from his or her own sexuality; but too little control will result in impulsiveness, a low tolerance for frustration, a preoccupation with self-indulgence, and guilt.

Coping with Adolescent Sexuality

sublimation: Unconsciously channeling socially unacceptable urges into acceptable behaviors.

Both Sigmund and Anna Freud held that the urges and conflicts of this age arouse so much anxiety that they inevitably evoke certain ego defenses within the individual. For the most part these defenses operate automatically and unconsciously, though they may be influenced by learning. Because civilized living requires a certain degree of *repression,* in which impulses are blocked unconsciously, adolescents, like adults, are not fully aware of their sexual urges and conflicts in the first place. But the intensity of the sex drive requires

displacement: Unconsciously redirecting threatening ideas or impulses onto less threatening objects.

other defenses as well. A major defense is *sublimation,* which is the unconscious redirection of sexual energy towards more socially acceptable goals. Common examples would be the increased curiosity, creativity, and pursuit of intellectual activity at adolescence. Another defensive strategy is *displacement,* in which sexual impulses are displaced onto other, emotionally safer objects or persons. A girl's preoccupation with physical beauty or clothes or a boy's interest in sports cars or motorcycles are examples of displacement.

identification: The unconscious process of attributing characteristics to ourselves which we perceive in others we admire.

New *identification,* with parental substitutes, such as a teacher, coach, or peers, also help adolescents cope with their sexual impulses with less anxiety. The fact that such adolescent crushes are relatively short-lived probably reflects their defensive function rather than the formation of genuine object relations with others, which usually require more maturity.

intellectualization: Unconsciously reducing anxiety by analyzing threatening issues in an emotionally detached way.

Anna Freud (1958) held that two defense mechanisms especially characteristic of adolescence are *intellectualization* and *asceticism.* In intellectualization adolescents use their newly developed powers of abstract thinking to discuss sex in a relatively impersonal manner to create a certain amount of emotional distance between ideas and impulses. An example would be the typical adolescent debate on such subjects as living together versus marriage, or homosexuality versus heterosexuality. The ideas in these discussions often represent opposite sides of an adolescent's inner conflicts. In asceticism, an adolescent unconsciously denies his or her sexual urges and

---

ADULT AMNESIA OF ADOLESCENCE

Anna Freud once observed that her adult patients suffered from an amnesia, or loss of memory, of their adolescence; she hastened to add that this did not resemble in extent or depth the amnesia of their early childhood. Memories of adolescence are ordinarily retained in consciousness and told to the psychotherapist without apparent difficulty. Adults may readily recall their sexual awakening—either their attempts at masturbation or their first moves toward sexual intercourse. However, such memories are usually no more than the bare facts, divorced from the feelings that accompanied them at the time.

What we fail to recover, as a result, is the atmosphere in which the adolescent lives, his anxieties, the height of elation or depth of despair, the quickly rising enthusiasms,

the utter hopelessness, the burning—or at times sterile—intellectual and philosophical preoccupations, the yearning for freedom, the sense of loneliness, the feeling of oppression by the parents, the impotent rages or active hates directed against the adult world, the erotic crushes—whether homosexually or heterosexually directed—the suicidal fantasies, etc.

It is these transitory and elusive mood swings of adolescence that are easily forgotten and often lost to the adult consciousness. Perhaps it is best this way. Otherwise, our memories of adolescence might be more painful than they are.

Anna Freud, "Adolescence," in R. S. Eissler et al., eds. *The Psychoanalytic Study of the Child*, vol. 13 (New York: International Universities Press, 1958), pp. 259–260.

---

asceticism: The unconscious mechanism by which adolescents may deny their sexual urges and avoid pleasurable associations with sex.

avoids pleasurable associations with sex. Unlike other defense mechanisms, which often express a compromise between the urge and its constraint, the ascetic response aims for a total control. An example would be the adolescent who joins a religious group that prohibits premarital sex as well as other related pleasures. Yet, because of the defensive character of asceticism, periods of restraint are often broken by periods of abandonment and sexual indulgence.

Newer Developments

A major criticism of the classical psychoanalytical view of adolescence is its preoccupation with the unconscious strivings of the id and the defensive strategies of the ego in coping with the sexual conflicts of this period. As a result, many of those who followed Freud have given greater emphasis to the functioning of the ego and positive relationships with the environment. Erikson's psychosocial account of adolescence, which will be dealt with in a subsequent chapter, is sometimes labeled *ego-psychology*. The more recent versions of analytical thought are often classified in terms of *self psychology* and *objects-relation theory*.

Heinz Kohut (1984) holds that adolescence is set in motion by the changes in the self accompanying maturation and independence of the parents, rather than puberty itself. Basically, the adolescent's growing disillusionment with his or her parents leads to changes in the ego-ideal—the part of the superego

ego psychology: The psychoanalytical approach that emphasizes the autonomous, adaptive functions of the ego in relation to the environment, rather than as something which developed solely to avoid psychic conflict.

self psychology: In psychoanalytical theory, the approach that emphasizes the origin, transformation, and organization functions of the self in personality development.

object-relations theory: In the psychoanalytical view, the idea that early self/other patterns decisively influence later interpersonal relationships.

which consists of positive ideals and values. The resulting search for new ideals and values leads to an extensive reorganization of the self, the central task of adolescence. Peer groups encourage this process through endorsing new values and also provide support for the development of self. At the same time, Kohut maintains that autonomy consists in a transformation rather than a severance of parental ties. Thus, healthy adolescents and adults are both independent and attached to the parents, but in a mature way. As a result, they are neither compulsively dependent on others, nor fearful of forming close relationships with them.

Peter Blos (1979), one of the most influential writers in the psychodynamic tradition, has given us a detailed account of adolescence as a temporal process. Blos explains that the passage of adolescence occurs throughout various phases, extending from the latency period to young adulthood. Blos begins his discussion of adolescence with a consideration of latency because he feels that the relative stability of this period is necessary for the turbulence which follows. That is, the increased control of the instinctual life by the ego and superego so characteristic of latency helps adolescents cope with the sexual struggles brought on by puberty. During early adolescence and midadolescence—or adolescence proper—the central issue is the problem of object relations. In early adolescence the first sign of psychosexual growth can be seen in the adolescent's preference for close relationships with same-sex friends. Entry into adolescence is marked by the transition to heterosexual object-finding and preference for opposite-sex friends. At the same time, many of these relationships involve people who bear a marked resemblance to the adolescent's parents in the sense of physical or mental similarity.

There is a characteristic increase in narcissism during adolescence, seen in adolescents' touchiness, self-absorption, and inflated ideas about themselves. Also, periods of regression are not only normal but an important part of growing up during adolescence. Reflecting on the lingering dependencies, anxieties, and needs in one's makeup, but now with a more mature ego and cognitive ability, provides the adolescent with an opportunity to resolve inappropriate ways of thinking, feeling, and behaving that may impede his or her development. By late adolescence, most individuals have achieved greater self-acceptance and a firmer sexual identity, and thus a new stability in their psychosexual development. The final phase of postadolescence, or young adulthood, is marked by the establishment of mature object relations, as seen in adult friendships, love relationships, marriage, and parenthood.

THE PSYCHOSOCIAL-CULTURAL PERSPECTIVE

An increasing number of professionals who study or work with adolescents now recognize that adolescence is considerably more complex than puberty. As a result, they put great emphasis on the mediating effects of the adolescent's learning and environment. Some, like Erikson, Piaget, and Kohlberg—whose views will be discussed in subsequent chapters—and Havighurst work within a developmental perspective, stressing the adolescent's readiness for mastering certain developmental tasks. Others, like Bandura, adopt a social-learning view, putting even greater emphasis on the role of learning and the environment. Still others, like Margaret Mead, point out how the experience of adolescence is largely shaped by patterns of socialization that vary from one culture to another.

Havighurst's Developmental Theory

Robert Havighurst, like Erikson, holds that personality develops through a series of stages with appropriate developmental tasks to be learned at each stage. The readiness for growth is set by biological *maturation,* or what Havighurst calls "teachable moments." The developmental tasks consist of certain skills, functions, or attitudes essential for adult competency.

The eight developmental tasks for adolescence (Havighurst, 1972) are:

1. Accepting one's body and using it effectively.

2. Achieving new and more mature relations with age mates of both sexes.

3. Achieving a masculine or feminine social role.

4. Achieving emotional independence of parents and other adults.

5. Preparing for an occupation and economic career.

6. Preparing for marriage and family life.

7. Desiring and achieving socially responsible behavior.

8. Acquiring a set of values as a guide to behavior.

Because these developmental tasks are significantly shaped by culture, they vary somewhat from one society to another, as well as among different social classes within the United States. Furthermore, the content of these tasks changes with the time in which one grows up. For example, acquiring a set of values continues to be an important developmental task at adolescence. Yet Havighurst contends that the specific values of youth are changing. Whereas youth in the 1950s were oriented towards the future and strove for an identity based on occupational and marriage roles, youth in the 1970s and 1980s are more oriented to the present and place a greater value on expressive activities. That is, they may value a meaningful relationship with someone of the opposite sex more than whether or not this leads to marriage. Havighurst feels that this shift from instrumental to expressive values may result in youth who are less sure of their identities than in previous generations.

Although Havighurst's developmental tasks have been criticized as being somewhat mechanical and age related, he has pointed out that the emphasis should be on the adolescent's active pursuit rather than on the final achievement of these tasks, because the latter occurs well beyond adolescence.

Bandura's Social-Learning Theory

social learning theory: The view that learning and the environment play a primary role in human behavior.

socialization: The process by which people acquire the attitudes, behaviors, and values that define themselves both as individuals and members of groups.

Albert Bandura, like other proponents of social-learning theory, holds that adolescent development is shaped primarily by the individual's reciprocal interaction with his or her environment. That is, individuals are not simply driven by inner urges or shaped by environmental factors; they also act to shape and control these other forces. As a result, human development is viewed as a life-long process of socialization in which biological drives are decisively shaped by environmental experiences—mediated by cognitive processes such as *modeling.*

The process of *modeling* helps to explain the acquisition of language, habits, attitudes, and values. Bandura and his colleagues paid special attention to the importance of modeling in aggressive behavior. In his work on adolescent aggression, Bandura (1973) pointed out that fathers of overly

Albert Bandura, a leading social learning theorist.
(Albert Bandura, Psychology Dept., Stanford University)

modeling: The process by which individuals observe and emulate the behavior of admired others.

aggressive boys were more rejecting of their sons, so that their sons became less dependent on them and spent less time with them. Parents of aggressive boys were also more likely to use harsh, physical punishment, so that their sons were more apt to imitate their parent's aggressive behavior than their verbal warnings to the contrary. Parents of aggressive boys also tended to encourage their sons in aggressive behavior, such as standing up for their rights or leading with their fists. In contrast, parents of the better adjusted boys showed more accepting attitudes towards their sons, explained their discipline and demands, and were less likely to use physical punishment. As a result, these boys tended to develop more inner controls of aggression, such as an adequate conscience and sense of guilt.

The consequences of a behavior also affect the extent to which it is followed. That is, when aggressive behavior is seen as rewarded, or *reinforced*, rather than punished, young persons are more likely to imitate it. In this connection, Bandura stressed the importance of vicarious reinforcement that comes from watching socially approved aggression—for instance, TV crime shows or violent sporting events. Spectators who see the advantages of aggression are more likely to engage in aggressive behavior themselves. Then, as they discover the personal payoff value of aggression in their own lives, they may administer their own reinforcement; such self-reinforcement is as effective as external reinforcement in influencing behavior. Accordingly, much of the aggressive behavior among adolescents and youth, whether in their driving behavior or delinquency, largely reflects the behavior and values of American society as a whole.

Adolescence is a time when individuals are exposed to different models and environmental influences—for example, folk heroes, hair styles, clothing, and changing social values. Therefore, it is not necessarily the stormy time depicted by the critical-stage theorists. In fact, Bandura maintains that the storm-and-stress view characterizes only the deviant minority. Most adolescents tend to develop in a positive direction, with a gradual increase in personal freedom accompanied by a corresponding decrease in parental guidance. When there is rebelliousness, it is more likely the result of changes in the adolescent's family, school, or social life.

The importance of models and reinforcement in Bandura's view reminds us that parents and teachers probably influence adolescents more by personal example and values than by spoken words.

Mead's Cultural Theory

anthropologist: A person who specializes in the study of the physical and cultural characteristics and institutions of humans.

Margaret Mead began her career as an anthropologist by studying adolescent girls in Samoa, in the South Pacific islands. There, Mead observed that adolescent behavior largely reflects the society in which one grows up, leading her to conclude that the experience of adolescence varies significantly

Margaret Mead, an anthropologist who studied adolescents.

culture: The pattern of attitudes, beliefs, values, and behaviors that are shared and passed on from one generation to another in a particular group of people.

from one culture to another. Her findings are published in two books *Coming of Age in Samoa* (1950) and *Growing up in New Guinea* (1953).

Mead found that adolescence among Samoan girls was an uneventful time of life, without the emotional turbulence associated with puberty in Western societies. Adolescent girls differed from their younger sisters mostly in one respect—namely, their more mature bodies. In growing up, Samoan girls encountered no taboo and separation at the onset of menstruation because they had learned the facts earlier. No new way of life was suddenly thrust upon them because they had been expected to share in the family work and obey their parents throughout the transition from childhood to adulthood. In fact, girls were taught dominance early in life, and 6 or 7-year-old girls dominated their younger sisters just as they were dominated by their older sisters. The older the girl became, the less she was disciplined by others and the more she disciplined others. Thus, Samoan girls passed painlessly from childhood to womanhood, so that adolescence was one of the pleasantest times in their lives.

Mead explained that the tranquil adolescence of Samoan girls was due to the cultural patterns in that country. For one thing, the young passed through a succession of well-marked periods of responsibility so that they always knew what was expected of them. There was also a more casual way of life in Samoa than in Western cultures. Emotional bonds were looser, both in the family and community. There was also less conflict and competitiveness in

Samoan culture. Nor were adolescents expected to make personal choices to establish their vocation or place in society. Mead pointed out that such cultural differences have their disadvantages too, especially the absence of deep feelings and little appreciation of individual differences.

Although some of these observations were later disputed by Derek Freeman (1983), who found the Samoans to be more aggressive and sexually inhibited than Mead did, her major findings remain intact. Lowell Holmes (1986), who did a restudy of Tau, the same village where Mead had worked earlier, concluded that Margaret Mead was essentially correct in her depiction of coming of age in Samoa.

Just before her death, Mead wrote about some of the cultural influences which are shaping adolescents' lives in American society today, especially the changes in family life (Mead, 1978–1979). First, a large proportion of youth grow up in cities and suburbs and are isolated from their grandparents and the extended family. In many homes both parents work, providing less supervision for adolescents during the day. Families are also more mobile than in the past, causing adolescents to be uprooted from their friends and schools. Young people in the inner cities face additional problems. Many of them grow up in single-parent families that are dependent on financial support from the government, and suffer from inferior schools and high-crime neighborhoods. Mead says that today's adolescents also suffer from the increased restlessness in society. With less guidance from their elders during the early years of marriage, young people tend to divorce at the first sign of trouble and then quickly remarry. Yet, the cumulative effect of the high divorce rate, along with the rise in single-parent families and remarried couples with children, compounds the difficulties of growing up today.

Mead points out that television has become a major factor in the changing American way of life. Children and adolescents with busy parents spend large amounts of time watching TV, often in place of interacting with other adolescents in the neighborhood. Television also exposes adolescents to different ways of life than their own. Poor, inner-city youth get a glimpse of what affluent, suburban life is like, thereby experiencing even greater dissatisfaction with their way of life. Margaret Mead suggests that television can become a more positive influence in society by providing more positive role models, portraying the problems of adolescents more realistically, and encouraging families to provide more emotional support for adolescents in a rapidly changing society.

Adolescence as a Cultural Phenomenon

The views of adolescence discussed above stress the role of social and cultural factors in adolescence. A closely related view suggests that adolescence can be explained entirely in socio-cultural terms. Ann Sieg holds that adolescence is not a necessary stage in human development, despite puberty or the age-related events associated with it. Instead, she defines adolescence solely in terms of the sociocultural process. Sieg says: "Adolescence is the period of development in human beings that begins when the individual feels that adult privileges are due him which are not being accorded him, and that ends when the full power and social status of the adult world are accorded to the individual by his society" (1975, p. 40). That is, adolescence depends primarily on the discrepancy between one's actual and desired status in society, and the attainment of adulthood on a satisfactory resolution of the same. Yet, as intriguing as this view is, it fails to take into account other sociocultural influences that affect adolescence as well as physical maturation and the adolescent's sometimes shaky readiness for exercising the responsibilities that accompany adult rights.

At the same time, such a view helps to explain why people may continue to behave in an immature manner long after they have attained physical maturity. Sieg cites the example of blacks in the United States, whose lack of adult status and power has contributed to the characteristic reaction of frustration, rebellion, irresponsibility, and crime. It's as if our society had assigned a status

---

MARGARET MEAD'S ADOLESCENCE

According to her autobiography, Margaret Mead was a lively, intense, and in some ways unconventional adolescent.

She was a precocious adolescent partly because of her unusual family. Both her parents had gone to graduate school and expected her to use her mind. They insisted on treating their children in a democratic way, so that Mead always felt treated as a person, even as a child. Her mother had a strong concern for social justice and actively participated in community affairs. As a result, Mead says that "in many ways I was brought up within my own culture two generations ahead of my time" (p. 2).

Mead's paternal grandmother exercised a decisive influence on her life. She had graduated from college, become a teacher, and later a school principal, in a day when these were exceptional accomplishments for women. Grandmother lived in the Mead home, and when frequent family moves made regular school attendance impossible, she taught her granddaughter at home. Later, Mead said

that it was the influence of her grandmother and mother as positive role models that inspired her to become both a professional person and a mother.

Margaret Mead became a superior student despite frequent changes in homes and schools. She not only made good grades, but was intellectually curious. She loved to read and write poetry; she kept a diary, and even began a novel. She was also physically active, playing vigorous games well beyond the age when girls were supposed to. Throughout her high-school years, Mead says, "I wanted to live out every experience that went with schooling, and so I made a best friend out of the most likely candidate, fell sentimentally in love with one of the boys, attached myself to a teacher, and organized as far as it was possible to do so, every kind of game, play, performance, May Day dance, Valentine party, and, together with Julian Gardy, a succession of clubs, in one of which we debated such subjects as 'Who was greater, Washington or Lincoln' " (p. 80). Yet, she felt she was

of permanent adolescence to blacks, to which the civil rights movement of the 1950s and 1960s was a corrective. The same reasoning may help to explain the youth movements in the past few decades, as well as the push for women's rights and drives for equality by various minority groups since the 1970s. You may recall from the opening chapter that social changes such as compulsory education, child-labor legislation, and the establishment of juvenile justice helped to prolong adolescence well beyond puberty. It just may be that the unrest among college-aged youth since the 1960s has been in part an attempt to reclaim a more realistic status for themselves in society. As Richard Braungart (1980) reminds us, when society fails to meet the legitimate needs of various groups and a significant number of people become aware of their common plight—and begin feeling something can be done to alleviate their dissatisfaction—social movements may appear. More specifically, as he points out, youth movements arise out of the discrepancy between individual needs and aspirations and the existing social and political conditions. However, the demographic changes of the 1960s—during which the first wave of the baby boom years surfaced as a disproportionately large number of youth—

demography: The statistical science dealing with the density, distribution, and vital statistics of populations.

"set apart" in school because of the educational advantages given by her family. As a result, she searched with a greater intensity than others, speculating at different times that she would become "a lawyer, a nun, a writer, or a minister's wife with six children" (p. 81).

Despite her academic excellence, Mead almost missed going to college because her father felt she should get married or become a nurse like other girls. She responded with "one of the few fits of feminist rage I have ever had" (p. 85). Her mother came to the rescue with a compromise choice, and persuaded her husband to send Mead to his own alma mater, De-Pauw College in Indiana. But it was a disastrous decision, because Mead's life style, with her self-designed dresses, serious books and posters, was at odds with the prevailing social atmosphere at the college. Consequently, she was rejected by the sororities and was deeply hurt. Feeling like an "alien," she transferred to Barnard College after her first year.

Life at Barnard College in New York City was an exciting and rewarding experience for Mead. Surrounded by stimulating teachers and other bright women like herself, she became active in literary and theatrical affairs. She debated and took part in student demonstrations. She also had a mischievous streak, and ran around with a group of friends known as the "Ash Can Cats," who remained her friends for life. Although at 17 she became engaged to a seminary student with similar interests, they did not marry until several years later when Mead was 22 years old.

Like many adolescents, Margaret Mead felt unsure of herself and her career at times. She had begun college as an English major, intending to become a writer. Later, she added a minor in psychology. Then, during her senior year she took a course in anthropology under Franz Boas, and acquired a serious interest in anthropology. She also began a life-long friendship with Boas's assistant, anthropologist Ruth Benedict.

Margaret Mead, *Blackberry Winter* (New York: William Morrow and Company, Inc., 1972). Used by permission.

combined with the accelerated rate of social change to make the youth movements of the 1960s different from those of earlier eras. Since then, young people have become an important source of social change, assuming a bigger role in defining their own place in society.

ADOLESCENT STORM AND STRESS IN PERSPECTIVE

As you may recall from the beginning of the chapter, one of the most persistent ideas about adolescence is that it is a special time of storm and stress. The phrase "storm and stress" is a translation of the German words "sturm und drang," taken from the works of the eighteenth-century writers Goethe and Schiller. In the novel *The Sorrows of Young Werther*, Goethe tells about a tormented young man who is beset by many desires and dreams, not unlike the author himself. But young Werther is unable to accomplish anything. To complicate matters, the young man falls hopelessly in love with a married woman. When he is rejected, he kills himself. Thousands of teenagers were influenced by this book at the time. Some of them, reminded of their own youthful despair, committed suicide clutching the book to their breasts. Ever since, some observers have insisted that the period of adolescence is a special time of turbulence.

Reevaluating Adolescent Storm and Stress

The major support for the storm-and-stress hypothesis has been psychoanalytical theory, especially the highly influential views of Anna Freud. Given the Freudian conflict theory of personality, it comes as no surprise that proponents of this view have seen adolescence as marked by extreme inner turbulence. Furthermore, much of the research on adolescents has been based on explicitly psychodynamic assumptions and has focused on disturbed youth to the neglect of normal youngsters. Yet, as Joseph Adelson and Margery Doehrman (1980) point out, studies of normal adolescents usually do not support the storm-and-stress bias. Instead, such studies have shown that the view of adolescence as a time of emotional upheaval has been based mostly on the articulate upper-middle-class adolescent. The average adolescent does not experience such turmoil, partly because of premature identity formation, ego constriction, and a general unwillingness to take risks. As a result, the stereotype of the stormy adolescent may continue partly because most research is done on upper-middle-class and lower-class adolescents rather than on working-class and middle-class youth, on the emotionally disturbed rather than on the well-adjusted, on the politically radical or alienated rather than on the conventional, and on males rather than on females. A survey of studies that included a wide variety of adolescents tends to cast doubt on the stereotype of the stormy adolescent. As Joseph Adelson says: "Taken as a whole, adolescents are *not* in turmoil, *not* deeply disturbed, *not* at the mercy of their impulses, *not* resistant to parental values, *not* politically active, and *not* rebellious" (1979, p. 37).

Patterns of Adolescent Growth

If most adolescents are not as beset by emotional upheaval as popularly thought, how may we best characterize adolescent growth? A partial answer comes from a follow-up study by the Offers (Offer & Offer, 1975). They found that 80 percent of their subjects demonstrated one of three types of growth: continuous growth (23 percent), surgent growth (35 percent), or tumultuous growth (22 percent). Although the remaining 20 percent of the subjects were not easily classified, their growth was more like the continuous or surgent growth patterns. Adolescents with a continuous growth pattern showed little evidence of an identity crisis or significant emotional turmoil. Instead, they had a stable ego development, adequate coping skills, and were reasonably well adjusted. Those with a surgent pattern—that is, characterized by a sudden or abrupt onset of growth—were also reasonably well adjusted, but they had greater difficulty coping with the changes of adolescence. They were especially prone to projecting their anger on others or regressing in behavior before consolidating their gains in growth. One reason for this may be the greater parental difficulties evident in this group; there were greater value conflicts between mothers and fathers in this group, and mothers had more trouble letting go of their adolescents. Only those with the tumultuous growth pattern showed the degree of emotional turmoil associated with the stormy adolescent. At least one-third of the adolescents in this group had received some counseling or therapy. Yet their problems were not wholly their own. For example, parents of this group were less sure of their values, and therefore were less apt models. Also, many of these parents expressed difficulty in separating from their adolescents.

It appears that adolescents who experience greater emotional stress in growing up may do so because of home problems rather than because of the developmental process itself. The study as a whole reminds us that there are several typical patterns of adolescent growth, and that the turbulent pattern is the exception rather than the rule.

The Impact of Social Change

empirical: Based on observation and experimental data.

Empirical studies are also confirming that the experience of adolescence is significantly affected by social change; thus, adolescents in one era are shaped by different forces than those in other eras. For instance, a longitudinal study of men 17 years of age and older showed that the later-born cohorts were lower in restraint and higher in assertiveness than those born in an earlier era, reflecting changing social patterns and values (Douglas & Arenberg, 1978).

social change: Changes in the structure of society, its institutions, and social patterns.

longitudinal study: A study in which researchers follow the same group of subjects over a period of time and make repeated observations of them at selected intervals of time.

cohort: A group composed of people born in the same year.

The impact of social change on adolescents is also evident in a study by Daniel Offer, Eric Ostrov, and Kenneth Howard (1981), which compared adolescents' self-perceptions in the 1960s, the 1970s, and in 1980. The vast majority of adolescents held positive self-images in all periods of the study, again refuting the stereotype of the stormy adolescent. However, the authors concluded that over an 18-year period, the self-perceptions of American adolescents have become decidedly less positive. Compared to adolescents in the 1960s, those in the late 1970s and 1980 were more emotional and less self-controlled. They were more sensitive and easily hurt, especially by criticism, even when justified. They also had less self-confidence. While they generally felt at ease in small groups, they were more fearful of meeting new people and of making friends in larger settings. They were also less positive about their families. Although sexual attitudes generally remained the same, more of the current adolescents felt "sexually behind"; 1 out of 5 felt this way in the recent periods, versus only 1 out of 10 during the 1960s.

The finding that most adolescents in each era show little evidence of the emotional upheaval implied in the psychoanalytic view does not mean adolescence is a tranquil experience. Instead, it implies that the characteristic storm and stress of adolescence has been exaggerated. We have also seen that the experience of adolescence is vitally affected by social and cultural factors, so that particular adolescents may become unduly rebellious more due to their personalities, families, or peers than to the growth process itself. Furthermore, the varied findings of empirical studies remind us of the importance of individual differences. Few, if any, adolescents conform to a given theory or group average. Both theories and empirical studies serve as a means to the end of more fully understanding real-life adolescents themselves.

SUMMARY

1. We began the chapter by citing the need for theories of adolescence which can serve as general explanatory statements for observed behavior among adolescents.

The Biological Perspective

2. The modern scientific study of adolescence began with the work of Stanley Hall, who held that adolescence is a critical stage of development during which the dominant biological factors are more readily modified by one's environment.

3. Gesell also regarded the biological growth process as the main cause of adolescent development and observed that it proceeds in an uneven manner with an alternating sequence of stable and unstable periods of growth.

4. Although Hall and Gesell's views assume a direct, causal link between puberty and the adolescent's psychological development, most current views of adolescence are based on a mediated-effect approach, which recognizes that the impact of puberty on the adolescent's psychological development is mediated by other variables.

The Psychoanalytical Perspective

5. Sigmund Freud viewed individual development as a sequence of psychosexual stages, with the personality dynamics at later stages being heavily dependent upon one's earlier development.

6. Anna Freud maintained that adolescence is a special period of turbulence because of the sexual conflicts brought on by puberty.

7. The intensified sex drive and resulting sexual conflicts arouse considerable anxiety, which in turn evokes a variety of ego defense mechanisms such as intellectualization and asceticism for coping with the stress of adolescence.

8. Recent psychoanalytic theory emphasizes the functions of the ego and more deliberate problem-solving strategies in relation to the adolescent's environment.

The Psychosocial-Cultural Perspective

9. Havighurst emphasized the varied developmental tasks of adolescence, which combine the individual's readiness for learning with sociocultural factors.

10. Bandura has gone a step further and holds that adolescence consists of a relatively continuous process of socialization in which cognitive and environmental factors outweigh the importance of biological drives.

11. Based on her studies of adolescents in the South Pacific islands, Margaret Mead believed that the experience of adolescence largely reflects the society in which one grows up, and thus varies somewhat from one culture to another.

12. Mead also pointed out that the way adolescents are growing up today is vitally affected by the changes in American family life and society.

13. A more extreme view regards adolescence as a wholly cultural process which begins in the perceived discrepancy between one's actual and desired status in society, with the attainment of adulthood dependent on a satisfactory resolution of the same.

Adolescent Storm and Stress in Perspective

14. Despite the persistent notion of adolescent storm and stress, studies of a wide variety of adolescents cast doubt on the stereotype of the stormy adolescent.

15. It appears that there are several typical patterns of adolescent development, with the turbulent pattern being the exception rather than the rule.

16. Empirical studies have also shown that the experience of adolescence is significantly affected by social change; thus, adolescents growing up in the 1980s will be somewhat unlike those in earlier eras.

17. Finally, it has been pointed out that few, if any, adolescents conform to a given theory or group average: Both theories and empirical studies serve as a means to the end of more fully understanding adolescents themselves.

REVIEW QUESTIONS

1. In what sense does Stanley Hall regard adolescence as a critical stage of development?

2. Explain the difference between the direct-effect and mediated-effect theories of adolescence, giving an example of each type.

3. How does Anna Freud explain adolescent storm and stress?

4. Explain the characteristic ego defense mechanisms of intellectualization and asceticism.

5. Describe Havighurst's developmental tasks of adolescence.

6. What are the most important determinants of adolescent development according to Bandura's social-learning view?

7. Identify some of the emerging sociocultural factors which are shaping adolescence in the 1980s.

8. Discuss the pros and cons of adolescence as a cultural phenomenon.

9. Describe the three characteristic patterns of adolescent development explained in the text.

10. Select one theorist who supports the storm-and-stress hypothesis and one who does not, then compare and contrast their views of adolescence.

3 Physical Development

- **PUBERTY**
  The Process of Puberty
  Hormonal Changes
  Trend toward Earlier Maturation

- **PHYSICAL CHANGES**
  Height and Weight
  Skeletal and Muscular Changes
  Physical Changes and Body Image

- **SEXUAL MATURATION**
  Girls
  Menarche
  Boys
  First Seminal Emission

- **EARLY VERSUS LATE MATURERS**
  Early Versus Late-Maturing Girls
  Early Versus Late-Maturing Boys
  Individual Differences

- **SUMMARY**

- **REVIEW QUESTIONS**

Adolescents have an intense desire not to be too different from others their age. Unfortunately, nature does not always cooperate with such a desire during puberty. Not only do girls begin to grow taller before boys their own age, but some adolescents can be nearly finished with their growth spurt before others their own sex even begin. Many adolescents don't understand what is happening to their bodies and feel awkward about being different. "Being so small I'm usually the last one chosen when we're dividing up into teams," complained a ninth-grade boy. "When I got my first period," said an eighth-grade girl, "I was so upset I didn't want to talk about it. I stayed in my room and went to bed for a couple of days." Such experiences are not only upsetting to adolescents themselves, but they also may become a source of concern to their parents, if and when they learn of their teenager's distress. Even then, providing factual information is less important than creating an attitude that adolescents can talk to their parents about such matters in trust and confidence. Parents who speak in a straightforward manner with an "I'm-glad-you-asked" tone are usually more helpful than those who laugh, tease, or use slang terms.

PUBERTY

Although *puberty* in the technical sense of the term means the attainment of sexual maturity, the word is generally used more broadly to refer to the entire process of glandular and bodily changes that accompanies sexual maturation. In both usages of the term, though, puberty is more accurately understood as a process rather than as a single marker event, and is more complex than ordinarily portrayed.

The Process of Puberty

growth spurt: The period of rapid physical growth preceding sexual maturation during adolescence, sometimes known as *pubescence.*

pubescence: See *growth spurt.*

menstruation: The part of the monthly menstrual cycle in which the bloody lining of the uterus is discharged through the vagina.

first seminal emission: The male's initial ejaculation of semen at adolescence.

masturbation: The self-manipulation of one's sex organ to produce pleasure.

Puberty is part of the unfolding growth process that begins at conception. That is, all of the components for sexual maturity are present in the developing fetus, and thus the newborn infant, but are immature. The rapid maturation of reproductive capacity at puberty is part of the overall growth process. One of the major factors distinguishing puberty from the rest of the biological life cycle is the rapid *rate* of growth along with the *magnitude* of bodily changes that accompany it. The rate of growth during infancy is even more rapid than at adolescence, but it may be less important because the infant does not experience change the same as an adolescent does. Major changes also occur in late adulthood, but they are usually slower and more gradual (Petersen & Taylor, 1980).

At some time in late childhood or prepuberty there is a rapid rise in the secretion of certain hormones, especially the sex-specific hormones, which signals the onset of puberty. This occurs at about 10 or 11 years of age in girls and at about 12 or 13 in boys. The rise in hormonal activity triggers a gradual increase in the size and secretion of the sex glands themselves, such as the ovaries in girls and the testes in boys. But because these changes are not immediately evident, the onset of puberty is often associated with the outward manifestations of puberty, such as the growth spurt in height and weight, the appearance of pubic hair, breast buds in girls, and the enlargement of the penis and testes in boys. This phase of puberty is sometimes referred to as *pubescence* (covered with hair) because of the pubic hair which appears on the pubic or genital area about the time the reproductive organs and secondary sex characteristics are reaching maturity.

Puberty in the narrower sense of the term refers to the culmination of sexual maturation or the attainment of reproductive powers. The first sign of this is the onset of *menstruation* in girls at about 12 or 13 years of age and the first seminal emission in boys (usually by *masturbation*) at about 13 or 14 years of age. However, it may take a year or more before some girls and boys are capable of sexual reproduction on a regular basis. The latter occurs as the girl's ovaries begin discharging a mature egg every 28 days or so in regular menstrual cycles and the boy's testes begin producing a greater proportion of mature sperm, thus increasing fertility in both sexes. Most adolescents are well into this culmination phase of puberty by their midteens, even though their body growth continues a few years longer.

Although the sequence of growth at puberty is similar for most adolescents, the age of onset, as well as the rate and duration of growth, varies somewhat from one adolescent to another. Futhermore, the psychological impact of puberty does not necessarily parallel these physical changes. Psychological development often lags somewhat, so that while adolescence may begin with puberty it usually extends beyond it.

Hormonal Changes

The onset and rate of growth at puberty is largely a function of hormonal changes. *Hormones* (to "stir up") are chemical substances secreted directly into the bloodstream by the endocrine glands, producing changes in the development of the body. The puberty-producing hormones are either secreted for the first time or in higher amounts at adolescence, because many of these glands have been functioning thoughout childhood. It is also important to note that the effect of hormones is less specific in humans than in animals, so that the changes of puberty characteristically result from the combined secretions of many glands (Tanner, 1978).

The hormonal changes are triggered by the hypothalamus, located at the base of the brain. The hypothalamus, which governs many aspects of behavior such as eating, drinking, and sleeping, stimulates the pituitary gland, which in turn stimulates the other glands.

hypothalamus: A small structure in the brain that regulates changes in the body's activities such as eating and sleeping.

The pituitary gland, located adjacent to the hypothalamus, secretes so many hormones that affect other glands that it is known as the "master gland." For example, the pituitary gland secretes thyroid-stimulating hormones, adrenal cortex–stimulating hormones, sex-stimulating hormones, and the human growth hormone. Except for the human growth hormone, which acts directly on certain body structures, these hormones act indirectly through other glands. As a result, the newly stimulated glands begin secreting their own hormones—for example, the male sex hormone testosterone is produced by the testes and the female sex hormones estrogen and progesterone are produced by the ovaries. These hormones, in turn, produce specific changes, such as the increase in size of the testes and penis, and hair growth in boys, and the increase in size of the ovaries, uterus, and vagina, and hair growth in girls. Actually, the sex glands play a dual role. They not only produce the sex hormones but they also produce ova (eggs) in girls and sperm in boys. The whole sequence of hormonal secretions operates on a negative feedback system. Once a specific growth function has been completed, the higher level of hormones in the bloodstream signals the hypothalamus, which then inhibits further production of a given hormone.

Although it was previously thought that females produced only female sex hormones and males produced only male sex hormones, recent research has shown that both sexes produce both sex hormones. At puberty the proportion of the respective sex hormones changes; from this point on female sex hormones dominate over male sex hormones in girls, and male sex hormones dominate over female sex hormones in boys, thus accentuating the distinctive characteristics of each sex (Money, 1980).

The combined effect of all these hormonal changes produces several types of growth: changes in height, weight, bone structure, muscles, sex glands, and reproductive organs, and the growth of the secondary sex characteristics. These changes will be discussed in greater detail in the rest of this chapter.

Trend toward Earlier Maturation

menarche: The developmental onset of the girl's first menstrual period in adolescence.

In recent decades, there has been a trend towards earlier maturation for boys and girls. It is called the *secular* (or "present") *trend* because marked evidence of it has appeared only in the last one hundred years, since about 1880. The first and most important change is that girls and boys are now beginning puberty at earlier ages. Girls now reach *menarche* at an average age of 12.8 years, which is 2 or 3 years earlier than was typical 100 years ago (Figure 3–1). Boys are also attaining their voice change at an earlier age, in contrast to the days when boys sang soprano in Bach's choir until their late teens. During the same period, there has been an upward trend in adult height and weight, with boys and girls growing taller and heavier. Adult males now reach an average height of 5 feet 9 inches, and adult females an average height of 5 feet 5 inches (Tanner, 1978).

Several explanations have been offered for this trend. One is that it is part of a long-term evolutionary trend, so that we may expect adolescents to mature physically at increasingly earlier ages; yet the evidence for this is contradictory. A more plausible explanation is that the greater social mobility of recent years has resulted in more genetically varied parents producing taller children. A dramatic example is that the first generation offspring of English and Polynesian parents grew a full 2 ½ inches taller than their parents (*New York Times,* June 13, 1976).

critical body weight: For young girls, a body weight of approximately 106 pounds that triggers the menarche.

The most widely accepted explanation for the trend toward earlier maturation is the improvement in nutrition along with better disease control which has occurred over the years. Better nutrition, in particular, has helped to lower the age at which boys and girls reach the critical body weight regarded by some, but not all, observers as a triggering event in the adolescent's growth spurt (Frisch, 1983). At the same time, it is now believed that the differences in maturation among youth in the present and past may not be as great as once thought. Instead, girls and boys may be maturing only 2 or 3 years earlier than a century ago, rather than the 6 years or so found by Tanner (Bullough, 1981).

In any event, the trend toward earlier maturation is slowing down, so that

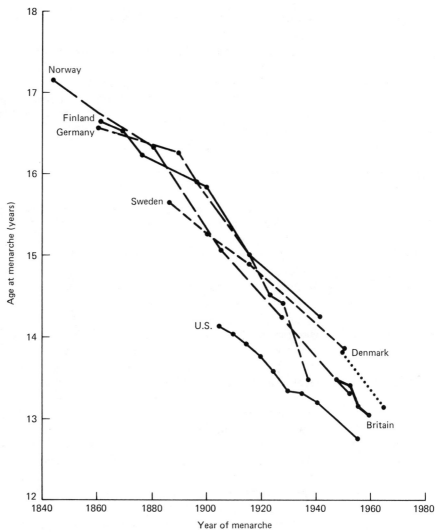

FIGURE 3–1 DECLINING AGE OF MENARCHE. The age of menarche, or the girl's first menstrual period, has declined in Western Europe and the United States so that today girls begin menstruation 2 to 3 years earlier than they did a century ago.

Source: J. M. Tanner, "Growing Up," *Scientific American* (September 1973): 43. Copyright © 1973 by *Scientific American.* All rights reserved.

we may expect smaller increases in earlier maturation during the coming years. Boys born in recent years have shown no sigificant increase in height over those born 5 to 10 years earlier. Also, mothers and daughters are now reaching their menarche at roughly the same chronological age (*New York Times,* March 28, 1976).

PHYSICAL CHANGES

Among the earliest signs that puberty has begun are dramatic changes in an adolescent's height and weight. These changes occur so quickly, for the most part, that they are referred to as the growth spurt. As we mentioned earlier, the general pattern of growth is similar for most adolescents, although the age of onset and the rate and duration of growth vary somewhat from one individual to another.

Height and Weight

The growth spurt in girls occurs, on the average, between 10 and 14 years of age. It usually begins after 10 years of age, peaks at about 11 to 12, and then declines to prepuberty rates from about 13 to 14 years. Girls continue growing at a slower rate, reaching their adult height and weight at about 15 to 16 years of age. A small proportion of girls do not reach their adult weight and height until several years later (Tanner, 1978).

The growth spurt in boys occurs, on the average, between 12 and 16 years of age. It usually begins after 12 years of age, peaks at about 13 to 14, and then declines to prepuberty rates about 15 to 16 years. Boys also continue growing at a slower rate for several more years, reaching their adult height and weight at about 17 to 18 years of age. A small proportion of boys do not reach their adult height and weight until several years later. Most growth in height at this age is due to the increase of trunk length rather than leg length, so that adolescent boys keep outgrowing their jackets after they have stopped outgrowing their pants (Tanner, 1978.)

Increases in height and weight tend to follow the same growth curve, though more so for girls than boys. That is, girls who grow taller than their friends also tend to be heavier, whereas shorter than average girls remain lighter. There is a greater variation among boys; some boys become tall and skinny, whereas others remain short and fat.

There is also a high correlation between prepuberty and adult height, so that adolescents who are shorter or taller than average tend to remain that way. Because the growth spurt at puberty is controlled by a different combination of hormonal secretions than in the preceding years, however, variations in the height patterns may occur among some adolescents (Figure 3–2). For example, a boy may be the runt of his class throughout childhood and then suddenly catch up with his peers at puberty; or a girl may be taller than her peers throughout childhood, but then grow at a slower rate until her friends catch up with her (Tanner, 1978).

Deviations in height are usually more significant than deviations in weight. The main reason is that although both may be affected by environmental factors such as malnutrition and disease, deviations in height are more likely

malnutrition: Faulty or inadequate nutrition.

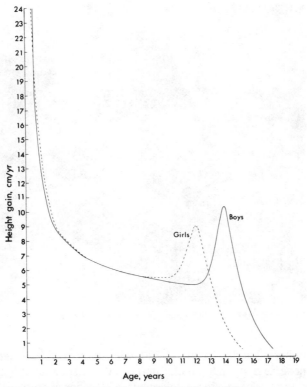

FIGURE 3–2 RATE OF GROWTH IN HEIGHT. The peak rate of growth in height occurs, on the average, about 2 years earlier in girls than in boys. In the peak years, girls gain about 9 centimeters (3.5 inches) and boys gain about 10 centimeters (4 inches) a year in height.

Source: J. M. Tanner, *Fetus into Man* (Cambridge, MA: Harvard University Press, 1978), p. 14. Used by permission.

to result from genetic or hormonal factors. In contrast, deviations in weight usually reflect other variables, such as nutrition and exercise.

Skeletal and Muscular Changes

The growth spurt is accompanied by skeletal and muscular changes because both are caused by the same hormones. Because adolescents of the same chronological age may vary as much as 5 or 6 years in their maturational age, skeletal age (determined by x-ray pictures of the hands and feet) is one of the most accurate measures of maturation. Although changes in the adolescent's bone structure and muscles follow the same pattern as the growth spurt, they follow different timetables, depending on the part of the body. For example, the head is the first to reach adult size; approximately 95 percent of the adult brain weight is attained before the growth spurt. Most changes in the head at

Girls begin their growth spurt sooner than boys. (Ken Karp)

puberty are due to growth of the jawbones and facial features, especially the nose. Growth of the hands and feet peaks soon after that of the head, leading some to observe that adolescents are "all hands and feet." As a matter of fact, arms and legs increase greatly in length during childhood and adolescence, with the arms increasing to four times their birth length and the legs to five times. Adult trunk size is achieved last, and grows longer and thicker, mostly during adolescence (Tanner, 1978).

One of the most noticeable differences between girls and boys during this period is the changing ratio of shoulders to hips. Even though girls reach the peak of their shoulder growth before their height spurt, the peak of their hip growth occurs after this. But just the opposite occurs in boys; the peak of their shoulder growth occurs after their height spurt. As a result, older adolescent girls have larger pelvic regions to facilitate childbearing, but older boys have larger shoulders and upper torsos necessary for the traditional male activities such as manual labor (Faust, 1983).

There is also a spurt in muscle growth during this period, though the timing is different for each sex. Although girls attain their peak muscle growth during the height spurt, boys reach their peak muscle growth about 1 year after the height spurt. In between, there is a short period in which girls generally have larger muscles than boys of the same age. But boys gradually develop larger muscles and greater intensity of force, so that grown boys, on the average, exert greater muscular strength than girls. Boys also develop larger hearts and lungs, greater capacity for carrying oxygen to the lungs, and greater power for neutralizing the chemical products of exercise. As a result, fully grown males are usually more adapted to the vigorous physical activity traditionally expected of them (Tanner, 1978).

But what about the effects of changing sex roles? Because muscle development benefits from strenuous exercise, the increasing participation of girls and women in sports today promises to lessen the differences in muscular strength between the sexes. Meanwhile, bear in mind that there is considerable overlap between the sexes, such that the stronger girls are as strong as or stronger than the weaker boys. Although older boys are usually stronger than girls, they also tire more quickly, mostly because boys exert greater force. Consequently, girls may have an advantage in tasks requiring endurance—that is, moderate strength over somewhat longer periods of time.

The spurt in skeletal and muscle growth is normally accompanied by a loss of body fat, especially in boys' arms and legs; girls generally have a slower growth of body fat. But as the adolescent growth spurt ends, fat accumulation again increases in both sexes, though more so in girls.

Physical Changes and Body Image

body image: The mental picture we have of our bodies, often taken to mean how we feel about our bodies.

The rapid physical changes at puberty vitally affect the way adolescents feel about themselves, especially their *body image.* Not surprisingly, when adolescents are asked what they most like or dislike about themselves, they mention physical characteristics far more frequently than intellectual or social ones. The most liked or disliked features tend to be those of the face and head, general appearance, size, and weight, and to a lesser extent body build. Generally, adolescents' feelings about their bodies, especially their general appearance, become more favorable as they move from junior high school to college (Jersild, 1978). Table 3–1 shows the results of a study in which college students were asked to rate the characteristics most important for their physical attractiveness.

early maturers: Adolescents who reach their physical and sexual maturity earlier than average for their sex.

Girls tend to express greater dissatisfaction with their bodies, mostly because physical attractiveness is more important for females in our culture. For example, in one study, college students were asked to rate the attractiveness of 24 body parts and then to rate these as effective or ineffective. The results showed that in girls the attractiveness rating was more closely related to the self-concept, whereas in boys the effectiveness rating was more related to the self-concept. This suggests that the meaning of one's body image differs somewhat for each sex. Girls are generally more concerned with the social appeal of their appearance. But boys tend to emphasize physical competence or what they can do with their bodies as a means of influencing their environment (Lerner, Orlos, & Knapp, 1976).

The relationship between maturation and body image appears to be more complex for girls than for boys, as shown in a longitudinal study by Blyth,

**Table 3—1   THE IMPORTANCE OF SELECTED BODY CHARACTERISTICS
FOR PHYSICAL ATTRACTIVENESS, BASED ON
THE SELF-RATINGS OF COLLEGE STUDENTS**

Body Characteristics	Female's Own Importance*	Male's Own Importance*
General appearance	1.3	1.5
Face	1.4	1.5
Facial complexion	1.6	1.8
Distribution of weight	1.7	2.0
Body build	1.7	1.9
Teeth	1.9	2.0
Eyes	1.9	2.4
Shape of legs	2.2	2.8
Hips	2.2	2.8
Hair texture	2.3	2.3
Waist	2.3	2.4
Chest	2.4	2.6
Nose	2.4	2.4
Mouth	2.4	2.4
Profile	2.5	2.3
Thighs	2.5	2.9
Height	2.9	2.7
Chin	3.1	2.8
Arms	3.1	3.0
Hair color	3.2	3.2
Neck	3.2	2.8
Width of shoulders	3.4	2.9
Ears	3.9	3.5
Ankles	4.1	4.2

*Scores range between 1 (very important) and 5 (very unimportant).

Source: Adapted from R. M. Lerner and S. A. Karabenick, "Physical Attractiveness, Body Attitudes, and Self-Concept in Late Adolescents," *Journal of Youth and Adolescence, 3* (1974):307–316. Used by permission.

late maturers: Adolescents who reach their physical and sexual maturity later than average for their sex.

Bulcroft, and Simmons (1981). In the sixth grade, early maturing girls reported greater satisfaction with their figures than the late-maturing girls. But by the tenth grade a different pattern had emerged. At this stage, when most of the girls had achieved physical maturity, it was the late-maturing girls who were the most satisfied with their height, weight, and figures. The most plausible explanation is that the late-maturing girls are usually taller and slimmer than early maturing girls, thereby more closely resembling the cultural ideal of feminine beauty. Maturation had considerably less effect on body image among the boys, and when differences appeared they usually favored the early maturing boys.

Teenagers are especially concerned with their looks during adolescence. (Ken Karp)

SEXUAL MATURATION

The increase in amount of sex hormones at the onset of puberty also brings about the maturation of the reproductive organs and secondary sex characteristics. In girls, the secretion of estrogen leads to an increase in the size of the ovaries, uterus, and vagina. In boys, the secretion of testosterone leads to an increase in the testes, scrotum, and penis. Secondary sex characteristics such as body hair and breasts also mature about the same time.

Girls

The typical developmental sequence in girls is as follows: Nonpigmented pubic hair appears soon after the beginning of the growth spurt; the appearance of breast buds, the increase of uterus and vagina size, and the menarche—or onset of menstruation—usually occur after the girl has reached her peak rate of growth in height. At the same time, it should be clear that this pattern may vary somewhat from one girl to another (Tanner, 1978).

Breast buds appear, on the average, at about 10 or 11 years of age and take about 3 to 4 years to full breast development (Figure 3-3). Girls with rapid growth may take only 1 ½ years to pass through all the stages, whereas those with slow growth may take 5 years or even more to reach their full *breast* development. Sometimes girls worry because their breasts may develop unevenly. One girl reported, "My mother took me to the doctor because it only happened on one side at first." Some girls also worry because one breast grows slightly larger than the other or because their nipples look different from those of other girls. But, as in other parts of the body, nature rarely achieves perfect symmetry. Furthermore, breasts not only vary in size from one adolescent to another, but also in shape.

pubic hair: Hair growth in the region of the sex organs.

Pubic hair begins growing about the same time as the appearance of breast buds. Fine and straight at first, pubic hair gradually becomes coarser, kinkier, and more extensive. Hair soon appears on other parts of the body, especially under the arms and legs (Figure 3–4). The color, texture, and extent of a girl's body hair is mostly determined by heredity.

uterus: Pear-shaped organ inside the female's pelvis connecting the vagina with the fallopian tubes; the normal site in which the fetus develops.

The *uterus* and vaginal canal continue growing in size throughout this period, along with the *clitoris* and labia or lips of the *vagina*. The *hymen,* or skin covering the vaginal opening, is often broken by vigorous exercise or the use of sanitary tampons before the first sexual intercourse.

vagina: The stretchable canal in the female that opens at the vulva and extends several inches into the pelvis to the uterus.

The menarche occurs relatively late in the developmental sequence, at an average age of 12.8 years in the United States. It rarely occurs before 10 or after 16 years of age. Menarche usually occurs after the peak of height growth, so that the tall girl may rest assured about future growth if her periods have begun. The menstrual cycle tends to be irregular at first, without the shedding of an ovum (egg) in some cases. However, it is usually a year or so before a girl is capable of reproduction on a regular basis (Tanner, 1978).

hymen: Tissue that partially covers the vaginal opening.

It is important to remember that there is considerable variation in the ages at which these changes occur, as well as in the rate and extent of growth. The fact that we have cited figures for the average pattern of growth means that as many girls experience the changes earlier than these ages as later (Figure 3–5).

vulva: External genitals of the female.

Menarche

A major marker event in the girl's passage to womanhood is the menarche or first menstrual period. The beliefs and expectations girls hold prior to the menarche and their subsequent reactions to this event can be seen in a study of 639 girls in selected grade levels from the fifth through the twelfth grades. In addition, 120 of the fifth- and sixth-grade girls who had not begun the menarche were followed longitudinally. As early as the fifth grade, the

marker events: Particular occasions or periods that signify important changes in an adolescent's development.

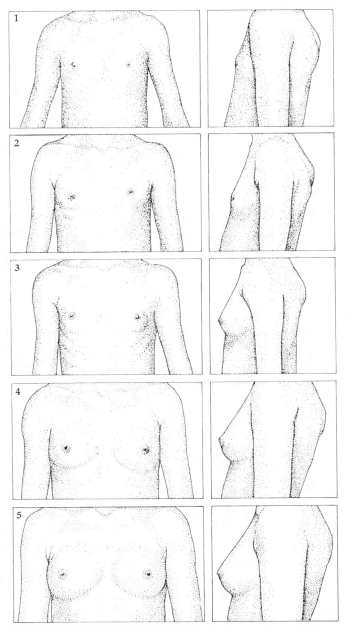

FIGURE 3–3 STAGES OF BREAST DEVELOPMENT. Stage 1 represents preadolescence; stage 2, the appearance of breast buds; stage 3, the small pubescent breasts; stage 4, projection of the areola (pigmented area around the nipple) beyond the contour of the breast; and stage 5, the mature prepregnant breasts. Stage 4 is skipped by some girls and persists for several years in a few girls.

Source: H. Katchadourian, *The Biology of Adolescence* (San Francisco: W. M. Freeman, 1977), p. 55. Adapted from J. M. Tanner, *Growth at Adolescence,* 2nd ed. (Oxford: Blackwell Scientific Publications, Ltd., 1962).

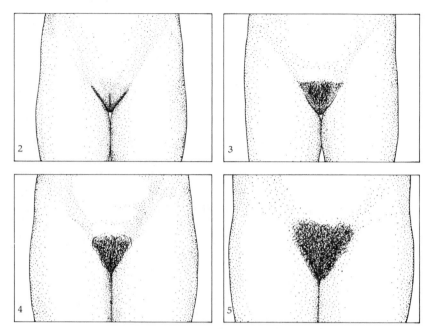

FIGURE 3—4 STAGES OF PUBIC-HAIR GROWTH. Stage 1 (not shown) represents prepuberty prior to the growth of pubic hair; stage 2, growth of the textured hair mostly at the sides of the labia; stage 3, more and darker, coarser hair; stage 4, adult hair; and stage 5, more extensive adult hair with horizontal upper border.

Source: H. Katchadourian, *The Biology of Adolescence.* (San Francisco: W. H. Freeman, 1977), p. 57. Adapted from J. M. Tanner, *Growth at Adolescence,* 2nd ed. (Oxford: Blackwell Scientific Publications, Ltd., 1962).

premenarcheal girls had clear expectations about the menstrual cycle which were similar to the reactions of the older adolescent girls. Yet, girls who had begun to menstruate usually experienced less severe menstrual distress, such as pain and water retention, than the younger girls expected to experience. At the same time, there was a positive relationship between a girl's earlier expectations of menstrual distress and symptoms later reported. Although there was little change in the amount of information learned from the various sources as the girls reached menarche, there were significant differences in regard to the sources of information. Girls who received most of their information from women related menstruation less negatively and reported less severe symptoms of menstrual distress than girls who learned mostly from male sources (Brooks-Gunn & Ruble, 1982).

The girls' reaction to menarche itself is generally mixed. Initially, there is a concern for secrecy, with moderate symptoms of menstrual distress. Girls who are the least prepared or who reach menarche early tend to report more negative reactions. However, the findings suggest that while menarche is initially experienced as inconvenient and somewhat confusing, the experience

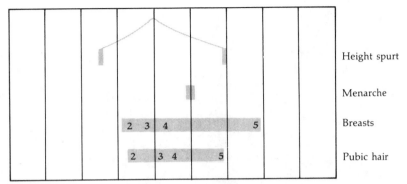

Height spurt

Menarche

Breasts

Pubic hair

FIGURE 3—5 THE SEQUENCE OF PHYSICAL CHANGES DURING A GIRL'S PU-BERTY. The average ages of onset and maturation of a given change are designated here by the two ends of the bar lines. The numbers shown within the breast bar line represent the stages of development using the pictorial rating scales in Figure 3—3 from 1 (prepuberty) to 5 (full development).
Source: J. M. Tanner, *Fetus into Man* (Cambridge, MA: Harvard University Press, 1978), p. 62. Used by permission.

for most girls is not as negative or traumatic as often depicted (Ruble & Brooks-Gunn, 1982).

Generally, it takes a year or more before the girl's menstrual periods occur at regular intervals, which may be disconcerting to those who have received little or no instruction about such matters. Also, ovulation may not occur for several years, at least on a regular basis. In one study of 200 normal girls, Apter and Vihko (1977) found that over half the menstrual cycles were not accompanied by ovulation in the first 2 years following menarche and only in 20 percent or less of such cycles by 5 years. Consequently, it is usually several years after menarche before the majority of girls are fertile on a regular basis.

amenorrhea: The absence of menstruation.

In addition, some girls are bothered by such dysfunctions as the absence of menses or painful periods. *Amenorrhea* may be primary, meaning that it has never occurred, or secondary, meaning that the girl's periods may have begun but ceased. Either condition may be caused by a variety of congenital, hormonal, and physical conditions. Secondary amenorrhea may also be brought on by a change of climate, overwork, or emotional stress. Furthermore, the fact that adolescent anorexics who starve themselves and female athletes in training may cease menstruation suggests that such factors as critical body weight and metabolism may also be involved (Frisch, 1983). A more common problem is *dysmenorrhea.* A study of a national sample of 7000 adolescents between 12 and 17 years of age found that 6 out of 10 girls experienced cramps before or during menstruation. Many girls reported that the pain of menstruation increased somewhat throughout the first 5 years after menarche. Yet, only 14 percent of them described such pain as severe. There are now several drugs that provide relief from menstrual distress by inhibiting the primary substance that produces cramps (Klein & Litt, 1983).

dysmenorrhea: Pain or discomfort before and during menstruation.

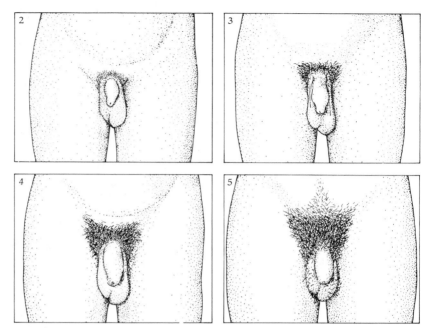

FIGURE 3–6 STAGES OF PUBIC HAIR GROWTH IN THE MALE. Stage 1 (not shown) represents prepuberty prior to pubic hair; stage 2, sparse growth of the textured hair, mostly at the basis of the penis; stage 3, more and coarser, curly hair; stage 4, adult hair; and stage 5, more extensive adult hair spreading upwards and outwards to the thighs

Source: H. Katchadourian, *The Biology of Adolescence* (San Francisco: W.H. Freeman, 1977), p. 67. Adapted from J. M. Tanner, *Growth at Adolescence*, 2nd ed. (Oxford: Blackwell Scientific Publications, Ltd., 1962).

Boys

testes: Male sex glands inside the scrotum that produce sperm and sex hormones.

scrotum: The bag-like pouch of skin that encloses the male testes.

penis: The male sex organ consisting of the internal root and external shaft and glans.

The typical developmental sequence in boys is as follows: The testes and scrotum begin to increase in size, pubic hair appears, the penis begins to enlarge about the same time the growth spurt starts, the voice begins to deepen as the larynx grows, and increasing sperm production leads to the initial seminal emission. At the same time, this pattern may vary somewhat from one boy to another, while remaining well within the normal sequence of events.

The *testes* begin increasing in size relatively early in the sequence of puberty, as you can see in Figure 3–6. Because the testes increase several times in size and even more in weight, such growth remains in progress over a period of several years. The *scrotum* also darkens and thickens with the growth of blood vessels in this area. During puberty the mass threadlike *seminiferous tubules* in the testes become differentiated into mature male *sperm*. From puberty on, the testes generate sperm more or less continually throughout adulthood. About a year after the testes have begun increasing in

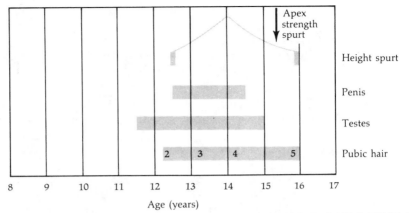

FIGURE 3–7 THE SEQUENCE OF PHYSICAL CHANGES DURING A BOY'S PU-
BERTY. The average ages of onset and maturation of a given change are designated by
the two ends of the bar lines.
Source: J. M. Tanner, *Fetus into Man,* (Cambridge, MA: Harvard University Press, 1978), p. 62. Used by
permission.

sperm: The male reproductive cell.

circumcision: Surgical removal of the foreskin
from the penis.

size, the *penis* begins increasing in length and circumference, reaching adult
size in several years.

Since urine and dirt may accumulate under the foreskin of the penis and
lead to irritation or infection, *circumcision* is commonly performed in hospitals
throughout the country. Yet there are arguments against circumcision on the
grounds that the foreskin may have some unknown but important function;
for example, the lymphoidal tissue present in these structures may contribute
to an individual's ability to fight disease (Crooks & Baur, 1983).

Although the difference in penis size among boys may be great in early
adolescence, this difference becomes less so by the late teens. Some authorities contend that differences in penis sizes tend to diminish in the erect state.
But contrary to folklore, the length and circumference of a male's penis is not
related to his physique or ability to give or receive pleasure (Masters, Johnson
& Kolodny, 1985).

Pubic hair appears soon after the onset of growth in the testes and penis.
Facial hair first appears at the corners of the upper lip, then over the entire
moustache area, and finally on the sides of the face and chin. Body hair also
appears on the arms, legs, and chest area. The color, texture, and amount of
hair growth is determined largely by heredity. As a result, some boys have
extensive body and facial hair, whereas others have only slight hair growth
(Figure 3–7).

A marked voice change occurs toward the end of the developmental sequence. The larynx (Adam's apple) enlarges and the vocal cords double in
length, causing the voice to drop about an octave in pitch. The quality of a
boy's voice also changes, mostly because of the enlargement of the resonat-
ing spaces above the larynx. The breaking of pitch may occur abruptly in

some boys, very gradually in others, but it usually takes several years before boys gain control of the lower range of notes (Tanner, 1978).

Other changes also occur during this stage, such as the increase in size of a boy's sweat glands and skin pores. This usually leads to an increase in perspiration and the likelihood of facial acne in some boys. As with girls, there is considerable variation in the age at which these changes occur among boys. Early and late maturers are especially likely to attain their growth at ages different from those given for the average developmental pattern.

First Seminal Emission

The erection of the penis, which is basically a reflex action, may occur anytime from birth on, usually as the result of local stimulation such as bathing or a full bladder. Even though genital stimulation is pleasurable throughout childhood, it rarely carries the degree of excitability or urgency it does for adolescents. However, after the onset of puberty the penis becomes tumescent or pleasurably erect more readily because of the growing intensity of the sex drive and maturation of the genitals. Boys may experience erections spontaneously or in response to a wide variety of erotic sights and sounds. They take pleasure in their erections and may begin to masturbate, but without ejaculation. Boys generally feel excited and proud of their erections as a sign of their emerging manhood. But spontaneous erections at unwanted times may also become a source of embarrassment.

semen: A white-colored fluid ejaculated through the penis that contains sperm mixed with other seminal fluids.

The first ejaculation of semen is often regarded as a marker event for male puberty similar to the menarche in girls. This usually occurs about a year after the onset of puberty, at 14 years or so on the average, though it may occur as early as 11 or as late as 16. The first seminal emission generally occurs during masturbation or in a nocturnal emission. The latter is frequently, though not always, accompanied by an erotic dream. Although today's adolescents are better informed and less likely to worry about such matters, younger boys and those who have not received any parental instruction may suffer from needless fears. Even the boy who has enjoyed the pleasurable sensations of masturbation without ejaculation may view his first seminal emission as a sign that something is physically wrong with him. Girls may also experience the equivalent of wet dreams—nocturnal dreams accompanied by orgasm—but only a small number of them do so and usually at an older age. At first, the boy's sperm contains so few mobile semen that he is sterile for all practical purposes. It usually takes anywhere from 1 to 3 years before the boy's ejaculate has sufficient sperm to be fertile (Money, 1980).

nocturnal emission: The involuntary ejaculation of semen during sleep, also known as a "wet dream."

EARLY VERSUS LATE MATURERS

Adolescents who reach their physical maturity earlier or later than the average tend to experience puberty somewhat differently than their peers. However, it

Being an early or late maturer makes teenagers
feel awkward in relation to their peers. (Ken Karp)

is not always clear what impact this has on their psychological development,
nor how long-lasting the effects will be. According to one view, much of the
positive or negative consequences of growing up sooner or later than one's
peers is due to the stress brought about by the bodily changes themselves. In
another view, the timing of maturation is a critical factor, so that adolescents
who reach physical maturity before their peers or lag behind them are regarded as members of a deviant group, which itself generates considerable
stress. In both instances, the consequences of being an early or late maturer
tend to be relatively short-lived. Yet in some instances individuals may experience special consequences of early or late maturation which may persist into
early adulthood.

Early Versus Late-Maturing Girls

In the longitudinal study conducted by Dale Blyth (1981) and his colleagues,
already cited, early maturation had more mixed effects for girls than for boys.
In the sixth and seventh grades, the early maturing girls felt better about their
figures, were more popular with the boys, and dated more frequently than the
late-maturing girls. They were also more independent than late-maturing girls

and were more likely to be left alone while their parents were out and to be allowed to babysit. Yet, early maturing girls were less likely to get good grades and were more apt to get into trouble at school than their peers. However, by the tenth grade, many of the differences between early and late-maturing girls had disappeared.

Generally, early maturing girls experience both advantages and disadvantages in their development. First, because of their physical development, early maturers are treated as more mature by adults and peers alike, so they tend to act that way. Then too, the greater acceptance from boys leads to a more active social life, which may help a girl to grow up socially, as long as she is emotionally ready for it. At the same time, some authorities point out that while early maturing girls enjoy a well-ordered external adjustment, they also suffer from more internal crisis and confusion, mainly because of their shorter period of preparation for the changes of puberty. Yet followup studies have shown that by adulthood, early maturing girls exhibit a high level of cognitive mastery and coping skills, partly because of their richer experiences throughout puberty (Livson & Peskin, 1980).

Late-maturing girls also have the characteristic advantages and disadvantages. Because they lag behind other girls in physical development, late-maturing girls are more likely to suffer from anxiety and self-doubt. They worry about whether their bodies will ever develop properly or whether they will be as well endowed sexually as those around them. At the same time, late-maturing girls may enjoy the advantage of growing up with less social pressure than early maturing girls. Given the 2 year lead of girls over boys, late-maturing girls may emerge from puberty at about the same time as their same-age male peers, giving them a more favorable opportunity to develop emotionally and socially at the same rate as boys. Late-maturing girls have been described as gregarious, socially poised, and assertive. Yet by the time they reach adulthood, late-maturing girls may have failed to achieve the high level of cognitive mastery and coping skills exhibited by early maturing girls because of a relatively "safe" adolescence—a longer period of preparation, coupled with a less hazardous sense of sexuality in relation to boys (Tanner, 1978).

AN EARLY MATURING GIRL

"I began to fill out early and looked awful. Throughout grade school I was the tallest girl in the class.

I got my first bra in the fifth grade. I was humiliated to death when I was the first of my peers to start my period in the fifth grade. Actually, I was thrilled at entering womanhood since I had seen a film on menstruation at school. But I was disappointed because I couldn't share my experience with my friends.

Luckily by the seventh grade, most of my friends had reached the same level of physical maturity."

A LATE-MATURING BOY

"Looking back, it seems to me that I was a late maturer. I never did get really big. It wasn't until college that I grew up.

I wasn't very good in sports or things like that which were very important in high school. The only thing I could do was study, which I did.

Being a late maturer made me more self-conscious and shy, with less self-esteem. But I've overcome most of these in time."

Early Versus Late-Maturing Boys

The psychological effects of being an early or late maturer are generally more marked for boys than for girls; there are, in addition, more advantages for early maturing boys and more disadvantages for late-maturing ones.

Generally, early maturing boys enjoy several advantages. First, their physical maturity gives them a competitive edge in sports activities. They are also more attractive to girls their age, who ordinarily mature more rapidly than boys. Because early developing boys appear more mature, they are also more likely to be chosen as leaders by their peers and to be expected to behave in a mature way by adults. As a result, early maturing boys exhibit more positive personality traits than average- or late-maturing boys. At the same time, early maturing boys risk certain disadvantages. Because they have a shorter time in which to adjust to their physical maturity, early maturing boys sometimes fail to grow intellectually and socially, at least through midadolescence. They may appear more settled than others their age, but sometimes at the price of shutting out adolescent experimentation and the search for personal identity.

In contrast, late-maturing boys tend to suffer far more disadvantages than early or average-maturing boys. The lag in their physical development puts them at a disadvantage in sports and boy-girl relationships. They are also less popular with their peers and less likely to be chosen as leaders. As a result, late-maturing boys may be plagued by a negative self-concept and feelings of inadequacy, dependency, rejection, and rebelliousness. Yet the relative social neglect suffered by the late-maturing boy, together with the longer period of puberty adjustment, may also lead to greater cognitive mastery and coping skills. That is, having to struggle all the harder to cope with puberty, the late maturer tends to approach new situations with greater intellectual curiosity, social initiative, and exploratory behavior, whereas the early maturer may adopt a more restricted, conventional approach (Livson & Peskin, 1980).

Individual Differences

Some adolescents are more likely than others to be early or late maturers. Generally, those with an athletic build tend to be early maturers and those with a thin build late maturers. Each individual's growth pattern ordinarily remains consistent, so that a girl who reaches her growth spurt early will also have an earlier-than-average menarche (Tanner, 1978). Yet, there is considerable variation within individual patterns, as was shown in a study analyzing the peak growth of five skeletal parts, including height, arm and leg length, and shoulder and hip width. The results showed that three-fourths of the boys and girls had unique development in some aspect of their physical growth (Faust, 1983). As one girl said, "I matured early in tallness, but got my period at the average age and was actually a bit late in my breast development."

Much of the psychological effect of being an early or late maturer depends on the individual, his or her family, peer support, and life experiences. For example, an early maturing girl with an understanding family may develop emotionally and socially all the sooner because of her early growth; yet an early maturing girl with a lot of stress at home and school may rely unduly on her physical assets as a way of compensating for her insecurity. Likewise, a late-maturing girl from a problem family and without close friends may suffer intensely, whereas a late-maturing girl with a supportive family and friends may actually benefit from a slower emotional, social development. Much depends on the adolescent's particular life experiences and how he or she copes with them.

SUMMARY

Puberty

1. Technically speaking, puberty refers to the attainment of sexual maturity. But the term is also used more broadly to include the entire process of glandular and bodily changes that accompany sexual maturation.

2. The sequence of growth at puberty is similar for most adolescents, though the age of onset as well as the rate and duration of growth varies somewhat from one adolescent to another.

3. The combined effect of the increased hormonal secretions at puberty results in various bodily changes, including changes in the adolescent's height, weight, bone structure, muscles, sex glands, reproductive organs, and secondary sex characteristics.

4. The secular trend toward earlier maturation is thought to be due mainly to better nutrition and appears to be slowing down in recent years.

Physical Changes

5. The growth spurt, which consists of dramatic changes in the adolescent's height and weight, occurs, on the average, between 10 and 14 years of age in girls and 12 and 16 years of age in boys.

6. The growth spurt is accompanied by skeletal and muscular changes, including the changing ratio of shoulder to hips so noticeable in boys and girls at this age.

7. The rapid changes at puberty vitally affect the adolescent's body image, with the relationship between maturation and body image being somewhat more complex among girls than boys.

Sexual Maturation

8. The usual sequence of sexual maturation in girls is the appearance of nonpigmented pubic hair, breast buds, accelerated growth of the uterus and vagina, culminating in the menarche—the first menstrual period.

9. Although the menarche is initially felt to be inconvenient and somewhat confusing, most girls do not find it as negative or traumatic as it is ordinarily portrayed.

10. The usual sequence of sexual maturation in boys is the growth of the testes and scrotum, appearance of pubic hair, and enlargement of the penis; the voice begins to deepen, and increasing sperm production leads to the first seminal emission.

11. Although the boy's first spontaneous ejaculation usually occurs a year or so after the onset of puberty, it normally takes anywhere from 1 to 3 years before the boy's ejaculate has sufficient sperm to be fertile.

Early Versus Late Maturers

12. Earlier maturation tends to have more mixed effects for girls than boys, with many of the differences between early and late-maturing adolescents disappearing by late adolescence.

13. There are both advantages and disadvantages to being an early or late maturer, though early maturing girls often exhibit greater psychological maturity because of their richer experience throughout puberty, and late-maturing girls often fail to attain the same level of maturity because of their relatively "safe" adolescence.

14. Generally, early maturing boys enjoy more advantages than disadvantages in their psychological development. At the same time, some late-maturing boys may, because of their longer period of puberty adjustment, acquire greater cognitive mastery and coping skills by early adulthood.

15. Much of the effect of being an early or late maturer depends on the individual adolescent, the support provided by family and peers, and his or her life experiences.

REVIEW QUESTIONS

1. Explain the meaning of puberty.

2. What is the relationship between puberty and adolescence?

3. Describe the major types of bodily changes that occur in puberty.

4. What is the secular trend?

5. Describe the usual sequence of bodily changes that occur during the boy's sexual maturation.

6. Describe the usual sequence of bodily changes that occur during the girl's sexual maturation.

7. What are some of the factors which influence a girl's experience of the menarche?

8. Explain the significance of nocturnal emissions in puberty.

9. What are some of the advantages and disadvantages experienced by early and late-maturing girls?

10. What are some of the advantages and disadvantages experienced by early and late-maturing boys?

4 Cognitive Development

- **PIAGET'S VIEW OF COGNITIVE DEVELOPMENT**
  Piaget's Concept of Intelligence
  Childhood Thought
  Adolescent Thought
- **ADOLESCENT THOUGHT AND LEARNING**
  Academic Studies
  The Developmental Gap
  Facilitating Formal Thought
- **ADOLESCENT THOUGHT AND SOCIALIZATION**
  Socialization
  Egocentrism
  Cognitive Maturity
- **CREATIVITY**
  Types of Creativity
  Characteristics of Creative Adolescents
  Creativity, Intelligence, and Achievement
- **SUMMARY**
- **REVIEW QUESTIONS**

At the same time adolescents are changing in their physical appearance, they're also going through important changes in their thinking. The impressive cognitive gains accompanying puberty make it possible for adolescents to think in a more adultlike way. An example is the different ways individuals can play the familiar game "Twenty Questions." As you may recall, the purpose of the game is to determine what one person is thinking of, using as few questions as possible. At the outset, younger children are apt to ask random questions such as, "Is it a car?" or "Is it an airplane?" Adolescents and adults are likely to adopt a more fruitful, problem-solving strategy involving a sequence of increasingly specific questions. That is, a player may initially ask, "Is this an animal, mineral, or vegetable?" Once the general category is determined, the players may ask more specific but relevant questions until they get the correct answer. As individuals reach adolescence, this type of thinking becomes increasingly important as they face more challenging subjects in school, crucial life choices, and intricate problems in human relationships.

We shall begin by explaining Piaget's concept of intelligence and cognitive development. Then we will explore the wide-ranging implications of the characteristic thought which emerges during adolescence. Finally, we shall examine adolescent creativity, an important but often neglected aspect of cognitive development.

PIAGET'S VIEW OF COGNITIVE DEVELOPMENT

cognitive development: Changes in mental activity such as attention, thinking, and memory that accompany the adolescent's maturation and experience.

Jean Piaget (1972), more than anyone else, has pioneered in exploring the qualitative changes in intelligence that accompany maturation. Essentially, Piaget holds that lifelong cognitive development results from the interaction of biological and environmental factors and unfolds through a sequence of four stages. Although the genetic and biological factors inherent in the growth process determine the individual's readiness for learning at each stage, the individual's learning experiences and environment affect the rate and extent to which one's intelligence is developed. Piaget's explanation of cognitive development is especially helpful in understanding the distinctive characteristics of adolescent thought—such as the increased ability to think logically and the growing interest in the hypothetical—as well as much of the characteristic behavior at this age.

Piaget's Concept of Intelligence

intelligence: The capacity for acquiring and applying knowledge.

One of Piaget's basic premises is that intelligence is more a process of understanding reality than a fixed trait. According to Piaget, the most essential characteristic of intelligence is the organizational properties of mental processes as a whole. Through extensive investigation with children and adolescents. Piaget discovered that the organization of mental processes changes predictably with maturation, in such a way that children think differently at the various stages of their development.

Cognitive development consists of a progressive reorganization of mental processes as a result of maturation and experience. That is, no sooner have children constructed a meaningful understanding of reality than they begin experiencing discrepancies between what they know and what their environment is presenting to them. It is the continuing process of resolving such discrepancies that transforms the child's intelligence into the more mature understanding of the adolescent and adult.

assimilation: The incorporation of new information into one's existing cognitive structures.

accommodation: The changes in one's existing cognitive structures in the process of adapting to an unfamiliar environment.

equilibration: The inherent regulatory process that facilitates cognitive growth through maintaining a functional balance between assimilation and accommodation.

Cognitive growth proceeds through the interaction of two complementary processes—*assimilation* and *accommodation*. When children and adolescents encounter something that is reasonably similar to what they already know, it is assimilated to their existing knowledge. When they encounter something that is quite dissimilar from what is already known, they either totally ignore it or change their way of thinking to accommodate their knowledge to the new and unfamiliar. For example, sexual maturation in adolescence brings about considerable accommodation in the way adolescents see themselves and relate to members of the opposite sex. The relationship between these two processes is governed by *equilibration,* a regulatory process inherent in the human organism that facilitates cognitive growth through maintaining a functional balance between assimilation and accommodation. Accordingly, individuals tend to

Jean Piaget based many of his findings on clinical interviews with children and adolescents. (Wayne Behling)

be attracted to situations that are interesting enough to warrant assimilation but familiar enough to justify accommodation. In this way cognitive growth proceeds through a progressive reorganization of mental processes that facilitates adaptation to one's environment.

Piaget holds that the process of cognitive development unfolds in an unchanging sequence of stages, mostly because of the primary role played by heredity or maturation. The respective stages are: (1) the *sensorimotor stage,* 0 to 2 years; (2) the *preoperational stage,* 2 to 7 years; (3) the *concrete operational stage,* 7 to 11 years; and (4) the *formal operational stage,* from 11 to 12 years and up. The ages stated are those at which three-fourths of a given age group can master the relevant stage-related tasks; they have been determined through extensive testing by Piaget and his followers. However, the rate at which a person progresses from one stage to another varies from one individual to another, mostly because of the different factors that affect the interaction between individuals and their environment including cultural, social class, IQ, and sex differences as well as the individual's learning experiences and social interactions.

Although we are interested primarily in the stage of formal operational thought, it needs to be considered in relation to the overall process of cognitive development. Consequently, we'll take a brief look at the first three stages of cognitive development in childhood before discussing the final stage of logical, abstract thought that emerges in adolescence.

Childhood Thought

sensorimotor stage: The earliest stage of cognitive development consisting mostly of trial and error learning through use of the senses and muscles, roughly from birth to 2 years of age.

Piaget characterizes the infant's cognitive development from birth to about 2 years of age as the sensorimotor stage. At this time, the infant possesses only rudimentary intelligence and functions mostly through trial-and-error learning. During the first 6 months or so, the infant learns through the senses, by sucking, touching, hearing, and seeing things. In the latter part of the first year, the infant begins to use the muscles more, and learns through crawling, climbing, and hitting things. Then, during the second year, the child begins coordinating these sensorimotor abilities better and actively experiments with his or her environment.

preoperational stage: The second stage of cognitive development in which the child acquires language and symbolic functions but uses these in a perception-bound way, roughly 2 to 7 years of age.

The maturation of mental processes and the increasing use of language enables the child to enter the preoperational stage of thought, roughly during the preschool years from about 2 to 7 years of age. "Preoperational" here means essentially prerational thought; that is, the child is thinking in the general sense of using symbols to represent external events in internal, mental processes, but it is not rational, directed thought like that of adults. Instead, throughout this period, childhood thought is characteristically rigid and perception bound, such that objects are equated with their surface appearances. Psychological realities like fantasies and dreams are easily substituted for physical realities. The child also sees events related in an associational rather than a causal way; that is, two events that accidentally occur at the same time are seen as causally related, even though they are not. Consequently throughout this period, the child's thinking remains highly egocentric and lacking in concepts, though less so with the gradual transition to the next stage of thinking.

concrete operational stage: The third stage of cognitive development in which the child can think in terms of basic concepts such as time and number, roughly 7 to 11 years of age.

With the further maturation of the brain and nervous system, aided by learning experiences, the child enters the concrete operational stage of thought, roughly from about 7 to 11 years of age. At this stage, the child is able to think in a more flexible, abstract way, at least in relation to concrete objects and situations. For example, the child is also able to engage in what Piaget calls *conservation,* the ability to see that certain properties remain constant despite changes in appearance. This results in the emergence of simple concepts or elementary abstractions, first in relation to subjects like number, and then gradually in relation to more complex subjects like space, time, and volume. All of this makes it possible for the older child to take part in more sophisticated learning both at school and home.

conservation: The recognition that a substance's properties such as weight and mass remain the same despite changes in its appearance.

Adolescent Thought

Sometime in early adolescence, beginning at 11 to 12 years of age on the average, individuals enter the formal operational stage of thinking. Here "for-

formal operational stage: The fourth and final stage of cognitive development in which the adolescent can think in an abstract, logical manner, beginning about 11 or 12 years of age.

some outcomes

mal" means entirely mental or capable of pure abstraction, as contrasted with the physically oriented concepts of the concrete operational stage. Adolescents can now manipulate ideas and symbols in a more abstract way, as in formal scientific and logical thought, from which the label for this stage is derived.

Two essential characteristics of thought at this stage are <u>increased possibility</u> and <u>flexibility</u> of thinking. In other words, adolescents become more hypothetical in their thinking. They can now distinguish more clearly between their thoughts about reality and reality itself. Consequently, they can engage in abstract, imaginative thinking, with more of the "if-then" quality characteristic of creative and scientific endeavors. Their thinking also becomes more flexible, so that they can grasp complex relationships. This, in turn, means that many individuals at this stage can comprehend all aspects of a problem and deal with it more systematically. (For an illustration of formal thought, see Figure 4–1).

The percentage of adolescents demonstrating formal thinking tends to increase with age, but not everyone reaches the formal stage of thought. In a comprehensive study of junior and senior high-school students, Renner and Stafford (1976) found that by the senior year of high school, two-thirds of the students remained at the concrete level of thought. A study of females with above-average intelligence showed that even though there was a steady improvement in formal thinking with each grade, not even the oldest group of subjects demonstrated formal thought on all tasks (Martorano, 1977). Further studies have also shown that only about 50 percent of the students entering college are ready to handle formal thought (McKinnon, 1976).

Even the effects of social class and IQ differences that are so noticeable up through the attainment of concrete thought tend to be negligible on formal thought. Neimark (1975) found that socioeconomic status had little or no effect on the development of formal thought. And Kuhn and Angelev (1976) found no correlation between measures of IQ and progression to formal thought. Yet, the finding that American adolescents aged 13 to 15 are more advanced in formal thought than the same-aged youth in Hong Kong suggests that cultural differences play a part in cognitive development (Douglas & Wong, 1977). Possible reasons for such differences may have to do with the contrast between rural and urban settings and the kinds of learning experiences fostered by the schools (Youniss & Dean, 1974). Indeed, Piaget (1972) himself became aware of these differences and eventually formulated the view that formal thought is not necessarily reached without specific learning experiences—a theme we'll explore further later in this chapter.

ADOLESCENT THOUGHT AND LEARNING

The emergence of formal operational thought at this stage of development has many implications for adolescent learning. Let us look briefly here at how

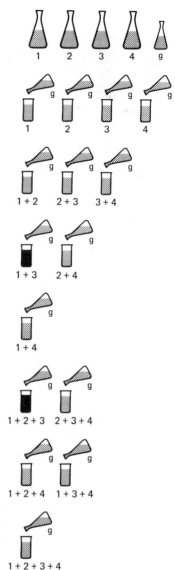

FIGURE 4–1 A PROBLEM REQUIRING SYSTEMATIC EXPERIMENTA-TION. Subjects are given four beakers of colorless, odorless liquids labeled 1,2,3 and 4, plus a smaller bottle, labeled g, which also contains a colorless, odorless liquid. Then they are given empty glasses and are asked to find the combination of liquids that will turn yellow when a few drops from bottle g are added to it. The combinatin that produces yellow is 1 plus 3 plus g. Because the liquid in 2 is plain water producing no effect on the reaction and the liquid in 4 prevents the yellow from appearing, subjects must try all the possibilities before discovering the correct combination. Elementary-aged children usually begin by systematically trying out all the single possibilities, like 1 plus g, then 2 plus g, and so forth. When none of these combinations turns yellow, they may say, "None of them works." Adolescents who have reached the stage of formal thought, however, can systematically consider all possible combinations of the four liquids. They may need the help of pencil and paper to keep track of the complicated procedure. But they are capable of conducting the systematic experimentation which is required in science courses.

Adapted from an exercise in B. Inhelder and J. Piaget, The Growth of Logical Thinking (New York: Basic Books, Inc., 1958).

the high school curriculum assumes that students all have the capacity for abstract thought. Then we shall examine some studies that confirm what many teachers and students have long suspected—that a significant proportion of high school students have not yet attained the capacity for abstract thought. Finally, we shall consider the implication of these findings for enhancing the intellectual development of adolescents.

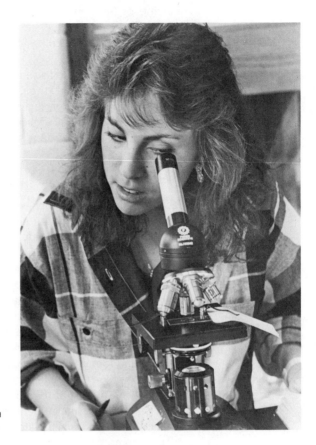

Science courses require the ability to think in a systematic manner. (Laimute Druskis)

Academic Studies

Much of what is studied in high school presupposes the capacity for abstract thought, even though this was not a deliberate policy of curriculum developers. Language courses invariably include some study of grammar, which requires one to think in terms of rules or principles of language usage. Diagramming a sentence involves isolating and separating the different parts of the sentence according to these principles. Foreign languages are often taught from a grammatical frame of reference, thus requiring an understanding of the structure of the language as a whole, an exercise in abstraction. Teaching foreign languages as a means of verbal communication is often more successful because this demands mostly concrete operational thought.

Courses in social studies and mathematics invariably require propositional thinking—that is, the ability to manipulate ideas and think in a logical manner. Courses in history, psychology and international relations involve an understanding of the general principles of behavior and the relation of specific events. The typical sequence of math courses—algebra, trigonometry, analytic geometry, and calculus—requires progressively higher levels of abstract thought.

hypothetical-deductive reasoning: Reasoning from known principles to assumptions.

Courses in the natural sciences demand the use of hypothetical-deductive reasoning, including the formation of hypotheses and the systematic examination of evidence through logic and experimental methods. The progression from general science, biology, chemistry, and physics calls for increasing ability to think in a systematic and scientific manner.

The Developmental Gap

Because not everyone reaches the stage of formal operational thinking at the same age, and some not at all, we would expect to find many high-school students not sufficiently ready for high school–level work. That such is the case becomes abundantly clear in the following discussion.

A comprehensive study of secondary-school students' ability to utilize abstract thought was conducted by John Renner and Donald Stafford (1976). Their subjects were 588 students in grades 7 through 12 in 25 public schools in Oklahoma. Schools were randomly selected within different sectors of the state to include rural, urban, and ghetto populations. Students were interviewed on six types of Piagetian tests designed to assess their abilities for concrete and formal operational thought. The findings are shown in Table 4–1. If you combine the first two categories, you can see that three-fourths of the students were at best at the concrete operational level of thinking. Furthermore, only 14 students in the formal category, or 2.4 percent of all students, had fully attained formal thinking.

The percentages of the students at each grade level who were at the

Table 4–1 CLASSIFICATION OF STUDENTS ACCORDING TO PIAGET'S STAGES OF COGNITIVE DEVELOPMENT

Score	Classification	Number of Students
0–5	Early concrete operational	20
6–11	Fully concrete operational	423
12–13	Transitional	87
14–16	Formal operational	58
		588

Source: J. W. Renner and D. G. Stafford, "The Operational Levels of Secondary School Students," in J. W. Renner et al., eds. *Research, Teaching, and Learning with the Piaget Model* (Norman, OK: University of Oklahoma Press, 1976), p. 95. Copyright 1976 by the University of Oklahoma Press, Publishing Division of the University.

Table 4–2 DISTRIBUTION OF STUDENTS AT THE LEVEL OF CONCRETE OPERATIONAL THOUGHT

Grade	Sample Size	Percentage of Sample at Level of Concrete Operational Thought
7	96	83
8	108	77
9	94	82
10	94	73
11	99	72
12	97	66

Source: J. W. Renner and D. G. Stafford, "The Operational Levels of Secondary School Students," in J. W. Renner/et al., eds. *Research, Teaching, and Learning with the Piaget Model* (Norman, OK: University of Oklahoma Press, 1976), p. 97, Copyright 1976 University of Oklahoma Press, Publishing Division of the University.

concrete level of thought or below are shown in Table 4–2. When compared to the secondary-school curricula, the results are devastating. For instance, sentence diagraming, which requires formal thought, is usually taught to seventh graders. Yet 83 percent of those seventh graders have not reached formal thought. Euclidean geometry, which also requires formal thought, is usually taught in the tenth grade. But 73 percent of the tenth graders are not sufficiently ready for such material. Even by the twelfth grade, fully two-thirds of the students are not ready for many of the courses in math and the natural sciences, such as chemistry and physics.

Similar studies of college students have also shown that a large proportion of them are not ready to handle the abstract thought usually required in college work. One study involved students in different types of both private and public schools ranging from 2-year colleges to universities. First-year English classes were randomly sampled within each school, and the students were tested on their abilities for concrete and operational thinking. The results showed that about 50 percent of the students entering college are not ready to handle abstract thought. The percentage of students who demonstrated the ability for abstract thinking varied from a low of 12 percent at the 2-year private colleges to a high of 61 percent at the 4-year private universities. The percentage figures for most public colleges fell somewhere in between, with 56 percent of the students at a 2-year public college and 46 percent of the students at a 5-year teachers' college demonstrating the ability for formal thought (McKinnon, 1976).

At the same time, a growing number of educators believe that the developmental gap between students' cognitive ability and the school curriculum may be partly attributable to Piaget's theory rather than the students themselves. More specifically, they contend that Piaget's theory does not make sufficient allowance for individual differences and environmental influences in the adolescent's cognitive development. Instead, when individuals reach adolescence, their thinking develops in a differential way depending upon their environment and learning experiences. Thus, one student may handle the higher level abstractions more readily with numbers, while another student may progress more rapidly toward formal thought in language skills. Much depends on the learning environment to which students are exposed as well as the manner in which the subjects are taught. In short, educators are discovering that the adolescent's cognitive development may be shaped more significantly by instruction than previously acknowledged by Piaget (Neimark, 1982).

Facilitating Formal Thought

There are encouraging signs that the use of strategies such as coaching and discovery learning may facilitate the development of formal thinking. In one study, students aged 10, 13, and 17 were coached if they failed to demonstrate thought on the first and second of three problems. Without prompting, none of the 10 year olds showed formal operations on the first problem, compared to about half of the 13 and 17 year olds. Coaching the students produced differential effects by age. Only one-fourth of the 10 year olds demonstrated formal thought even after two promptings on the first two problems. Yet, three-fourths of the 13 year olds and all of the 17 year olds exhibited formal thinking after two promptings. Such results confirm that formal thinking is more readily available to older students, even though they may not spontaneously demonstrate it. At the same time, these results suggest that studies which evaluate formal reasoning without suitable coaching may underestimate the adolescent's ability to reason at the formal operational level (Danner & Day, 1977).

College courses can also be taught in a way that will promote the development of formal thought. One example is an experimental approach to the introductory science courses for nonscience majors in college. The subjects were 143 first-semester students in a private university who were randomly divided into two groups. The experimental group was taught in an inquiry-oriented method, with the emphasis on understanding scientific concepts, problem solving, and learning to think independently. The control group consisted of students enrolled in the usual college courses in which the emphasis tended to be on the mastery of content as demonstrated on tests. Students in both groups were given pre- and posttests to determine the number of them at the concrete and formal operational levels of thought. The results showed a net gain of more than 50 percent in the experimental group; significantly more students in the experimental group than in the control group progressed from the concrete and transitional stages to formal operational thinking (McKinnon, 1976).

ADOLESCENT THOUGHT AND SOCIALIZATION

The emergence of abstract thought at this age helps to explain much characteristic adolescent behavior. For one thing, it helps to explain why adolescents become so idealistic and romantic on the one hand, yet more critical and cynical on the other. It also explains why adolescents are so busy clarifying their own identity, yet suffer from the accompanying conflicts and confusion. The newly discovered capacity for abstract thought also may explain the characteristic self-centeredness of this stage of life.

Socialization

The heightened ability for abstract thought has a marked effect on adolescents' *socialization*—the process by which they learn the attitudes, skills, and values necessary to function adequately in society. This is especially evident in matters such as role fulfillment, competition, decision making, and defense mechanisms.

For one thing, adolescents are now better able to understand and fulfill their social roles—the appropriate attitudes and behavior expected by society. On the positive side, adolescents may fulfill a variety of roles, shifting easily from one to another, whether as students, workers, or friends. They may also choose and experiment with their personal identity, especially that of their sex role. On the negative side, however, the heightened ability for abstract thought also generates a certain degree of conflict and confusion about one's identity. Rebellious adolescents may conspire to do the opposite of what is expected of them, and they may adopt a negative identity—for example, as a repeated delinquent or school dropout.

Competition also becomes more intense at this age, whether in the classroom or on the playing field. Unlike preadolescents, who are only intent on their tangible answers given on the test or their skill in hitting the ball, adolescents are better able to understand abstract competitive rankings like academic grades and competitive team sports. Although this enables them to participate more meaningfully in competitive adult society, it also makes them more vulnerable to the ills of competitiveness—for example, anxiety over grades or fear of failure. Because of their personal values, many of today's adolescents disapprove of the excessive emphasis on competition in society, but at least with the emergence of formal thought they are able to understand it.

The heightened capacity for abstract thought also promotes the adolescent's problem-solving, decision-making abilities. At this point, they can gather information more systematically, conceive of alternatives, and weigh the consequences of each course of action before making a decision. By the same token, adolescents become more anxious and aware of the conflicts between alternatives, especially when they are uncertain of their underlying values or goals. Consequently, decision making becomes more agonizing but realistic, which is fortunate, considering that some of the most difficult life decisions are made at this stage of life: choice of a college major or of a career, or even of a marriage partner.

Adolescents also use their newly discovered ability for abstract thought to cope with anxiety, which results in the increased use of *defense mechanisms*—unconscious, automatic ways of defending ourselves against psychological threat. A common defense mechanism at this age is *intellectualization,* in which one's unacceptable feelings are hidden behind a highly intellectual but heated discussion on the subject in question. Another example is *rationalization,* seen in the adolescent's readiness to "explain" away behavior like failing a test or being late for work. Although fantasy, the tendency to relieve unfulfilled desires through imaginative action is not in itself defensive; it easily becomes so when unrealistic aspirations and romantic longings are substituted for action, as they often are at this age.

rationalization: Justifying one's unacceptable behavior through "good" reasons.

Egocentrism

egocentricism: According to Piaget, the inadequate differentiation of one's thoughts and feelings from those of others, especially evident during preoperational thought and again in the early phases of formal thinking.

The emergence of formal thought at early to midadolescence is accompanied by a marked increase in *egocentrism*—an inadequate differentiation between one's own thoughts and feelings and those of others. Being temporarily overwhelmed by the heightened self-consciousness and abstract thought at this stage, adolescents become noticeably more self-centered. The effects of adolescent egocentrism are spelled out in two related concepts—the *imaginary audience* and the *personal fable*—which account for a wide variety of typical adolescent behaviors (Elkind, 1978).

imaginary audience: The adolescent's heightened self-consciousness and preoccupation with the anticipated reactions of others.

One consequence of adolescent egocentrism is the failure to differentiate between one's own thoughts and those of others, which can lead to the belief that others are preoccupied with one's thoughts and behaviors. As a result, adolescents feel as if they are constantly on stage anticipating the reactions of an "imaginary audience." It is an audience in that adolescents feel they are the focus of attention, but it is imaginary in the sense that others are actually not that concerned with the adolescent. Anticipating the reactions of this imaginary audience helps to account for the heightened self-consciousness and excessive concern for one's appearance at this age. The boy or girl who stands in front of the mirror combing his or her hair for an hour is probably thinking of how he or she will look to others. Preoccupation with others' reactions also helps to explain why *shame* rather than *guilt* is the characteristic emotional reaction—that is, reaction to an audience—at adolescence. This also explains adolescents' intense desire for privacy and their shyness or reluctance to reveal themselves. Intense awareness of others' reactions also causes adolescents to overreact to criticism and blame, as every parent has discovered. One of the most common manifestations of the imaginary audi-

ROLE TAKING AND PERSPECTIVE TAKING

Increasing attention is being given to social cognition—how people think about themselves and others—and how this changes throughout development. A crucial part of social cognition is social role taking, or perspective taking, which includes the ability to understand one's self and others as subjects, to react to others as like the self, and to react to one's own behavior from another person's viewpoint. Robert Selman (1980) has formulated a stage theory of social cognition, along with the approximate ages for each stage, as follows:

Stage 0. *Egocentric, undifferentiated social perspective taking* (3 to 6). At this stage children cannot make a clear distinction between their own interpretation of a situation and another's point of view. Nor can they realize that their own view may not be the correct one.

Stage 1. *Differentiated, subjective perspective taking* (5 to 9). Children begin to realize that others may have different views than their own, but they are unable to understand such views accurately.

Stage 2. *Self-reflective, reciprocal perspective taking* (7 to 12). Older children and preadolescents can reflect on their own thoughts and feelings from another person's perspective. But they cannot hold both perspectives simultaneously as in the next stage.

Stage 3. *Third-person or mutual perspective taking* (10 to 15). Adolescents can step outside their own perspective and those of others and assume the perspective of a neutral third person. Thus, friendships become more than mutual back-scratching, and conflicts may be viewed in terms of mutual differences.

Stage 4. *In-depth and societal perspective taking* (adolescence to adulthood). Individuals become aware that motives, thoughts, and actions are shaped by psychological factors, and may move to a more abstract level of social perspective taking which includes that of a generalized, societal perspective. The increase in role-taking skills and empathy makes possible more mature self-understanding, friendships, peer group participation, and parent-adolescent relationships. Furthermore, the realization that each person can consider the societal perspective fosters more accurate communication and problem solving with others, as well as grasping the idea of law and morality as social systems of thought. However, since stage 4 corresponds to Piaget's level of formal thought, Selman points out that not all adolescents or adults reach this final stage of social cognition.

From R. L. Selman, *The Growth of Interpersonal Understanding* (New York: Academic Press, 1980).

ence is the adolescents' anticipation of how others will react to their deaths—hopefully with a belated recognition of their good qualities.

A complementary process to the imaginary audience is the *overdifferentiation* of one's feelings from those of others, leading adolescents to exaggerate their own uniqueness. Perhaps partly because they feel of such importance to so many people, adolescents tend to regard themselves as more special and unique than they really are. This is known as the personal fable, a subjective story they tell themselves but which is not true. The concept of the personal fable helps to account for the typical adolescent phrase, "You don't understand," which *every* parent has heard at one time or another.

personal fable: The adolescent's exaggerated sense of personal uniqueness.

Evidence of the personal fable is also prominent in adolescent diaries, in which they express feelings that "nobody understands me" and that their loves and frustrations are of universal significance. The belief in their uniqueness often becomes a conviction that they will not die, leading some adolescents to feel they are not subject to the dangers or fate commonly suffered by others. Consequently, they may fail to use their seat belts, drive too fast, or dispense with contraceptives out of the conviction that "it won't happen to me." The personal fable is also illustrated in the adolescents' search for superhuman powers, whether in ESP, romantic love, technology, superstitions, or religion.

Cognitive Maturity

maturity: Fully developed.

According to Piaget, the egocentrism of early adolescence declines as formal operational thought becomes firmly established, on the average at about 15 or 16 years of age. At this point, adolescents tend to modify their imaginary audience in the direction of the real audience, increasingly recognizing the difference between their own thoughts and the thoughts and reactions of others. The personal fable, though perhaps never entirely overcome, is also progressively modified through intimate exchanges with others. By taking the role of others and sharing mutual confidences, adolescents discover that others have also had similar experiences to theirs. As a matter of fact, there is a postiive association between cognitive and social maturity, and stimulation in one area facilitates growth in the other (Nissim-Sabat, 1978). The rate at which adolescent egocentricism is outgrown varies somewhat from one person to another, and depends on factors such as individual experience and maturity. But much of the growth at this age comes with the transition to young adulthood with all its practical decisions about school, jobs, and personal relationships.

An empirical study by Robert Enright, Daniel Lapsley, and Diane Slukla (1979) has shown that although certain aspects of adolescent egocentrism decline with age, the overall picture of cognitive and social maturity is somewhat more complex. The subjects for their study were 60 students, 20 each in the sixth grade, eighth grade, and college, with an equal number of males and females at each age. The students were given instruments designed to assess different aspects of adolescent egocentrism, including the imaginary audience, personal fable, and a general focusing on the self. The results showed that different aspects of egocentrism dominate at different periods of adolescence. That is, whereas the notion of the imaginary audience declined significantly at each age level, the notion of personal fable declined in a much more gradual way. But contrary to expectations, the researchers found that the general focusing on self showed a significant increase with age. Part of the explanation for the latter is the adolescent's increasing concern for understanding and solving conflicts that come with late adolescence. This increase in self-awareness without the self-conscious expectation of others' reactions (imaginary audience) and with less insistence on one's own uniqueness (personal fable) most likely represents an *unself*-conscious striving for betterment of self in the college years, which tends to overshadow other concerns.

CREATIVITY

The ability to think in a more abstract manner along with the heightened self-consciousness that occurs at adolescence makes possible greater creativity at this stage of life. Yet, adolescents often demonstrate less creativity in their thinking than children. One explanation is that while adolescents have greater potential for creativity, they may be more inhibited in expressing it because of their peers' judgment (Wolf & Larson, 1981). Another reason is that schools put greater stress on academic achievement and grades rather than creativity; thus, creativity often goes unrecognized and undeveloped during adolescence.

Types of Creativity

creativity: That which is associated with original or unique contributions to society.

Creativity may be defined in different ways, though most meanings of creativity follow one of three approaches. A common approach is to focus on the *creative process*, which involves combining ideas or things into novel arrangements. Such a process not only requires original thinking and flexibility, but also the use of logical reasoning and problem-solving skills. A second approach focuses on the characteristics of the *creative product* and on whether or not it conforms to criteria such as unusualness, appropriateness, and simplicity. A third approach to creativity is to describe the salient personality characteristics of highly *creative people*—for example, independence, risk taking, unconventionality, and childlike playfulness.

self-actualizing people: Those who have reached a healthier, more optimal level of functioning than the average person.

After studying many types of highly creative people, Abraham Maslow (1971) made a helpful distinction between special-talent and *self-actualizing creativeness*. Special-talent creativeness involves the ability for a specific type of creativity, like painting, music, writing, or acting, which varies widely from one person to another. Self-actualizing creativeness, on the other hand, refers to a quality of personality that is potentially present in everyone, though in different degrees, but may be used in any type of activity. Self-actualizing creativeness has more to do with personal qualities such as spontaneity, flexibility, courage, and perceptiveness, which make one creative in anything he or she undertakes. In this sense, creativeness is an integral part of psychological health or human fulfillment, rather than something linked with madness or illness, as in the popular notion of the creative but eccentric genius. Although self-actualizing creativeness may be more highly developed in the so-called self-actualizing people, it is something that can be developed

CREATIVITY

These three test items make use of "divergent" thinking—the ability to generate many unusual solutions for the same problem. Accordingly, there is no one right answer for each of the items. Instead, the creativeness of your response is evaluated by such norms as how many different responses you make (fluency), the number of shifts from one class of responses to another (flexibility), and the number of unusual responses (originality).

1. Ingenuity

A very rare wind storm destroyed the transmission tower of a TV station in a small town. The station was located in a flat prairie with no tall buildings. Its former 300-foot tower enabled it to serve a large farming community, and the management wanted to restore service while a new tower was being erected. The problem was temporarily solved by using a _____.

2. Unusual uses of things

Name as many uses as you can think of for each of the following things:

a. a brick
b. a paper clip
c. a coat hanger

3. Imagination

Complete this drawing:

Original   Ordinary response   Creative response

(1) J. C. Flanagan, "The Definition and Measurement of Ingenuity," in C. W. Taylor and F. Barron, eds. *Scientific Creativity* (New York: Wiley, 1963), p. 96. (2) J. P. Guilford, *The Nature of Human Intelligence* (New York: McGraw-Hill, 1967), p. 143. (3) F. Barron, "The Psychology of Imagination," *Scientific American, 199* (3) (1958): 154; used by permission of the author.

to varying degrees in all of us, and expressed in any aspect of life, not just in scientific discovery or the arts.

Characteristics of Creative Adolescents

Maslow's notion that self-actualizing creativeness is associated with certain personal qualities like spontaneity and courage is also supported by studies of highly creative individuals in various fields. For instance, young people judged to be especially creative either because of their accomplishments or scores on measures of creativity tend to have certain personality characteristics. For one thing, they are highly independent and self-assured and are not afraid to take risks. They also have a strong curiosity, which attracts them to new ventures. Creative young people tend to be somewhat introspective, reserved, and at times aloof. They are also more unconventional in their interests, tastes, moral values, and life styles (Getzels & Csikszentmihalyi, 1976).

Highly creative adolescents are usually characterized by moderate levels of anxiety. By comparison, adolescents with either too little or too much anxiety tend to exhibit reduced creativity. Thus, creative adolescents have a high tolerance of ambiguity and are able to entertain opposing ideas and values in a way that produces novel combinations (Lunzer, 1978).

There seem to be no consistent sex differences in creative ability among males and females (Maccoby & Jacklin, 1974). Yet, there is some evidence that males who demonstrate creative ability are more likely to do creative things (Forisha, 1978). An obvious explanation for this discrepancy is the conventional outlook in society that has limited the opportunities and recognition of creativity in women. As the opportunities and support for women increases in various fields of endeavor, we may expect to see a greater proportion of creative, outstanding females of all ages. Another reason for the gap in creative performances between the sexes is that many of the traits associated with creativity, such as independence, nonconformity, and adventuresomeness, have been more valued in men than in women. Here again, the recent trend toward less stereotyped sex roles may further diminish the gap in creative performance between the sexes. Growing evidence suggests that creative individuals of both sexes are likely to combine personality traits which traditionally have been regarded as masculine, such as independence and risk taking, and feminine, such as sensitivity and flexibility (Spence & Helmreich, 1979).

Creative accomplishments are better predictors of creativity than school grades.
(Laimute Druskis)

Creativity, Intelligence, and Achievement

Highly creative adolescents tend to be intelligent, though not necessarily extremely intelligent. But because the emphasis in high school is on academic achievement, highly creative adolescents often remain unappreciated as students. Most teachers prefer the highly intelligent student who makes good grades, rather than the more creative but less conforming student. As a result, academic performance is not a good predictor of creativity in later life. Real-life creative accomplishments in school, such as acting in school plays and performing in concerts, may be better predictors of creativity in one's career than grades or test scores.

Creativity and intelligence tend to be combined in different proportions from one adolescent to another, as seen in Welsh's (1977) study of 1000 gifted adolescents. On the basis of test results and the various combinations of these two dimensions, he defined four groups of gifted adolescents: (1) those relatively high in creativity compared to their intelligence; (2) those high on both dimensions; (3) those low on both dimensions; and (4) those relatively low in creativity compared to their intelligence. Adolescents who are highly creative compared to their intelligence tend to be sociable and impulsive, with fantasy-based thinking that inclines them toward the performing arts and sales occupations. Those who are high on both dimensions, tend to be introverted and self-sufficient, concerned with imposing intellectual order; they tend to select careers in art, journalism, and literature. Gifted adolescents who are relatively low on both dimensions are indiscriminately sociable and conforming, with factual thinking that inclines them toward business and service occupations. Finally, those who are relatively low in creativity compared to their intelligence are apt to be friendly but shy in social interaction; their thinking tends to be logical and oriented toward practical solutions, leading them into practical careers such as engineering, law, or medicine. Welsh's studies suggest that the variations in these two dimensions of cognitive style have important consequences for understanding adolescents, especially the gifted ones.

Adolescents and adults who achieve excellence in some field readily admit that it takes more than intelligence and talent, or even the advantages of being in a program for the gifted. In one survey, Bloom (1983) interviewed 120 individuals who had achieved fame in their respective fields, including concert pianists, sculptors, research mathematicians, research neurologists, tennis champions, and Olympic swimmers. Each person was asked to self-evaluate what was responsible for his or her unusual accomplishments. The replies suggested that these famous persons had achieved excellence largely because of the particular environmental support, superior teaching, special experience, and encouragement received throughout their development. As a result, all of them devoted an unusual amount of time in training and practice for the optimal development of their skills.

The achievement of excellence demands an unusual amount of training and practice. (Laimute Druskis)

Schools can and should do a better job of teaching and encouraging creativity, especially the self-actualizing creativity advocated by Maslow. Schools can foster creativity in several ways. First, the creative process can be explained to students. Examples of the way creative people in all fields, not just the arts, go about creating can also be shown. Second, creative teaching methods can be employed in the school. Use of brainstorming sessions in class, discovery learning, problem solving, and more imaginative assignments model the process of creativity itself. Finally, teachers and parents can encourage adolescents in their efforts to be creative. Above all, they should avoid being overly critical or harsh. Creative endeavors may help adolescents to use their imagination, curiosity, and adventuresomeness in a positive way.

SUMMARY

Piaget's View of Cognitive Development

1. We began the chapter by explaining that Piaget's view of cognitive development emphasizes the qualitative changes in intelligence accompanying maturation.

2. Piaget holds that cognitive development unfolds in an unchanging sequence of four stages, culminating in the attainment of formal operational thinking at adolescence.

3. Adolescents who have reached the stage of formal thought are capable of thinking in a more abstract, hypothetical, systematic, and problem-solving manner.

Adolescent Thought and Learning

4. Many of the courses in high school presuppose some degree of formal thought, especially those in the language arts, mathematics, and sciences.

5. At the same, a significant proportion of high school students have not reached formal thinking, with a notable gap between the average student's cognitive abilities and the typical school curriculum.

6. Recent studies suggest that the use of teaching strategies such as coaching and discovery learning may facilitate the attainment of formal thinking, especially among those who have reached midadolescence.

Adolescent Thought and Socialization

7. The wide-ranging implications of formal thought on adolescent socialization were also explored, including interpersonal relationships, competitiveness, decision making, and the use of defense mechanisms.

8. We also discussed the effects of the characteristic egocentricism that occurs at this stage of cognitive development, as seen in the adolescent's experience of the imaginary audience and personal fable.

9. Although egocentricism declines as formal thought becomes firmly established, the notion of the imaginary audience seems to disappear more rapidly than the exaggeration of one's uniqueness (personal fable).

Creativity

10. Despite the greater potential for creativity that accompanies formal thought, adolescents often fail to demonstrate creativity in their thinking, partly because of the fear of their peers' judgment and the emphasis on academic achievement by parents and schools.

11. Creativity may be defined in different ways, including the creative ˉˉˉˉ the creative product, and the characteristics of the creative person.

12. Maslow also distinguishes between special talent and self-actualizing creativity, with the latter referring to a quality of personality that is potentially present in everyone in different degrees.

13. Highly creative adolescents tend to be independent, self-assured, and adventuresome, with a moderate level of anxiety and a combination of personal traits traditionally regarded as masculine and feminine.

14. Outstanding achievement usually involves a combination of creativity and intelligence as well as the advantages of special environmental support throughout the individual's development.

REVIEW QUESTIONS

1. Explain Piaget's concept of intelligence.

2. What is formal operational reasoning?

3. How does the capacity for formal thinking help the adolescent in school?

4. To what extent does the attainment of formal thought depend on the adolescent's environment and learning experiences?

5. How would you account for the gap between the typical adolescent's cognitive abilities and the high school curriculum?

6. What are some teaching strategies schools might adopt to foster the development of formal thinking?

7. In retrospect, can you recall some manifestations of "the imaginary audience" during your adolescence?

8. How would you define creativity?

9. Would you agree that creativity requires moderate but not superior intelligence?

10. What are some ways schools may encourage creativity among adolescents and youth?

5 The Family

- **CHANGES IN THE AMERICAN FAMILY**
  Smaller, More Mobile Families
  Changing Functions of the Family
  More Flexible Roles
- **DUAL-CAREER FAMILIES**
  Working Mothers
  Household Chores
  Latchkey Teens
- **DIVORCE**
  Adolescents in Strife-Ridden Homes
  How Adolescents Are Affected by Divorce
  The Long-Term Effects of Divorce
- **SINGLE-PARENT AND REMARRIED FAMILIES**
  Single-Mother Families
  Single-Father Families
  Remarried Families
- **SUMMARY**
- **REVIEW QUESTIONS**

Each of us grows up in some sort of family—a group of two or more persons related by birth, marriage, or adoption who share a common household. Some people are fortunate enough to spend their entire childhood and youth with both of their natural parents along with any brothers or sisters in the family. Others may grow up in strife-ridden homes that are eventually dissolved by divorce and will spend much of their youth in a single-parent or remarried family. Either way, the family is a major influence in the adolescent's life. Just take a look at the teenagers who excel in school and get along well with their peers. More often than not, they receive a lot of love and support at home. On the other hand, those who are doing poorly at school and tend to get into trouble are likely to be abused or neglected at home. In each case, it is the *quality* of family life which exerts the decisive influence on the adolescent's development. But today, the quality of family life is being vitally affected by a variety of changes occurring in the American family. In this chapter, we'll examine some of the changes in American family life, such as the trend toward smaller, more mobile families; more mothers working outside the home; and the high rate of divorce, resulting in a greater proportion of adolescents growing up in single-parent and remarried families. In the next chapter, we'll focus on the parent-adolescent relationship and how this affects the adolescent's achievement of autonomy.

CHANGES IN THE AMERICAN FAMILY

Much is being said about the changes in American family life. Yet, a look at the past shows that the family has always been changing. In previous eras, the family has been modified by such events as the influx of immigrant families from other countries and the migration of families from the farm to the cities. Today, family life continues to change because of rapid social change throughout society. Some of the changes, such as the increase in two-child families, may have a predominantly positive impact on the quality of family life. Other changes, such as the increase in single-parent and remarried families, probably have a more mixed impact on family life and are still being evaluated.

immigrant families: Families who move to the United States from another country.

Smaller, More Mobile Families

First of all, families are becoming smaller than they were in the past. During colonial times, families had an average of eight children, partly because children were needed to run the farms. But by the 1950s, families were having an average of only four children. The major explanation was that most people lived in cities and had no need for a large family. In recent years, people have been marrying more for companionship and personal fulfillment and are having fewer children. The typical couple with children now has only one or two children. (See Figure 5–1.) When people were asked to give the ideal number of children in a family, 51 percent said two children. Only 16 percent of those questioned favored families with four or more children, down from the 41 percent who favored large families in 1968. (*New York Times,* April 27, 1980).

Families also move around more than they used to. It has been estimated that one-half of all families move once every 5 years (U.S. Bureau of the Census, 1984). The most frequent moves are made by adolescents and young adults as they pull up stakes, go to college, enter the job market, and get married. Most of the long-distance moves are made by those at the upper end of the socioeconomic scale, like professionals and executives, and those at the lower end of the scale, like farm laborers and the unemployed.

All this means that adolescents tend to grow up in smaller, more mobile families. They encounter a larger number of people than their parents did as adolescents, but they form fewer lasting ties with any of them. Relationships tend to become more transient and less intimate. Adolescents see their grandparents less often, sometimes not at all. As a result of these changes, adolescents tend to become even more dependent on their parents and ambivalent about leaving home. They are likely to be plagued by a sense of rootlessness and loneliness, which may in part account for the high value contemporary adolescents and youth place on personal relationships.

Changing Functions of the Family

Families are smaller and more mobile largely because the family fulfills different functions than it did in the past. For instance, in the early days of our country, the family was an economically productive unit. Families built their

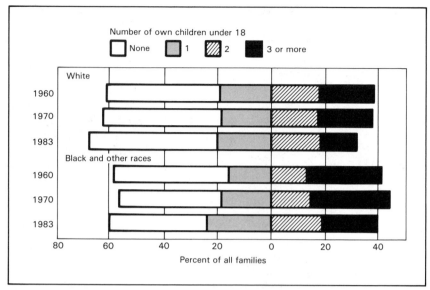

Figure 5—1 FAMILIES, BY RACE AND NUMBER OF OWN CHILDREN UNDER 18: 1960, 1970, AND 1983.

Source: Adapted from U.S. Bureau of the Census, Statistical Abstract of the United States, 1986, 106th ed. (Washington, DC: U.S. Government Printing Office, 1985), p. 45.

own homes, raised their own food, and sewed their own clothes. They were economically self-sufficient. With the rise of manufacturing and modern labor-saving devices, however, families have become primarily consumers. Family members are economically dependent on builders and merchants for their houses, food, and clothes.

Along with these economic changes, the family has also surrendered many of its traditional functions to other agencies. Education of the young is now the primary reponsibility of the schools. Moral and religious instruction have become the function of churches and synagogues. Care of the sick and aged is no longer the sole responsibility of the family, but is shared with physicians and hospitals. Protection and security have become something we expect of the city and state.

At the same time, the family has retained some vital functions and strengthened others. The family remains the approved group for the fulfillment of affectional and sexual needs, despite the rise in number of unmarried couples living together. The family also retains the biological function of having children. Responsibility for the support and socialization of children remains with the family, though it is shared with other agencies. The family also provides for the financial and psychological support of the adolescent's education,

Today's families are smaller than in the past. (Laimute Druskis)

which has become increasingly important in recent decades. Furthermore, life in an individualistic, competitive, and rapidly changing society accentuates the adolescent's need for acceptance and security within the family. As you will discover throughout this book, many aspects of adolescent development—academic achievement, the attainment of independence, vocational aspirations, or future marital success—depend in large measure on the quality of emotional nurture adolescents receive within the family.

nurture: The act or process of rearing an individual in a way that promotes his or her development.

More Flexible Roles

Another change is toward more flexible roles within the family—that is, what is expected of parents and adolescents respectively. In the first place, parental roles are becoming more functional and shared between the sexes than in the past. Although fathers and mothers play distinctively different roles in socializing their young, they do so in a way that is less determined by stereotyped sex roles. Traditionally, fathers have fulfilled an instrumental role, and mothers have fulfilled an expressive role. Today, more mothers are working outside the home and helping with the financial support of the family. And in many homes, fathers are taking a greater responsibility for childrearing. Young adult males are especially apt to expect fathers to be more involved in childrearing and family recreation and to share the provider role (Eversoll, 1979). Yet, such changes in the father's role are not as widespread as we would expect, as we'll discuss in the next section of this chapter. The major reason may be, as Joseph Pleck (1981) points out, that women's roles have been changing

instrumental role: The expected behavior associated with supporting the family and modeling the assertive behavior needed in a competitive society.

expressive role: The expected behavior associated with caring for the young and fostering their emotional and social development.

CHARACTERISTICS OF TODAY'S PARENTS

According to a national survey, there is a new breed of parent emerging, one that is more self-centered, more permissive, and less authoritarian than the traditional parent. The new breed makes up 43 percent of parents with children under 13, with the traditionalists making up the remaining 57 percent.

Although children of both types of parents agree on a number of matters, they differ on at least one important matter. Children of the new breed parents are more willing (58 percent) to see their parents separate if they are unhappy than are children of traditional parents (47 percent).

The New Breed—43%

Believe having children is an opportunity, not a social obligation

Self-oriented, not ready to sacrifice for their children

Do not push their children

Permissive with their children

Believe children should be free to make their own decisions

Parents question authority

Believe boys and girls should be raised alike

The Traditionalists—57%

Believe having children is a very important value

Ready to sacrifice for their children

Want their children to be outstanding

Not permissive with their children

Believe parents should be in charge of their children

Parents respect authority

Believe boys and girls should be raised differently

The General Mills American Family Report, 1976–1977: Raising Children in a Changing Society (Minneapolis: General Mills Consumer Center, 1977), p. 28.

faster than men's roles have. As a result, in the typical dual-career family, mothers are under great stress, mostly because they continue to bear primary responsibility for childrearing and running the home. At the same time, adolescents reared in these homes tend to acquire less stereotyped sex roles than those in one-wage-earner homes, so that family roles are indeed changing gradually.

Teaching-learning roles within the family are also becoming more mutual. Living in a society of rapid social change, young people face a future that is largely uncharted by their parents. They must necessarily learn many things for themselves, and look more to their peers and less to their parents for guidance. It is not that parents have nothing to teach. But parents must adopt a more flexible role, teaching their young how to learn and how to make decisions, rather than automatically attempting to pass on their own learning and values to them. Parents must also be willing to learn from the young, especially in those areas in which young people are more knowledgeable and experienced. But parents who habitually dominate their adolescents with simple assertions such as "because I say so" or "because it's right" not only fail to learn from their adolescents, but also undermine the latter's ability to cope with life in a rapidly changing society.

DUAL-CAREER FAMILIES

dual-career families:
Families in which both parents are employed in the workplace.

One of the most significant changes in the American family is the dramatic increase of families in which both husbands and wives work. More than half of all married women with children or adolescents are now in the work force, compared to one out of five in 1960. The proportion is even higher among women who are separated or divorced. (See Table 5–1.) Although the additional demands of time and effort required in dual-career families affect everyone involved, they are likely to be especially stressful for women who continue to bear primary responsibility for childrearing and running the home.

Working Mothers

How a mother's working outside the home affects a particular family depends on a variety of factors, including the mother's attitude toward her job, her husband's support, their marital happiness, the division of labor in the home, the time spent with young children and adolescents, and the latters' attitudes toward working mothers.

Table 5–1. WOMEN IN THE LABOR FORCE BY MARITAL STATUS AND AGE OF OWN CHILDREN, 1986 (in millions)*

	Total	Children Under 18 Years	Children 6–17 Years	Children Under 6 Years
Married	28.2	12.9	8.8	6.6
Separated	2.1	1.0	0.6	0.5
Divorced	6.2	3.5	2.0	0.7
	Total Percent	Percent Under 18 Years	Percent 6–17 Years	Percent Under 6 Years
Married	54.6	48.2	68.4	53.8
Separated	62.2	60.4	70.6	57.4
Divorced	76.0	72.1	84.7	73.8

*Although a greater *number* of women employed outside the home are married (many with children under 18 years of age), a higher *percentage* of separated and divorced women with children 6 to 17 years of age enter the labor force.

Source: U.S. Bureau of the Census, *Statistical Abstract of the United States: 1987,* 107th ed. (Washington, DC: Government Printing Office, 1986), p. 383.

The attitude of both partners toward working wives is an important factor. As you might expect, a large number of couples agree that both partners have a right to work. Yet, a substantial minority feel otherwise. In Blumstein and Schwartz's (1983) in-depth survey of couples, one-third of the husbands were opposed to both partners working, compared to only one-fourth of wives who felt this way. The survey also showed that married couples who disagree about the wife's right to work are less happy with their relationship. Furthermore, when the wife works, couples tend to fight more about how young children and adolescents are being raised. Much of this stems from the conventional wisdom that children grow up best with their mothers at home. Although there are studies demonstrating that this may not be true, many husbands and wives are not convinced. Interviews with such couples have disclosed two major reasons why men and, in some instances, women feel it is important for the woman to stay at home. One is the belief that women are better at "mothering" than men. The second is that since one parent is needed at home, it makes more sense that the woman be the one because she is likely to earn less than her husband. Unfortunately, the concentration of women in low-paying, service-oriented jobs tends to add fuel to such an argument.

The working mother in dual-career families bears a great deal of the responsibility for reconciling the couple's work schedule and family life. In one study of dual-career couples, neither the combined number of hours a couple works nor the amount of time one or both spouses work is strongly related to the quality of family life. However, the couple's work schedule tends to have stronger effects on the attitudes and behaviors of the wife than those of the husband. As the total time parents work increases, they spend less time with their young children and adolescents. Yet, women feel they lose more time with their children than men do, a perception shared by most husbands. Again, much of the reason for this is because in many homes women continue to bear the primary responsibility for childrearing and domestic tasks (Kingston & Nock, 1985).

One advantage of dual-career families is that adolescents in these homes tend to acquire less stereotyped sex roles than their counterparts in the more traditional, one-wage-earner family. One study comparing adolescents from dual-career families with those from traditional families included whites, blacks, and Mexican-Americans in grades 8 through 12. As expected, adolescents from the dual-career families expressed less traditional sex roles and were more favorable toward mothers working outside the home. Girls from dual-career families were more likely to aspire toward a career and want to combine work and family roles. Boys from the dual-career families were more apt to support a mother's working outside the home and to believe that husbands whose wives worked in the labor force have higher prestige than those whose wives don't work outside the home. In contrast, adolescents from the more traditional families aspire to raising a greater number of children and expect less help with child care from the father than do adolescents from dual-career families. Such findings suggest that dual-career families may foster an ever-increasing number of adolescents who, in turn, will produce dual-career families of their own. If this is true, we may also expect an increase in the number of individuals with egalitarian sex role attitudes and behavior (Stephan & Corder, 1985).

Household Chores

When the woman works outside the home there are usually adjustments made in the division of labor within the home. But the changes are not always as impressive as one would expect. Blumstein and Schwartz (1983) found that working wives do less housework than fulltime homemakers do, but they still do the bulk of what needs to be done at home. About 60 percent of working wives put in 10 or more hours of housework a week, while more than three-fourths of the husbands contribute 10 or less hours a week to household chores. (See Figure 5–2.) A few husbands help substantially with housework, but these seem to be in the minority. Men are most likely to share or take greater responsibility than their wives in such tasks as bedmaking, food shopping, cleaning, and doing the dishes. They are likely to help or take primary responsibility for doing such tasks as ironing, mending, and cooking. At the same time, the younger and better educated the couple, the more evenly household chores are shared (Haas, 1981).

How much do adolescents help out around the house? And is it true that

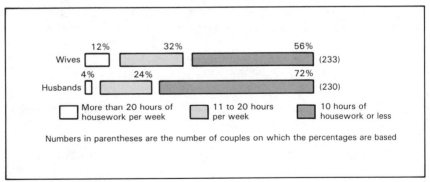

Figure 5–2 TIME SPENT ON HOUSEWORK AMONG MARRIED COUPLES IN WHICH BOTH ARE EMPLOYED FULL-TIME AND BOTH FEEL STRONGLY THAT HOUSEWORK SHOULD BE SHARED EQUALLY.

Source: From *American Couples* by Philip Blumstein, Ph.D., and Pepper W. Schwartz, Ph.D. Copyright © 1983 Philip Blumstein and Pepper W. Schwartz. By permission of William Morrow & Company, 1983, p. 145.

Teenagers are more likely to help with household chores when their mothers work full-time outside the home. (Marc Anderson)

girls continue to do more housework than boys? In an effort to find answers to such questions, Margaret Sanik and Kathryn Stafford (1985) studied 1031 adolescents in two-parent, two-child families. They found that adolescents of full-time employed mothers contributed the most housework, and adolescents of part-time employed mothers contributed the least. At the same time, daughters did more housework than sons, and first-born adolescents of both sexes performed more household chores than second-borns did. Thus, first-born daughters contributed 60 percent more housework than first-born sons. One of the most important but often overlooked factors affecting housework was the school day, with adolescents performing less housework on school days than on weekends. Generally, adolescents decrease the time spent in housework as they grow older. Yet, surprisingly, when adolescents take outside employment, they generally increase their contribution to housework. The most likely explanation is that parents give their permission conditionally—that is, "You can accept the job only if you do your home chores first."

The types of chores adolescents are willing to do at home depend on a number of factors. First, as we've already described, adolescents from dual-career families tend to have less stereotyped sex roles and are more flexible

siblings: Brothers or sisters.

about the particular tasks they perform at home than those in the more traditional, one-wage-earner homes. At the same time, the *sibling structure* of the family—or the number and proportion of boys and girls in the family—is also a significant factor. Generally, as the number of sons increases in a family, the sex-typing of traditionally feminine tasks decreases. But as the number of daughters increases in a family, the sex-typing of traditionally feminine tasks becomes more pronounced. Only in families with no sons does an increase in the number of daughters reduce the sex-typing of traditionally male tasks (Brody & Steelman, 1985). At the same time, there are probably numerous exceptions to these general patterns. Much depends on the individual adolescent and cooperation between family members.

Latchkey Teens

latchkey teens: Adolescents who let themselves in the home with their own key while the parents are at work.

The growing number of mothers who work outside the home has resulted in more young children and teenagers returning to empty houses every afternoon after school. These are often called "latchkey children" or "latchkey teens." There are now more than 2 million school children under the age of 13 left on their own after school, with the largest share in white, upper-income households. The proportion of latchkey teens among those 14 to 18 years of age is even higher. Furthermore, these numbers are expected to increase steadily in the coming years (Schmid, 1987).

Being forced to fend for yourself at such an early age can be a lonely and stressful experience. Many younger adolescents feel isolated because they are ordered to stay inside or have to limit the friends they can have over to the house. Then there's the stress of not having anyone to help when things go wrong. Physical and sexual abuse by older brothers and sisters is higher among unsupervised children and adolecents. And about one-fourth of the fires in homes are caused by unsupervised children (McCrary, 1984). Teenagers who are left to themselves in the afternoons are also more susceptible to peer pressure in regard to experimenting with alcohol and drugs. Sexual encounters between teens are more likely to happen at home while mom is away at work than was the case in the past. And parents often discover what is happening at home the hard way. For instance, a mother returned home unexpectedly one afternoon to find a full-scale party involving drugs and sex among 14 and 15 year olds. Shocked and disillusioned by the experience, she and her husband laid down firmer guidelines for their 15 year old son.

In some communities, parents and school administrators are cooperating to set up after-school programs for children and supervised activities for teens. Many parents are doing a better job preparing their teens to cope with this situation by providing emergency telephone numbers, designating backup places for teens to go if they lose their house keys, and setting ground rules about who may be invited over to the house. Parents may also find it helpful to keep in better touch with their teenagers by telephone during the afternoon. Few educators or parents advocate leaving teenagers completely on their own after school, but more of them are accepting it as a fact of life. Properly prepared, teenagers may learn a sense of responsibility and gain a sense of confidence in looking after themselves. Again, a lot depends on the particular teenager as well as his or her family. (McCrary, 1984).

DIVORCE

divorce: The legal dissolution of marriage, usually accompanied by emotional, social, and financial adjustments.

Returning from school to an empty house during the weekdays is one thing; living in a home with only one parent week after week, seeing the other parent only occasionally, if at all, is another. Yet, this has become a familiar experience to more and more teenagers because of the rising divorce rate. The United States now has the highest divorce rate in the world, with one out of every two marriages ending in divorce. More than half of the divorces involve children under 18. Although the majority of divorces involve only one or two children, over 1 million children and adolescents are affected by divorces each year (U.S. Bureau of the Census, 1985).

Adolescents in Strife-Ridden Homes

family dissension: The presence of a high degree of emotional conflicts and quarreling within a family.

emotional divorce: The emotional alienation between marriage partners.

Because of concern about divorce, we frequently overlook the fact that adolescents may suffer as much, if not more, from the family dissension or "emotional divorce" in the home that precedes the actual legal divorce. The latter usually comes only after a lengthy period of emotional strife in the home which inevitably affects adolescents to some degree, either directly or indirectly. In one study, the home environments of adolescents were observed as much as 6 years before the divorce occurred. As you might expect, these families were less stable and more conflict-ridden than other families. The parents expressed less warmth and concern for their adolescents and emphasized stricter discipline, creating a home atmosphere in which adolescents suffered from degrees of emotional deprivation (Morrison, Gjerde, & Block, 1983).

In a similar study, Deborah Luepnitz (1979) found that over half of adolescents in a divorcing family experienced the greatest stress in the years prior to the divorce. Adolescents described their homes as strife-ridden, with violence not uncommon. They would often avoid bringing friends home out of fear of being embarrassed by a family feud. Understandably, these adolescents viewed the eventual divorce with a sense of relief. At the same time, another one-fourth of the adolescents experienced the greatest stress in the year or so immediately after the divorce. For some, divorce meant losing a parent or ending up on welfare, a stressful experience in itself. For others, the stress came from having to adjust to a stepparent. A smaller number of teenagers experienced the greatest stress during the transition between the pre- and post-divorce phases of family life. The most common sources of stress were the increasing tension prior to one parent's departure, the pressure to take one parent's side, and feelings of rejection.

How Adolescents Are Affected by Divorce

Although divorce may come as a relief in strife-ridden families, the great majority of adolescents find their parents' separation and divorce a very painful experience. Divorce usually has a disruptive effect on family life, which at best requires several years to overcome. In a long-term study of 60 divorced families, Wallerstein and Kelly (1980) found that young children and adolescents were affected somewhat differently by divorce. Young children often felt they were to blame for their parents' divorce and became very anxious about the future. Teenagers felt more vulnerable to the emotional demands of their parents and were more likely to feel stress when parents put pressure on them to take sides in the marital dispute. However, because they were older and more socialized, teenagers were more apt to hide their feelings, thus intensifying their emotional suffering. In this study, adolescents reported the greatest stress during the year following the announcement of their parents' divorce. But a check on these adolescents a year or so later found that over half of them had made a reasonably good adjustment and had returned to a normal developmental pattern. One-fourth of the subjects who had shown trouble at home or school before the separation were about the same a year afterwards. However, another one-fourth of them, who had histories of long-standing difficulties in their families, became even more troubled after their parents' separation, exhibiting lowered self-esteem and superficial and unrewarding relationships with others.

dating: The practice by which individuals of opposite sexes agree to meet at a specific time and place—either alone or in a group—for a social engagement.

There is also a tendency for young people from broken homes to engage in somewhat greater dating activity, as seen in a study of college students involving parental divorce. Students of divorced parents were as likely as students from intact families to form long-term relationships with opposite-sex peers. But they were more likely to have dated, to have had sexual intercourse, or to be living with someone in the 2-week period prior to the administration of the survey questionnaire. Males and females were equally affected. The increase in dating activity was even greater if the parental divorce had been accompanied by acrimony during and after the divorce, by a deterioration of the parent-child relationship, and by the custodial parent remaining single (Booth, Brinkerhoff, & White, 1984).

custodial parent: The parent (or parents) who retains legal custody of his or her children.

Adolescents may cope with their parents' separation and divorce in a variety of ways. Luepnitz (1979) found that many adolescents find the attention and support lacking at home in community groups, such as those at school, church, or Boy or Girl Scouts. Adolescents may also derive emotional support from their friends, with an increasing number of young people now finding themselves in a similar situation. One of the most successful teenage strategies is keeping a safe emotional distance between one's self and the divorcing parents. Adolescents who rely on this approach learn not to get involved in parental disputes and resolve not to let their parents' divorce ruin their own lives. Wallerstein and Kelly (1980) also found that adolescents who coped most successfully with their parents' divorce kept their distance from the divorce proceedings and maintained strong ties with their peers. Adolescents who have great difficulty coping with their parents' divorce may experience significant problems at school, with alcohol or drug abuse, or precocious sexual behavior. In some instances, adolescents from divorced homes may need professional help. But conjoint therapy with the custodial parent and adolescent is often more successful than individual therapy with the teenager, with the latter often evolving out of the former arrangement (Kalter, 1984).

Long-Term Effects of Divorce

The increasing proportion of young children and adolescents from divorced families has prompted researchers to investigate the long-term effects of divorce. Are adults from divorced families more maladjusted and unhappy than those from intact families? Are they more apt to become divorced themselves? In a 10-year followup study, Judith Wallerstein (1984) interviewed 30 adolescents 12 to 18 years of age whose parents had divorced when they were young children. Forty of their parents were also interviewed. She found that adolescents had few conscious memories of the intact families or the marital rupture which ensued. Most of them had made a reasonably good personal and social adjustment and were doing well in school. However, a significant number spoke of the emotional and financial deprivation which had resulted from the divorce. Half of the adolescents reported fantasies that their parents would get back together. Relationships with custodial mothers often reflected closeness, and the relationship with the noncustodial father continued to be of central concern. The results suggest that children who were very young at the time of the marital breakup are considerably less burdened by the divorce in later years than those who were older.

Adults whose parents divorced when they were young may continue to bear emotional scars from the divorce, or they may have benefited from the experience, with a great deal depending on the individual and family involved. According to one view, divorce tends to have lasting, negative effects on the adult lives of those involved. Norval Glenn and Kathryn Kramer (1985) investigated the psychological well-being of adults whose parents had divorced when they were young and found the negative effects of parental divorce were consistently strong on several measures, especially for women. The latter reported greater negative responses on self-rated health, satisfaction from health, friendship, and community of residence. Since experiences of respondents who had lost parents through death were similar but did not have the same long-term negative effects, the authors suggest that the lasting negative effects of divorce may be due largely to the parental conflict before and after the divorce, the social stigma of divorce, and the emotional problems of the custodial parent, rather than the living arrangements in single-parent families.

Another view is that adults whose parents divorced when they were young may actually benefit from the experience in some ways. They may become more responsible, self-reliant, grow up faster, and work even harder at making their own marriages work, thus mitigating some of the negative effects of parental divorce. Partial evidence for this view can be seen in a study by Kulka and Weingarten (1979) showing that adults from divorced parents eventually achieve the same degree of personal and social adjustment as those from intact families. Adults whose parents divorced when they were young did not report any greater level of personal unhappiness or anxiety about their future. They were more apt to regard their youth as the most unhappy time in their lives. But they were also likely to report poorer physical health than individuals from intact families, though such reports tend to diminish with age. Unlike the findings of the study by Glenn and Kramer, described above, women from divorced families generally coped better with the anxiety and stress in their lives than men. However, both men and women from divorced families generally reported greater marital dissatisfaction and a higher incidence of separation and divorce than those from intact families. The most likely explanation for this is that such individuals have learned early in life that divorce is a legitimate solution to an unhappy family life and are even less hesitant to seek it themselves.

SINGLE-PARENT AND REMARRIED FAMILIES

cohabiting couples. Individuals living together as a couple, sharing bed and board.

"Typical" is no longer an appropriate adjective to describe the American family. Instead, a greater variety of living arrangements is emerging, with an ever-decreasing proportion of adolescents living in the traditional nuclear family. Although married couples made up three-fourths of all households in 1960, they constitute a little over half of such households today. The remaining households are composed of an increasing number of never-married singles, cohabiting couples, single-parent families, and remarried families.

Single-Mother Families

single-parent families: Families composed of one or more children or adolescents living with only one parent.

About one out every five children under 18 lives in a female-headed single-parent family, as shown in Figure 5–3. In the majority of these families, the mother is separated or divorced. A much smaller number of such families is headed by a widow. At the same time, the number of households headed by single mothers who have never married has doubled in the past few years,

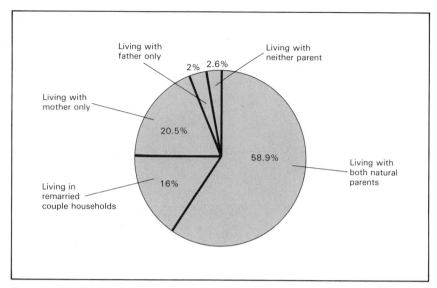

Figure 5—3 CHILDREN UNDER 18 YEARS OLD BY PRESENCE OF PARENTS.
Sources: U.S. Bureau of the Census, *Statistical Abstract of the United States:* 1985, 105th ed. (Washington, DC: U.S. Government Printing Office, 1984), p. 46. Andrew Cherlin and James McCarthy, "Remarried Couple Households: Data from the June 1980 Current Population Survey," *Journal of Marriage and the Family, 47* (1985): 23–30.

remarried families: Families in which one or both parents have been previously divorced, with one or more children and adolescents from an earlier marriage.

now constituting about half of all single-parent families headed by women among blacks.

Robert Weiss (1984) found that separation and divorce almost always brings drastic changes in the amounts and sources of income. As a result, the typical single-parent family tends to suffer from financial deprivation, though there is a wide range of income in these families. Single-parent families at higher incomes rely on earnings of the new household head, and, increasingly with time, on the earnings of others, supplemented in the early years by alimony and child support. However, single-parent families at lower-income levels rely much less on earnings of the new household head, much less on alimony and child support, and much more on public assistance.

Many single parents feel good about the way they are rearing their adolescents. But single parents of both sexes are somewhat more likely than their married counterparts to worry about the job they are doing in rearing their young. (See Figure 5-4.)

sex-role model: Someone who exemplifies the appropriate behavior associated with a male or female sex role.

Boys are generally affected more adversely than girls by the absence of a father, mostly because of the lack of an appropriate sex-role model. Boys in father-absent homes are more apt to suffer from emotional and social problems, drop out of school in greater numbers, and engage in delinquent behavior than boys in intact families. One study, drawing on a sample of 813 adolescents aged 12 to 18 in a stable middle-class suburb of a large city,

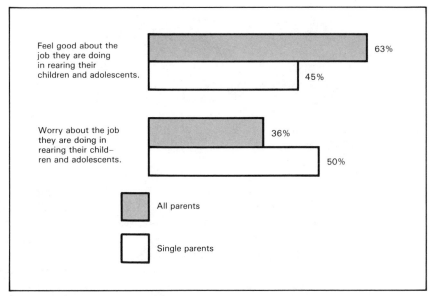

Figure 5—4 PARENTS' ASSESSMENT OF THEMSELVES AS PARENTS.*

One percent of all parents and 6 percent of single parents "not sure."

Source: The General Mills American Family Report, 1976–1977: Raising Children in a Changing Society. Conducted by Yankelovich, Skelly and White, Inc., © 1977 General Mills, Inc., p. 64.

found that boys are affected significantly more by the absence of the father than girls are. The absence of the father resulted in a greater number of problem behaviors, such as alcohol consumption, drug use, and sexual activity, for both sexes. But this was especially true for boys. Over half the boys in father-absent homes were high-rate alcohol users, compared to about one-third of the boys in father-present homes. In addition, about half the boys in father-absent homes were high-rate marijuana users, compared to less than a third of those in father-present homes. Furthermore, 54 percent of the boys in father-absent homes reported very high rates of sexual activity, compared to just 22 percent of the boys in father-present homes. The finding that boys from father-absent homes are at greater risk for such problems should be given greater consideration by community agencies (Stern, Northman, & Van Slyck, 1984).

Girls who grow up without fathers tend to have greater anxiety and difficulty in relating to boys, as was shown in an extensive study by Mavis Hetherington (1981). Three groups of 13- to 17-year-old girls were compared: those from homes with both parents, those who had lost their fathers because of death, and those who had lost their fathers because of divorce. Daughters of widows were tense and shy around boys. As a result, they began dating at a later age and remained more sexually inhibited. Even though daughters of divorced women were somewhat insecure around boys, they compensated for their loss by becoming more flirtatious and aggressive with the opposite sex. They usually began dating at an earlier age and became more sexually active than girls in the other two groups. Girls with divorced parents also had more conflicts with their mothers and held more negative attitudes toward their fathers.

As these girls reached young adulthood, a followup study focused on their sexual behavior, marital choices, and marital behavior. The daughters of divorced women generally marry at a younger age and tend to select marriage partners with inconsistent work histories as well as alcohol and drug problems. Daughters of widows are more apt to marry men with a stable but somewhat straight-laced makeup. Both daughters of widows and divorced women tend to have greater sexual adjustment problems, such as fewer orgasms. In contrast, daughters from two-parent families exhibited greater variation in their sex-role behavior and marital adjustment. They also seemed more at ease in their marital roles, suggesting that they had worked through their father-daughter relationships more successfully.

Black adolescents are especially likely to grow up in single-parent families, with the dramatic rise in female-headed homes one of the most significant changes in the black family in the past 30 years. In 1960, three-fourths of all black families with children had two parents. But by the mid-1980s, less than half of all black families with children had parents of both sexes. Today, 51 percent of black children under 18 live with their mothers in single-parent families (U.S. Bureau of the Census, 1985).

out-of-wedlock birth: Birth to an unmarried woman.

The biggest single reason for this change is the increase in out-of-wedlock births to blacks. About half of all children born to blacks are conceived out of wedlock. Most of these births are attributable to teenage pregnancies, with 40 percent of black women having given birth at least once by 20 years of age. The high rate of out-of-wedlock births combined with the greater proportion of black women who never marry and those who remain single after divorce all contribute to the increase in female-headed homes among blacks. But, as Robert Staples (1985) points out, the prevalence of female-headed homes among blacks is due to a combination of social, economic, and cultural factors. A major factor is the high rate of unemployment among black males, rendering many unable to assume the responsibilities of husband and father. In addition, many black women have achieved greater independence, thereby encouraging the trend toward single-parent families. Yet, many black women are in dead-end jobs, lack alimony and child support, and require public assistance. The median income of single-parent black families is about $7,500, barely a third of the average income of two-parent black families. By contrast, the black families who have achieved middle-class or upper-middle-class standards derive even greater satisfaction from family life than from their jobs. Black teenagers growing up in middle-class homes share many of the same opportunities and values as their white counterparts. Meanwhile, an

alimony: Court-ordered allowance paid to an individual by his or her spouse after a legal separation or divorce.

even larger number of black teenagers are growing up in father-absent, economically deprived homes, thereby setting the stage for problem behaviors such as dropping out of school, delinquency, alcohol and drug abuse, and teenage pregnancy. Male teenagers are especially at risk. If the present trend continues, Staples estimates that by 1990, of the black families with young children or adolescents, 6 out of 10 will be headed by women, with most of them living on incomes below the poverty level.

Single-Father Families

Another significant change in family life has been the dramatic increase in single fathers rearing children. There are now more than 600,000 fathers rearing young children and adolescents alone, triple the number since 1970. According to a survey by Geoffrey Greif (1985), single fathers tend to be white, well-educated, and successful in their careers. Most of them are in middle-level positions, such as managers, administrators, buyers, and store owners, and have higher than average incomes. More than three-fourths of them have one or two children. But a small proportion of single fathers are rearing three or more children. Male children outnumber females four to three, with an average age of 11 or 12 years. In most instances, the men didn't want their marriages to end, didn't initiate the divorce, and find their current situation very stressful. They usually gained custody through mutual consent or were picked by the children, mostly because the mothers were either unstable, had alcohol and drug problems, or were incapacitated. In some cases, the mothers had left the marriage. Men who were deserted by their wives had the least confidence in themselves, gave themselves the worst rating as fathers, and found the experience of being a single parent most stressful. Yet, two-thirds of the single fathers wanted custody very much, felt they were prepared for it, and experienced satisfaction in many areas of parenting.

custody: The legal arrangement following a divorce specifying which parent the children and adolescents will reside with until they reach legal age.

How well adolescents fare in single-father families depends on a variety of factors, such as the father's personality, his emotional involvement with the adolescents, his parenting skills, his use of support systems, the ongoing relationship between the ex-spouses, and how consistently they deal with their adolescents. A great deal depends on the particular adolescent and single-father family. But one extensive investigation of children of both sexes 6 to 11 years old growing up in father-custody, mother-custody, and intact homes showed that boys achieve more competent prosocial behaviors in father-custody homes, whereas girls seem to achieve better adjustment in mother-custody homes (Warshak & Santrock, 1983). A major reason boys benefit from single-father homes may be the father's greater tendency to use available support systems, like household help and relatives; use of such support systems is positively related to the youth's self-concept. Also, the

prosocial behavior: Actions intended to benefit others that have no apparent selfish motivation.

presence of a male model and greater use of authoritative methods of parenting—as contrasted with authoritarian or permissive methods—fosters the growth of social competence among boys.

Remarried Families

More than three-fourths of divorced people remarry, usually within a few years, so adolescents are as likely to grow up in remarried families as in single-parent homes. These families are labeled variously as "remarried families," "reconstituted families," "blended families," and "remarried couple households." An analysis of census data by Andrew Cherlin and James McCarthy (1985) has shown that there are slightly more than 9 million remarried families—in which one or both parents has been previously divorced. In about a third of these marriages only the wife has been divorced; in another third only the husband has been previously divorced; and in still another third of these families both partners have been previously divorced. The majority of partners in remarried families are still in their thirties, with many of their children just reaching preadolescence. Parents over forty are more likely to have adolescents. Almost 10 million children under 18 years of age, or 16 percent of the nation's children, now live in remarried families. About half of them are living with a stepparent, with stepfathers outnumbering stepmothers six to one. The other half of the children were born after the new marriage, with such families resembling intact families in many ways. Only a handful of remarried families have young children or adolescents from both previous marriages *and* the new marriage.

stepparent: The person who has married one's parent after the death or divorce of the other parent.

Some of the impact of divorce, father absence, and single parenting tends to be modified when the remaining parent remarries, especially so when fathers remarry (Parish & Dostal, 1980). At the same time, life in remarried families can be stressful for everyone concerned. In addition to the usual difficulties of marital adjustment, the new marriage partners experience special stresses, such as the possible interference of an ex-spouse. But much of the difficulty revolves around the presence of a stepparent, partly because of the lack of clear roles for stepparents and stepchildren. How do you relate to an adolescent whose development has already been heavily influenced by someone with different values? Or, how do you behave toward a stepfather or stepmother when you already have a biological father or mother? Three of the most common problems encountered by stepparents are adjusting to the habits and personalities of adolescents, gaining their acceptance, and disciplining them. Such problems generally demand an extensive period of adjustment on the part of parents, stepparents, and adolescents alike. Whenever stepparents and adolescents can discuss their problems in an open and trustful manner, the adjustment goes more smoothly. Stepmothers tend to have more difficulty making these adjustments than stepfathers, probably because mothers traditionally have been more emotionally involved with the children

(Rice, 1983). Also, the older the adolescent at the time of the divorce and remarriage, the more difficult the adjustment, mostly because of the longer emotional involvement in the former marriage.

Adolescents may react to the stress of stepparents in a variety of ways. Adolescents in remarried families with stepfathers tend to engage in more delinquent acts than those in intact families (Haney & Gold, 1977). Girls with stepfathers also exhibit higher levels of alcohol and drug use as well as sexual activity (Kalter, 1977). Adolescents in remarried families are also more likely to be involved in family violence and become victims of sexual abuse or neglect than those in intact families. It appears that adolescents are at high risk for abuse in families that have undergone a series of stressful changes in the preceding year—often the parents' divorce—have stepparents, and lack a flexible, appropriate family structure (Garbarino, Sebes, & Schellenbach, 1982).

Many of these problems can be minimized or avoided when parents and stepparents face up to the challenges and prepare for them appropriately. Generally, it is wise for prospective partners to give adolescents ample opportunity to get to know the future stepparent and to explore the changes remarriage will bring, such as new living arrangements and relationships with the noncustodial parent. It's also usually best not to force relationships with stepchildren but to let them develop gradually. Stepparents may find it more appropriate, especially at first, to become an "additional" parent rather than a replacement for an absent parent. Comparisons with the absent parent are inevitable, and stepparents should be prepared to be tested, manipulated, and challenged in their new role. At the same time, parents and stepparents should keep in mind that remarried families, like intact families, generally function as their members make them function and have vast resources for cooperation and growth. When parents and stepparents do their utmost to establish good communication in the family, show consideration and fairness for the young people involved, and model cooperation, remarried families may function as well, if not better, than many intact families.

SUMMARY

Changes in the American Family

1. We began the chapter by pointing out that the family is a major influence on the adolescent's development, and that the quality of family life is being affected by a variety of changes in the American family.

2. Adolescents are growing up in families that are smaller and more mobile than those in the past.

3. Although the family has surrendered many of its traditional functions to other agencies, it has strengthened the importance of the emotional nurture given adolescents within the family.

4. Parents and adolescents alike are adopting more flexible, functional roles within the family, reflecting our rapidly changing society.

Dual-career Families

5. The increase in the number of dual-career families has put disproportionate stress on mothers who work outside the home, mostly because they continue to bear primary responsibility for childrearing and running the home.

6. The extent to which husbands and adolescents help with household chores depends on a variety of factors, with adolescents from dual-career families less likely to have stereotyped sex-role notions about household chores.

7. Although latchkey teenagers are exposed to greater hazards than those with a parent at home, they may also learn a sense of responsibility and gain confidence in looking after themselves.

Divorce

8. More than half of the divorces involve children under 18, and more than 1 million young children and adolescents are affected by divorce each year.

9. Adolescents suffer as much from the conflicts and "emotional divorce" within strife-ridden families as from the separation and legal divorce which often follows.

10. Most adolescents find their parents' separation and divorce a very painful experience; such children are especially vulnerable to pressure from parents to take sides.

11. The extent to which adults suffer later from the long-term effects of parental divorce when they were young varies according to a number of factors, though adults from divorced homes tend to have a higher incidence of divorce than those from intact families.

Single-Parent and Remarried Families

12. Because of the rising divorce rate, an increasing proportion of adolescents are growing up in single-parent and remarried families.

13. The majority of single-parent families are headed by women, with boys being more adversely affected by the absence of a father than girls.

14. Daughters of divorced women tend to date at an earlier age, become more sexually active, and marry earlier than daughters of widows or married couples.

15. About half of black children under 18 live with their mothers in single-parent families, of which a major contributing factor is the high rate of out-of-wedlock births among teenage blacks.

16. Single-parent families headed by men have tripled in the past decade or so, with boys especially likely to benefit from the presence of a father.

17. Three-fourths of divorced people eventually remarry, and thus adolescents are as likely to grow up in remarried families as in single-parent families, with stepfathers greatly outnumbering stepmothers.

18. Much of the difficulty in remarried families revolves around the presence of a stepparent, but when parents and stepparents do their utmost to establish good communication in the family and show consideration for the young people involved, remarried families may function as well as intact families.

REVIEW QUESTIONS

1. How has American family life changed in the past two decades?

2. What are the major functions of the family in American society?

3. How are adolescents affected by the mother working outside the home?

4. To what extent are the household tasks performed by adolescents still sex-typed?

5. Discuss the growth opportunities as well as the hazards of being a latch-key teenager.

6. What are the characteristic ways adolescents are affected by parental divorce?

7. Describe some of the potential long-range effects of parental divorce.

8. In what ways are male adolescents more adversely affected by the absence of their fathers than females are?

9. Why do you think stepmothers tend to have greater difficulty than stepfathers in gaining the acceptance of adolescents in blended families?

10. What suggestions could you make to someone who is becoming a stepparent to an adolescent?

6 Family Relationships and Adolescent Autonomy

- **PARENT–ADOLESCENT RELATIONSHIPS**
  How Adolescents Feel about Their Parents
  Parental Control
  Adolescents' Perception of Parental Control
- **PARENT–ADOLESCENT COMMUNICATION**
  Attitudes and Values
  Parent–Adolescent Conflicts
  Improving Parent–Adolescent Communication
- **ADOLESCENT AUTONOMY**
  The Process of Achieving Autonomy
  Types of Autonomy
  Autonomous Decision Making
- **SUMMARY**
- **REVIEW QUESTIONS**

Mark and Cheryl, classmates in high school, have grown up in families that are quite different. Mark's parents are rather strict, dictating when he is to do his homework and which chores are his responsibility at home. Whenever Mark objects, his parents will say, "As long as you live in this house you'll do as you're told." In order to get along with his parents, Mark often feels he must deceive them, saying he has completed his homework when he hasn't and lying about where he's going on his evenings out. On the other hand, Cheryl's parents are less strict and more trusting of their daughter. When it comes to such matters as school work, household chores, and dating, they're more apt to talk things over with Cheryl rather than issuing orders. When Cheryl disagrees with her parents, as she does from time to time, she feels she can negotiate with them until they reach some sort of mutual agreement. For instance, once when Cheryl's parents asked what time she planned to return from a party, she replied, "Oh, I don't know—probably about 2 or 3 o'clock at the latest." When her parents protested that this was too late, they discussed the matter at length until they reached a mutual decision—Cheryl would be home by 1 o'clock.

PARENT—ADOLESCENT RELATIONSHIPS

Which of these families most resembles the one in which you grew up? Did your parents tend to be autocratic, always telling you what to do like Mark's parents? Or were they more reasonable and willing to talk things over like Cheryl's parents? Chances are the great majority of parents would fall somewhere in between, but probably would be more like Cheryl's parents. In each case, the way parents treat their child has a decisive influence on the latter's development of autonomy or self-direction. Individuals who grow up feeling they have little to say in family matters are more apt to rebel as a way of asserting their independence, though often with the opposite results. By contrast, adolescents encouraged to make their own decisions are more apt to acquire genuine autonomy at an earlier age, relating to their parents, other authorities, and their peers in a more adult manner.

How Adolescents Feel about Their Parents

If you were to travel across the country interviewing several hundred teenagers and tabulating questionnaire responses from another 160,000 adolescents, what do you think they would say about their parents? You might be pleasantly surprised to discover that the chief fear of America's teenagers is not a nuclear war or being critically injured in an accident. Instead, what they most fear is losing their parents, and not by divorce, but by death. Teenagers do care about their parents. They may get angry at their parents and occasionally rebel against them. But when all is said and done, the great majority of American teenagers feel positive about their parents (Norman & Harris, 1981).

Similar evidence comes from a longitudinal study by Offer, Ostrov, and Howard (1981), which includes a cross-section of adolescents 13 to 18 years of age from a wide variety of homes and schools in the 1960s, late 1970s, and 1980. Again, the researchers found that the great majority of adolescents feel close to their parents and do not perceive any major problems in their adolescent-parent relationship. About 9 out of 10 typical adolescents feel that their parents are satisfied with them most of the time. And more than 8 out of 10 feel they will become a source of pride to their parents in the future. The study did show modest changes between adolescents' attitudes in the 1960s and in the 1970s, the latter adolescents feeling slightly less satisfied with their parents and more distant with each other than did adolescents in the early 1960s. But the great majority of adolescents in both eras, male and female, older as well as younger, decribed their family as a harmonious and well-functioning system.

juvenile delinquent: Generally, someone under the legal age who has engaged in illegal behavior; technically, a person convicted of such a violation in juvenile court.

However, the juvenile delinquents and emotionally disturbed adolescents surveyed were much less positive about their parents. They were much less apt to share their parents' values as completely as the typical adolescents did, or to feel that they were an integral part of a cohesive family. Only 6 out of 10

Most teenagers feel close to their parents. (Laimute Druskis)

of the troubled adolescents felt their parents were satisfied with them. Further-more, the delinquents and disturbed adolescents were more likely to say they liked one parent better than the other, that their parents were ashamed of them, and that they had been carrying a grudge against their parents for years. Almost half the delinquent and disturbed adolescents tried to stay away from home most of the time, compared to only one-fourth of the normal adolescents. In short, while troubled teenagers tend to harbor bad feelings toward their parents, the great majority of typical teenagers have decidedly positive feelings toward their parents.

Parental Control

power assertion (authori-tarian): The method of parental guidance and dis-cipline that relies primar-ily on the appeal to power, allowing adoles-cents little or no participa-tion in such matters.

How close teenagers feel toward their parents depends largely on the way parents attempt to influence or control adolescents, especially in matters of discipline. After reviewing the research on this subject, Martin Hoffman (1980) has described three broad types of parental control or discipline: power assertion, love withdrawal, and induction. Parents using the *power assertion* or authoritarian method attempt to control their adolescents by relying on physical power. Generally, this includes the use of threats, physical punishment, and control over the adolescent's material resources. Parents relying on *love withdrawal* tend to give direct but nonphysical expression of

love withdrawal: The method of parental discipline that relies primarily on the threat of separation or rejection by parents.

induction (democratic): The method of parental guidance and discipline that relies primarily on a rational explanation of parental rules and encourages adolescents' choices and responsibility.

their anger or disapproval. They may refuse to listen to their adolescents, turn their backs on them, or explicitly express their negative feelings toward their adolescents. Although the withdrawal of love poses no immediate physical threat to the adolescent, it may be more punitive—because of its implicit threat of abandonment—and more prolonged than the authoritarian method, lasting for hours or days. Parents using the *inductive* or democratic method explain the reasons for their rules. Instead of relying on the fear of punishment or abandonment, these parents appeal to their adolescents' capacity to understand the situation and their desire for mastery and independence.

Each type of parental control tends to evoke certain behaviors in young people. Adolescents reared with the authoritarian approach may become dependent, submissive, and overly conforming on the one hand, or rebellious and hostile on the other. Such adolescents may act compliant and evasive in the presence of their parents or other authorities, but defiant and resentful of them behind their backs. In either case, they remain dependent on some external authority. Adolescents reared with the love-withdrawal approach may inhibit expressions of anger for fear of being rejected. They may mistakenly equate their misconduct with being inferior and go through life with a strong need for approval. Adolescents growing up in homes where parents use a more democratic, inductive form of control are fortunate in several ways. First, by having assumed privileges and responsibilities on appropriate occasions, they are more likely to achieve behavioral autonomy in age-appropriate tasks. Second, feeling loved and respected by their parents, they may have less need for the approval of authorities. Third, through regular participation in the decision-making process, these individuals learn the value of acting in a reasonable and responsible manner.

In practice, of course, the exercise of parental control usually includes a mixture of authoritarian, love-withdrawal, and democratic tendencies, with more varied results. Also, much depends on the parents' personalities, their consistency in the use of a particular approach, as well as the combination of affection and control used. For instance, warm, accepting parents who use authoritarian control will probably have a less restrictive impact on adolescents than would authoritarian parents who are distant and harsh. Conversely, undemonstrative parents using the democratic approach may be perceived as being permissive, making adolescents feel insecure and rejected. Furthermore, too much or too little control or affection usually has detrimental results. Instead, parents who are affectionate and warm but not afraid to exercise the appropriate authority when necessary will get their adolescents' cooperation most of the time.

In one study (Smith, 1983), mothers and fathers of teenagers in the sixth, eighth, and tenth grades were asked about their use of parental control. Most parents reported that they used a moderate level of parental control, with mothers being slightly more controlling than fathers. In the parents' eyes, most disagreements were resolved slightly in favor of the parents. Adolescent

compliance: The act or tendency to give in readily to others' requests.

compliance was greater when mothers used an intermediate level of control, rather than very high or low levels. But fathers got more compliance at intermediate and high levels of control, probably because of the authority traditionally associated with the father's role. Girls expressed more resentment of parental attempts to control them than boys did, but they were no more or less likely to comply with their parents' requests than boys.

Adolescents' Perception of Parental Control

How do teenagers feel about all this? To find out, Cay Kelly and Gail Goodwin (1983) asked adolescents how they perceived the use of parental control in the home. They obtained data from 100 high school students representing a variety of social and cultural backgrounds and racial groups. The investigators were surprised to find that 83 percent of these students felt their parents adopted a democratic type of parental control. Only 11 percent saw their parents using autocratic control, with another 6 percent having permissive parents. Adolescents with democratic parents responded favorably to two-thirds of all the questions asked, whereas those with autocratic parents did so for only half the questions and those with permissive parents for barely a third. The low percentage of favorable responses among students with permissive parents suggests that adolescents would prefer some sort of discipline to none at all. See Table 6–1 for students' responses to selected questions.

Of special note are the responses to two key questions, 1 and 8 respectively. When asked, "Would you like to be the kind of person your mother/father is?" the majority of adolescents with democratic parents said they would, whereas only a minority of those with autocratic or permissive parents responded favorably. Again, in response to the question, "Would you discipline your children in much the same way as your parents are disciplining you?" two-thirds of adolescents with democratic parents said yes, several times as many as those with autocratic parents. Such results confirm the widely accepted notion that adolescents who grow up in families using a democratic approach are more apt to have positive feelings toward their parents and conform with parental rules than those with autocratic or permissive parents. Much of the rationale here is that democratic parenting appeals to the adolescent's capacity for understanding and self-direction, thereby fostering a high degree of autonomy and cooperation with parents.

In actual practice, however, parents are probably much less consistent in their use of parental control than we have been led to believe. This was brought out in another study of adolescent perception of parental control by Mark deTurck and Gerald Miller (1983). These investigators found that parents' persuasive strategies are much more complex than implied in the traditional three broad categories—authoritarian, love withdrawal, and demo-

Table 6—1 ADOLESCENTS' PERCEPTION OF PARENTING STYLES

	Positive Response	Negative Response	Percentage of Positive Response
Autocratic			
1. Be Like Mother/Father	3	16	16%
2. Stop Seeing Friends for Mother/Father	4	15	21%
3. Change Girlfriend/Boyfriend	7	12	37%
4. Rebel in Other Ways	12	7	63%
5. Usually Does What Mother/Father Says	18	1	95%
6. Improve Grades for Mother/Father	18	1	95%
7. Cut Down on Outside Activities	13	6	68%
8. Discipline Like Mother/Father	3	16	16%
Permissive			
1. Be Like Mother/Father	2	9	18%
2. Stop Seeing Friends for Mother/Father	1	10	9%
3. Change Girlfriend/Boyfriend	3	8	27%
4. Rebel in Other Ways	2	9	18%
5. Usually Does What Mother/Father Says	5	6	45%
6. Improve Grades for Mother/Father	9	2	82%
7. Cut Down on Outside Activities	4	7	36%
8. Discipline Like Mother/Father	3	8	27%
Democratic			
1. Be Like Mother/Father	98	48	67%
2. Stop Seeing Friends for Mother/Father	72	74	49%
3. Change Girlfriend/Boyfriend	59	87	40%
4. Rebel in Other Ways	92	54	63%
5. Usually Does What Mother/Father Says	136	10	93%
6. Improve Grades for Mother/Father	139	7	95%
7. Cut Down on Outside Activities	121	25	83%
8. Discipline Like Mother/Father	93	53	64%

$N = 100$

Source: Cay Kelly and Gail E. Goodwin, "Adolescents' Perception of Three Styles of Parental Control," *Adolescence* (Fall 1983): 570.

cratic. They found that adolescents' perceptions of parental strategies varied considerably by age and sex of both participants as well as the particular situation involved. For instance, when parents were attempting to get their adolescents to help with spring cleaning, they used at least 16 different persuasive-message strategies, with significant age-by-sex interactions for the following three strategies:

Pregiving: "Since I raised your allowance not too long ago, I expect you to help out with the cleaning this Saturday."

Aversive stimulation: "If you don't help out this Saturday, you will have to do the dishes every night until you learn to do your share of the work around here."

Positive moral appeal: "As a member of this family, it is morally right for you to help out around here with the work."

Preadolescent boys reported that their mothers were more likely to use these strategies, while older adolescent boys said their fathers were more likely to use such strategies. Female perception was just the opposite. That is, preadolescent girls reported that their fathers were likely to use these strategies, while older adolescent girls said their mothers were more apt to use them. Furthermore, adolescents in their mid-teens indicated that their mothers were especially likely to use still another strategy, namely, the *debt appeal:* "Since I let you go out with your friends last weekend, you owe it to me to help clean the house this Saturday." All of this suggests that the process of parental influence is more complex than ordinarily depicted. It appears that parents use a wide range of persuasive strategies with their adolescents, varying them by the age and sex of both parent and adolescent as well as the particular context involved.

PARENT–ADOLESCENT COMMUNICATION

How well adolescents get along with their parents may be gauged by the communication within the family. But communication includes more than the verbal exchange of information. Even more important are attitudinal factors such as trust, being comfortable enough to express one's feelings, empathy, and the willingness to listen to another's feelings, especially during disagreements. Relatively normal, healthy adolescents tend to come from homes where there is reasonably good communication between themselves and their parents, to the extent that on occasions they can even "agree to disagree" without threatening family cohesiveness. By contrast, families with disturbed adolescents tend to have faulty communication. Parents and adolescents tend to talk past each other, without recognizing each other's feelings or needs, and thus failing to become responsive to each other's needs. Parents are especially apt to express harsh criticism, act intrusively, and induce guilt in their teenagers. And adolescents in such families express a greater frequency of negative or contrary responses, experience lowered self-esteem, and feel a sense of isolation (Fischer, 1980).

Attitudes and Values

Suppose you wanted to find out how well things were going in a particular family, who would be the best person to ask? Would you get a more accurate report if you asked the parents or the teenagers? Chances are their responses wouldn't be that different, depending, of course, on the issues involved. In a study involving almost 4000 secondary school students and their parents in New York State, Dorothy Jessop (1981) found a remarkable agreement in the perceptions of family life between parents and their teenagers, similar to reports in previous studies. Apparently, parents are not as inclined to "idealize" family life as popularly believed. Instead, both parents and adolescents have a tendency to enlarge their own power and influence in interpreting family matters. Thus, adolescents tend to exaggerate the actual differences between themselves and their parents, while parents tend to minimize them, with such misperception serving to justify the respective positions of each generation. But on the whole, Jessop found a remarkably high level of agreement between parents and their teenagers regarding the quality of relationships in the home, including the closeness between parents and teenagers, the decision-making process in family matters, and adolescents' dependence on parental guidance. At the same time, there was wider disagreement in their perceptions of certain areas of parent-adolescent relationships, especially the ease with which they could discuss the teenager's use of drugs or dating certain members of the opposite sex.

hypothetical: Conditional thinking based on that which is assumed or supposed.

The degree to which parents and teenagers agree or disagree on particular issues may be seen in a study involving 1002 people of all ages from 17 to 94, with 42 percent of them being parents. Respondents were asked whether they would back the parents or the adolescent in 42 hypothetical family situations. The majority of respondents strongly supported firm parental guidance in matters pertaining to the teenager's sex lives. For instance, 90 percent of them supported the mother who "refused to discuss the issue" with a 15-year-old girl who wanted to spend the weekend with a college student she had dated only twice. Similarly, 84 percent of the group supported the parents who insisted that a 15-year-old boy leave his bedroom door open when his girlfriend was visiting in the home. However, respondents of all ages were more inclined to grant teenagers greater autonomy in matters relating to educational choices, religion, and personal grooming. For instance, 65 percent of the group would permit a 10-year-old Jewish girl who is "uninterested" in her own faith to attend the services of a variety of other religions. And only 42 percent of the respondents supported parents who insisted a 12-year-old boy cut his hair to a length more acceptable to the parents (Bornstedt, Freeman, & Smith, 1981).

Similar levels of agreement and disagreement have been discovered in the values of parents and adolescents. In a survey of the literature, Norman Feather (1980) found that both parents and adolescents give high importance to values such as world peace, happiness, wisdom, honesty, and responsibility. Likewise, both generations give relatively low rankings to values such as pleasure, social recognition, salvation, and obedience. At the same time, however, parents tend to be the more conservative of the two groups and to give greater

Table 6–2 THE TRANSMISSION OF VALUES TO YOUTH
(in percents)*

	Believe and want youth to believe	Have doubts but still want to teach youth	Don't believe and don't want to pass on to youth
It's not important to win, it's how the game is played.	71	21	8
The only way to get ahead is hard work.	65	31	3
Duty before pleasure.	58	33	9
Any prejudice is morally wrong.	51	33	15
There is life after death.	51	27	21
Happiness is possible without money.	50	36	13
Having sex outside of marriage is morally wrong.	47	25	28
Everybody should save money even if it means doing without things right now.	42	37	20
People are basically honest.	37	47	16
My country right or wrong.	34	41	24
People in authority know best.	13	56	30

*"Not sure" responses not included.

Source: The General Mills American Family Report, 1976–1977: Raising Children in a Changing Society. Conducted by Yankelovich, Skelly and White, Inc., © 1977 General Mills, Inc., p. 82.

importance to family values and national security. Adolescents generally assign more importance to values such as true friendship, pleasure, freedom, excitement, equality, and happiness. Generally, daughters are more likely to hold values similar to their parents than sons. But it is important to recognize that adolescents are actively involved in sorting out their values, so that, as they grow in experience and acquire greater education, their values increasingly diverge from those of their parents. Then, too, parents are rethinking their values and which ones they want to pass onto their youth. (See Table 6-2.)

Parent—Adolescent Conflict

No matter how well adolescents get along with their parents, some degree of disagreement and conflict between the generations is inevitable for several reasons. First, parents have become so accustomed to socializing their young that they are reluctant to relinquish control. Yet it is normal for growing adolescents to assert their independence. Thus, some clash of wills is to be expected. Second, parents and adolescents are at different developmental stages of life. Physically and psychologically, adolescents are on the way up, while their parents are on the way down. Adolescents have boundless energy, budding sexuality, and are eager to explore their potentials. At the same time, parents have diminished energy, waning sexual attractiveness, and are more aware of the limitations and claims on their lives. Furthermore, the self-assessment that accompanies middle age makes parents especially sensitive to the complaints and criticisms of their adolescents. Finally, parents' lives have been shaped by a different era than that of their adolescents, thereby increasing the potential for disagreement and conflict between generations.

Parent—adolescent conflict generally increases during early adolescence, remains stable throughout middle adolescence, and then declines as adolescents leave home. Most of the conflicts during early adolescence revolve around everyday matters such as school work, home chores, dress, sibling conflicts, and disobedience. Older adolescents pose more serious conflicts involving the use of the car, alcohol, drugs, and sex. Parents are more likely to argue with their sons about taking care of family property or the use of the car. Parents tend to disagree with their daughters more about dating and sex. One study showed that adolescents had an average of one argument with their parents every three days, with each disagreement lasting about 11 minutes on

Parent—adolescent conflicts generally increase during early adolescence. (Laimute Druskis)

the average. Most of the conflicts were moderately upsetting to the teenagers. These conflicts were more likely to involve the mother rather than the father, with the majority of conflicts being between girls and their mothers (Montemayor, 1982).

A mild to moderate degree of conflict between parents and adolescents is to be expected as adolescents develop a firmer identity of their own choosing and assert their independence. In fact, lack of such disagreements may reflect an adolescent's undue dependence, anxiety over separation, and fear of independence. On the other hand, adolescents who are actively exploring their own identity and asserting their independence will normally evoke a certain degree of disagreement with their parents. However, frequent conflict in the home is detrimental to the adolescent's growth. Chronic or intense conflict between parents and adolescents is associated with a wide variety of problem behaviors, including academic underachievement, dropping out of school, delinquency, teen pregnancy, running away from home, alcohol and drug abuse, as well as physical and sexual abuse of teenagers.

Improving Parent—Adolescent Communication

Many problems at home stem from the failure to establish good parent-child communication during the child's formative years of development. Parents may unwittingly adopt habitual ways of expressing themselves that are judgmental, such as advising, warning, criticizing, praising, and questioning. They may rarely stop to realize that such statements put their children on the defensive. Parents are quick to complain about the disrespectful ways their teenagers talk back to them, without realizing that the latter have learned their basic style of communication through imitating their parents. Mothers and fathers who have repeatedly told their children to "shut up" or "do it because I said so" may be dismayed to hear such phrases being thrown back at them when their children reach adolescence. Such faulty communication habits make it all the harder for adolescents to establish good communication with their parents and other adults. (See Table 6–3.)

Parents may improve communication with adolescents through workshops such as Thomas Gordon's (1975) Parent Effectiveness Training (P.E.T.). Recently, adolescents have also begun receiving similar help in Gordon's Youth Effectiveness Training (Y.E.T.). Essentially, such training helps individuals establish honest but nonjudgmental communication which fosters cooperative problem solving. Parents and adolescents alike learn how to express themselves in a nonjudgmental way and to listen to others more accurately. Effective self-expression involves what Gordon calls the "I" message, by which individuals express their emotions nonjudgmentally. The "I" message includes four components: (1) a nonjudgmental description of the other person's objectionable behavior, (2) the concrete effects on me, (3) how I feel about it, and (4)

"I" messages: The expression of one's feelings about another person's behavior in a nonjudgmental way.

Table 6–3 PEOPLE WITH WHOM CHILDREN AND ADOLESCENTS FIND IT HARD TO COMMUNICATE (in percents)

Principals	46
Teachers	41
Doctors	34
Priests/Ministers/Rabbis	32
Fathers	27
Brothers	22
Sisters	18
Mothers	16
Friends	11

Source: *The General Mills Family Report, 1976–1977: Raising Children in a Changing Society.* Conducted by Yankelovich, Skelly and White, Inc., © 1977 General Mills, Inc., p. 138.

empathetic listening: Listening and telling people what you've heard them say to you in an accurate, nonjudgmental way.

what I'd prefer the person do about it. For instance, suppose you are a parent and your daughter comes home several hours later than promised. You may instinctively say angrily, "Where have you been?" Chances are your daughter may hear only the judgmental anger and feel she has to defend herself, with an unproductive argument ensuing. But suppose you were to say something like, "When you come home much later than you told us, we worry about you. We'd like to know when you're coming home." Your daughter is more likely to get your main message, which is, "We worry about you," rather than reacting to your anger at her.

Equally important is the skill of reflective or empathetic listening, which involves giving nonjudgmental feedback to the speaker as a way of checking on the accuracy of what has been heard. For instance, suppose your son complained, "I can't stand my math class." Instead of criticizing him or giving a lot of advice—which usually falls on deaf ears—you might give a more empathetic response such as, "You really dislike that class." The teenager who feels that you're really listening might continue to express his feelings, including the deeper reasons he doesn't like the class. "Yeah," he might say, "the teacher gives us too much homework to do." An understanding parent might then say, "Then it's more the excessive amount of homework rather than the subject of math itself you dislike." This type of communication usually leads to a more fruitful conversation in which both the parent and the adolescent may explore the real issues involved, thereby setting the stage for a more constructive solution to their differences.

Gordon's effectiveness training is a way of implementing the type of democratic parent–adolescent relationship explained earlier. Gordon rightly contends that much of the adolescent's rebelliousness is directed more at the parents' misuse of power than at the parents themselves. As adolescents become more mature and independent, they resent being ordered about and reprimanded in front of others like little children. And they become increasingly immune to parental threats of punishment. Parents who have relied on autocratic control since childhood may discover that their influence over the young wanes sooner than they expected. In contrast, parents who respect the rights of their adolescents, talk to them in nonjudgmental ways, and really listen to them not only promote good parent–adolescent communication but also the teenager's self-respect and autonomy. It should come as no surprise that adolescents who are satisfied with the communication and help received at home tend to have higher self-esteem than those who are dissatisfied at home (Burke & Weir, 1978).

ADOLESCENT AUTONOMY

autonomy: the state of being self-determining.

It's been said that parents should love their children in order that the children may become persons in their own right—and then should prepare to be jilted. Accordingly, as parents stand back and encourage adolescents to make their own decisions, they may not always be pleased with the outcome. Think of the parents with careers in the professions whose son or daughter chooses not to attend college right away, if at all, or think of the young person who declines to enter the family's business. Yet, these same parents, in time, may come to realize how fortunate they are to have such young people when they see other youths having difficulty breaking away from their parents. All this reminds us that one of the goals of rearing children is to help them eventually achieve autonomy, so that they in turn can move toward adulthood and, if they so choose, become parents in their own right.

The Process of Achieving Autonomy

satellization theory: The idea that socialization occurs through a process of initial dependence upon parents (satellization) and eventual breaking away from parents (desatellization).

The achievement of autonomy is more gradual and difficult than meets the eye largely because of the inherent discontinuity between dependence and independence in the developmental process. Ausubel's satellization theory of development (Ausubel, Montemayor, & Svajian, 1977) may be a helpful way of understanding this transition.

Young children are seen as satellites who, in the normal course of development, become dependent on their parents for an extensive period of socialization before becoming relatively autonomous at adolescence. Successful development implies that individuals eventually become self-governing and form families of their own. Yet the quality of the dependency relationship throughout childhood has a significant bearing on the eventual attainment of autonomy. Children growing up with warm, loving parents tend to internalize their parents' unconditional acceptance as a lasting basis of self-worth, thereby facilitating the eventual attainment of autonomy. But children growing up with cold, domineering or permissive, neglectful parents tend to feel undervalued in such a way that they remain unduly dependent on the approval of others, which impedes the eventual attainment of autonomy.

THE DEMAND FOR PRIVACY

At what age do young people start closing the bedroom door behind them for privacy? Ross Parke and Douglas Sawin (1979) investigated this among 112 middle-class children and adolescents aged 2 to 17. The demand for privacy was determined by how readily the subjects let other people into occupied bedrooms and bathrooms.

A big factor in the demand for privacy was family density—the ratio between the size of the house and the number of people living in it. The strongest desire for privacy occurred at the extremes: those living in smaller houses (less than 1600 square feet) with more than five people and those living in larger houses with four people or less. Children and adolescents in moderately crowded houses were less likely to keep bedroom and bathroom doors closed. The researchers explained that individuals who grow up in overly crowded homes may feel the deprivation more severely and insist on privacy whenever they can, whereas those who grow up in more spacious homes get accustomed to privacy and expect it.

The desire for privacy generally increases with age, with the biggest change occurring just before and during adolescence. Whereas boys and girls aged 2 to 5 closed bedroom doors behind them only 29 percent of the time and bathroom doors only 64 percent of the time, the 10-to-13-year olds closed bedroom doors 46 percent of the time and bathroom doors 96 percent of the time. Those aged 14 to 17 closed bedroom doors 58 percent of the time and bathroom doors 100 percent of the time.

The only group that stood out as an exception to the general rule was the 6-to-9-year olds—those in Freud's latency stage of sexual development. Boys and girls this age were less inhibited than at any time before or after. Most of them allowed anyone to come in regardless of what they were doing, whether dressing, bathing, or using the toilet. In contrast, adolescents demand more privacy partly because of their greater physical and sexual maturity, their growing independence from other family members, and their need for self-reflection.

R. D. Parke and D. B. Sawin, "Children's Privacy in the Home: Developmental, Ecological, and Child-Rearing Determinants," *Environment and Behavior, 11,* 1 (1979): 87–104. Copyright © 1979. Reprinted by permission of Sage Publications, Inc.

At adolescence, satellization is gradually replaced by desatellization. That is, adolescents normally break away and gain more distance and autonomy from their parents at this stage of life. But adolescence does not bring total self-rule. Instead, adolescent autonomy is an initial step in the eventual attainment of full autonomy. For instance, adolescent desatellization normally involves resatellization, or the transfer of dependency from parents to others. This can be seen in adolescent crushes on teachers or coaches or in the slavish conformity to peers at this age. It can also be seen in the intense emotional identification in adolescent friendships and romances. Although the element of dependency in these relationships is usually rather-obvious to parents, from the adolescent's point of view, such relationships help to diminish their dependence on their parents.

Desatellization is a gradual process involving the transformation rather than the severance of emotional ties with parents. Throughout this process the attitude and behavior of parents may either help or hurt the attainment of autonomy. Parents who are reasonably satisfied and fulfilled in their marriage and careers, especially if they are independent people themselves, may be more willing to let go of their young and encourage them in the achievement of autonomy. But parents who are frustrated and unfulfilled, especially if they are dependent people themselves, may attempt to hold on too tightly, thus thwarting the development of autonomy in their young.

It is important that parents continue to have an accepting attitude towards their adolescents and that they be willing to modify their discipline and supervision in a way that encourages autonomy. They must expect more responsible behavior from their young but within appropriate limits. That is, they must not tolerate childish excuses for unacceptable behavior nor demand adult behavior too quickly. Parents who abdicate their authority in the face of adolescent defiance may undermine the adolescents' attainment of autonomy by permitting immature or self-defeating behavior, however "adult" such behavior may seem to the adolescents themselves. Wise parents will realize that teenagers have a right to make their own mistakes, while at the same time maintaining reasonable limits in matters of health and personal safety.

Types of Autonomy

So far we have referred to autonomy as if it were a unitary trait signifying self-direction. Yet, upon closer inspection, autonomy appears to be a complex mixture of different traits and skills, and it follows that they may be more evident in some aspects of behavior than others within the same individual. For instance, one teenager readily learns how to drive a car but is still fearful of his parents' disapproval, while another teenager has not obtained her driver's license but feels free to make decisions without her parents' approval. We might say that the first teenager has moved toward behavioral autonomy faster than emotional autonomy, with the sequence being just the reverse for the second teenager. Accordingly, it is customary to distinguish various types of autonomy, such as behavioral autonomy, emotional autonomy, intellectual autonomy, as well as moral and value autonomy.

behavioral autonomy: The ability to perform age-appropriate tasks in a relatively independent manner.

The distinction between behavioral and emotional autonomy is especially important. *Behavioral autonomy* refers to the ability to perform age-appropriate tasks in a relatively autonomous manner. Such autonomous behavior is already evident by the second year of life, when small children begin taking a more active role in their eating and toilet habits. Even though children's ability to act in a self-reliant way generally increases with maturation, they continue to need help in mastering many of the skills necessary for everyday living. With the rapid growth at adolescence, however, there is a marked increase in behavioral autonomy between 11 and 18 years of age. Adolescents not only want to do more things for themselves but are increasingly able to do so. By the late teens, most adolescents can take care of themselves in terms of their daily needs, such as eating, dressing, and learning the appropriate skills needed in school, sports, and social activities. Adolescents who take jobs

emotional autonomy:
The ability to rely on one's own inner reserves of self-esteem and self-confidence in relation to disappointment, criticism, and rejection at the hands of others.

outside the home also learn how to handle responsibility in the workplace, manage money, and get along with people they dislike. At the same time, most parents recognize that on occasion adolescents are likely to need assistance in matters requiring more seasoned judgment, such as the choice of an appropriate college or career goal.

Emotional autonomy develops more gradually than behavioral autonomy, largely because it involves a transformation of the affectional ties which have been so important to both the child and parent. Basically, emotional autonomy is more relationship oriented, involving learning to rely more on one's own inner reserves of esteem and self-confidence in the give and take of relationships. This includes the capacity to form adultlike close ties with friends of one's own choosing and the ability to survive disappointment, criticism, failure, and rejection at the hands of others. Adolescents who feel they have received much love and acceptance at home tend to feel free to choose close friends and confidants outside the family, including those their parents might not approve of. On the other hand, teenagers who feel they have been neglected or unloved at home may remain unduly dependent on the approval and support of others, whether their parents or peers. Sometimes teenagers as well as adults mistakenly think adolescents have achieved their independence when they have done so only in a behavioral sense. For example, Brad has a job after school, has bought his own car, and gets along well with his friends. Yet he is very touchy whenever his work is criticized at school or work. He also becomes very moody and despondent whenever he is disappointed by his girlfriends or breaks up with them. Even though Brad has achieved behavioral autonomy in several areas, his low tolerance for criticism and rejection suggests he's lacking in emotional autonomy, which in the long run may become even more important than behavioral autonomy.

Autonomous Decision Making

autonomous decision making: The ability to make important life decisions with a minimum of assistance from others and without an undue need for their approval.

One of the most important aspects of autonomy is the ability to make decisions on one's own. This ability increases steadily during adolescence largely because of the cognitive changes that occur at this age. First, the increased ability to abstract enables adolescents to distinguish and compare various viewpoints simultaneously. The capacity to take the role of another person, which also increases during this period, permits adolescents to consider divergent views in light of another person's distinctive perspective. Then too, the adolescent's hypothetical reasoning enables him or her to consider the possible consequences of an anticipated course of action. The capacity for autonomous decision making is also aided by the emergence of a self-chosen identify, emotional maturity, value autonomy, as well as increased peer involvement and support.

The improvement in decision making skills during this period of development can be seen in a study of 100 adolescents in three age groups, including seventh, eighth, tenth-, and twelfth-graders (Lewis, 1981). The subjects were presented with a series of problem situation in which the teenager involved

Teenagers increasingly make more decisions on their own. (Bob Rashio, Click/Chicago)

faced a difficult decision he or she needed help with. The problems involved several factors of decision making, such as different types of risks, the possibility of differing with an opinion of a respected person, and reconciling advice from different experts. The adolescents' responses were evaluated according to five aspects of decision making, including whether they were aware of the future consequences; the potential risks; whether their views were revised in light of new information; whether parents, peers, or outside experts were recommended as consultants; and whether adolescents considered the self-interests of those giving advice. The results showed that teenagers' decision-making abilities improved steadily throughout adolescence. Older adolescents showed more refined decision-making abilities in four of the five areas, including awareness of possible risks, consideration of future consequences, the willingness to use an independent consultant, and consideration of the vested interests of those giving advice.

Some parents are more helpful than others in facilitating the development of autonomous decision making. Parents who adopt the democratic approach tend to respect the rights of their youth and encourage them to make their own decisions, intervening mostly in matters pertaining to health and safety. On the other hand, parents who are constantly dictating things like hair style and dress generally undermine the young person's opportunities to make his or her own decisions. Yet, as individuals move into late adolescence and early adulthood, they face some of the most important decisions in their lives, including whether or not to pursue further education, the choice of a career, and the selection of friends and a prospective marriage partner. Wise parents realize that individuals at this stage of life will make some mistakes but may also learn from their experience. Then too, going away to college often provides a valuable opportunity to make more decisions on one's own. In one study, 242 males from 12 high schools in several northeastern states were surveyed in high school and again in college (Sullivan & Sullivan, 1980). Their autonomy was measured by the extent to which their parents had encouraged them to make their own decisions or criticized them and tried to dictate things like hair styles. Generally, those who had gone away to college demonstrated a greater degree of autonomy than those who were still living at home. Apparently, being away from home provides more opportunities for autonomous decision making, thereby setting the stage for entry into adulthood and a greater degree of independence, as we'll discuss in the last chapter.

SUMMARY

Parent–Adolescent Relationships

1. We began the chapter by noting that most adolescents feel close to their parents and do not perceive any major problems between parents and themselves.

2. Parents' use of control is generally classified according to three broad patterns—the authoritarian, love withdrawal, and democratic methods—with most parents relying on a moderate level of control.

3. The majority of adolescents feel their parents use a democratic approach in parenting. Teenagers from such homes are more likely to identify with their parents and want to rear their own children the same way than are teenagers from other homes.

4. In actual practice, the use of parental persuasion and control is more complex than ordinarily portrayed and varies by age and sex of both participants as well as by the particular situation involved.

Parent–Adolescent Communication

5. The communication between parents and adolescents is a crucial factor in adolescents' development; thus, families with faulty communication are more likely to have troubled adolescents.

6. Parents and adolescents tend to share many of the same attitudes and values—though to a different degree—with adolescents being granted greater autonomy in matters such as educational choices, religion, and personal grooming.

7. Parent—adolescent conflict generally increases in early adolescence, though a moderate degree of conflict between generations is positively related to the adolescent's identity exploration.

8. A major way of improving family communication involves parents and adolescents alike learning how to express themselves in a nonjudgmental way and listen empathetically to each other.

Adolescent Autonomy

9. Adolescents tend to achieve autonomy gradually, usually through the transformation rather than the severance of emotional ties with parents.

10. There are different types of autonomy, including behavioral, emotional, intellectual, moral, and value autonomy.

11. An important distinction was made between behavioral- or task-related autonomy and emotional- or relationship-oriented autonomy, with the latter being especially important in the give and take of human relationships.

12. Finally, autonomous decision making improves steadily during adolescent development, with older adolescents being increasingly able to make decisions in a more mature way.

REVIEW QUESTIONS

1. Which adolescents are least likely to feel close to their parents and why?

2. What are some of the advantages and disadvantages of using authoritarian, love withdrawal, or democratic parenting?

3. How do adolescents tend to react to each of these types of parenting?

4. Why is the democratic method of parenting generally preferable?

5. To what extent has the generation gap between adolescents and their parents been exaggerated?

6. Would you agree that a moderate degree of conflict between the generations is not only inevitable but desirable for a young person's development?

7. What is meant by an "I" message?

8. Explain Ausubel's satellization theory of personality development.

9. Describe the difference between behavioral and emotional autonomy.

10. Explain what is meant by autonomous decision making among adolescents.

7 The Self-Concept and Identity

- ■ **THE SELF-CONCEPT**
  Changes during Adolescence
  Self-Esteem
  Change and Stability in Self-Esteem

- ■ **THE SEARCH FOR IDENTITY**
  Erikson's Psychosocial Theory
  The Adolescent Identity Crisis
  Problems in Identity Formation

- ■ **VARIATIONS IN IDENTITY STATUS**
  Types of Identity Status
  Changes during Adolescence
  Factors Facilitating Identity Achievement
  Sex Differences in Identity

- ■ **SUMMARY**

- ■ **REVIEW QUESTIONS**

In high school, Roy Baumaster (1984) was a straight, all-American boy. He says, "I did lots of homework, was on the swim team, and stayed away from drugs, and sex and other delights." Once in college, he began experimenting with other lifestyles. "I grew a beard, played in a rock band," he says, "and in general became quite a different person." Later, in graduate school, he changed again.

Like so many people at this stage of life, Roy Baumaster was busily engaged in exploring the kind of person he wanted to be. In the process of redefining themselves, young people like Baumaster tend to reject some past identities and affirm others, while retaining a definite connection between their past and present selves. They're less certain of the connection between their present and future selves, realizing full well that they continue to change. However, friends and acquaintances are not always so understanding. For instance, during graduate school, Baumaster found it annoying to interact with people who had known him in college and even more irritating to meet people who had known him in high school. He says, "These people expected me to act the way I had acted when they had known me." This observation eventually led Baumaster, now a psychologist, to a consideration of how the growth of one's inner, private self may be constricted by one's public self-image—the impressions others have of us.

THE SELF-CONCEPT

self-concept: The overall image we have of ourselves, including perceptions of "I" and "me," together with the feelings, beliefs, and values associated with them.

Baumaster's experience illustrates the importance of the self-concept and the struggle to achieve one's personal identity that takes place during adolescence. Among other things, the self-concept provides us with a sense of personal identity, reassuring us of who we are amid the changing circumstances and relationships of our lives.

In the early months of development, children's self-images emerge from the awareness of their bodies. But the contents of the self-concept are largely acquired through interaction with the "significant others" in children's lives. Parents are usually the most important people in children's lives, though peers become increasingly important as they reach adolescence. During this formative period, children tend to internalize the reaction of others into their self-concepts. For instance, if a parent repeatedly tells a child, "You're stupid," the child gradually sees himself or herself in this way. In a similar manner, children learn to anticipate what others expect of them and adopt the same attitude toward themselves. In this way children build up pictures of themselves as a "me" or an object to themselves. As they acquire greater self-consciousness and personal identity, they become aware of themselves more as an acting subject or an "I" rather than a "me." But much of this self-awareness comes from the internalized reactions and expectations of others. At the same time, the self-concept is not simply a mirrorlike reflection of external reality. Instead, the self-concept, like impressions of other people and the world, involves an integration and interpretation of a tremendous amount of information and continues to change with one's personal experience (Harter, 1983).

personal identity: The sense of who one is as distinct from others.

Changes during Adolescence

The cognitive changes accompanying puberty, especially the heightened self-consciousness and ability to think in an abstract way, lead to extensive changes in the adolescent's self-concept. First, adolescents perceive themselves in a more detailed, differentiated way than in the past, so that, for example, an adolescent may feel apprehensive about herself with her father but feel comfortable with herself in relation to her best friend. Second, adolescents acquire a more individuated view of themselves than they did as children, and are more apt to affirm themselves in terms of their differences or uniqueness in relation to others in addition to their similarities to them. Third, adolescents also acquire more perspective on the self and are better able to put themselves in the shoes of others and see themselves as others see them. As a result, adolescents become very much concerned with how others react to them.

Adolescents gradually learn to view themselves in a more abstract and differentiated way than do children, as was demonstrated in a study by

Montemayor and Eisen (1977). The subjects included students from preadolescence through late adolescence. Younger and older adolescents' responses to the question "Who am I?" were analyzed according to 30 different categories, such as age, sex, body image, social status, and career role. As expected, the younger adolescents were more likely to describe themselves in concrete ways, such as "I have blue eyes. I have two brothers. I like ice cream." In contrast, the older adolescents relied on more ideological statements and interpersonal characteristics, such as "I am an ambitious person. I believe in God. I'm someone other people like to tell their problems to." Such findings show that older adolescents use their growing cognitive skills to perceive themselves and the surrounding world in a more complex, differentiated way.

As a result, it may be more accurate to view the adolescent's self-concept in terms of multiple self-images than of one global self-concept. In one study an equal number of male and female students in a midwestern high school were administered questionnaires measuring both their global self-concepts and their self-image based on four roles: student, athlete, best friend, and son/daughter. The results showed that the students' global self-ratings differed considerably from each of the role-specific self-ratings. Girls made more significant distinctions than boys between self as "daughter/son" and self "with best friend." Boys made more significant distinctions than girls between self as "athlete" and self as "student." Such findings indicate that girls and boys differ in their perceptions of these roles and acquire different expectations of themselves. All of this suggests that it may be more realistic to approach the adolescent's self-concept in terms of multiple self-images than of a global self (Griffin, Chassin, & Young, 1981).

Self-Esteem

self-esteem: The personal judgment one makes of his or her own self-worth; the sense of personal worth associated with one's self-concept.

A critical aspect of one's self-concept is the sense of self-esteem associated with it. Adolescents' assessment of themselves is obviously affected by their outward successes and failures as well as their relationships with others. But, ultimately, adolescents judge themselves in light of their own aspirations and ideals. Thus, having experienced the same level of success at school, an individual with a modest self-ideal may feel reasonably satisfied with himself while another person with an unrealistically high ideal may remain dissatisfied.

global esteem: One's characteristic level of self-worth or self-esteem.

Although people customarily speak of self-esteem as a single entity—"global esteem"—there is a growing recognition that self-esteem, like the more inclusive self-concept, functions differently in the various areas of one's life. Thus, Mark may have high self-esteem in regard to playing certain sports, such as basketball and tennis. But he may lack confidence in himself as a student in many subjects in the classroom, especially those requiring a high level of verbal abilities. On the other hand, Lisa may feel good about herself as a student, though this also varies somewhat from subject to subject. At the same time, she may suffer from low esteem in regard to her looks, mostly because she is overweight for her age.

The way adolescents feel about themselves is also affected by the bodily changes that occur during puberty. The more favorably adolescents perceive their physical appearance, attractiveness, and physical competence, the better they tend to like themselves. In evaluating themselves, adolescents tend to compare their own physical development to that of their peers as well as the cultural body ideals for their sex (Peterson & Boxer, 1982). Individuals in their teens and twenties are generally more concerned about their appearance than any other age group, with females being somewhat more dissatisfied than males with their looks. Generally, there is a mutual relationship between body image and self-esteem, so that those who like their appearance also feel good about themselves as a person. But it works the other way too, with high-esteem individuals able to accept their bodies despite their physical shortcomings (Cash, Winstead, & Janda, 1986).

Adolescents' self-esteem is especially affected by family life. Since self-esteem is shaped by a combination of various factors, including family structure, family interaction, cohesiveness, and so forth, it is difficult to assess the impact of any one factor in relation to the others. However, some patterns seem fairly well established. For instance, it appears that parents' supportive behaviors—such as nurturance, warmth, and approval—tend to enhance adolescents' self-esteem. Similarly, parents' use of democratic methods of parental persuasion rather than coercive methods facilitates the development of positive self-esteem in adolescents (Openshaw, Rollins, & Thomas, 1984).

family cohesion: The degree of emotional bonding between family members, including the sense of mutual trust and cooperation.

It also appears that family cohesion is positively related to adolescents' self-esteem. That is, the stronger the emotional bonding between family members and the more autonomy they feel in relation to each other, the more likely adolescents will experience positive self-esteem (Openshaw & Thomas, 1986). At the same, the curvilinear hypothesis put forth by Olson, Russell, and Sprenkle (1983) suggests that adolescents may suffer a loss of esteem in families at the high or low end of family cohesion. Thus, adolescents from homes characterized by "emotional divorce" or disengagement may be affected adversely in their self-images because of the conflict and turmoil in the home. Yet, adolescents in highly cohesive families may lack sufficient freedom and the individuality normally needed to acquire positive self-esteem.

An important but controversial factor is the relationship between differences in family structure and adolescent self-esteem, especially differences between single-parent and two-parent families. Although there is substantial evidence that adolescents are often adversely affected by the loss of a parent through death or divorce, such factors work in combination with other influences. For instance, in an effort to measure the impact of such variables, Joycelyn and Thomas Parish (1983) studied adolescents from intact, single-parent, and remarried families. Although they found that family structure had an impact on

Males and females are equally likely to enjoy high levels of self-esteem. (Susan Rosenberg)

androgynous (psychological): The combination of desirable traits from both male and female sex roles, in contrast to traditional sex roles.

adolescents' self-esteem, this factor interacted with the family atmosphere, or how happy adolescents were in their respective families. Thus, an adolescent from a happy single-parent home might exhibit high esteem, whereas another adolescent of the same age and sex from an unhappy two-parent home might suffer from low esteem. Similarly, other studies have shown that the effect of family structure also depends on the age and sex of the parent and adolescent. Openshaw, Thomas, & Rollins (1984) found there was a negative relationship between self-esteem and the single-parent family resulting from divorce, but they also found in the same situation a positive relationship with self-efficacy—or sense of mastery in a particular situation. Thus, while adolescents may have been hurt by their parent's divorce, they may also gain more self-confidence in their own abilities to cope with life in such a situation. All of this reminds us that the connection between family structure and self-esteem is sufficiently complex to resist simplistic generalizations.

The impact of masculinity and femininity on adolescent self-esteem has

undifferentiated: Individuals who perceive themselves as neither masculine nor feminine in sex-role orientation.

traditional sex-typed: Possessing high levels of traits regarded as appropriate for one's biological sex and low levels of traits associated with the opposite sex.

cross-sexed typed: Possessing low levels of traits regarded as appropriate for one's biological sex and high levels of traits associated with the opposite sex.

been studied by James Rust and Anne McCraw (1984). Their sample included 195 adolescents of both sexes, with an age range of 14 to 18, from a mixed social and racial background in urban, suburban, and rural high schools. These investigators found no difference in the level of self-esteem among males and females. Androgynous students generally exhibited higher esteem than adolescents with a confused, undifferentiated sexual identity. However, androgynous females were higher, but not significantly higher, in esteem than the traditionally sex-typed and cross-sexed females, that is, those who exhibited low levels of femininity and relatively high levels of masculinity. But males with low levels of masculine traits coupled with high feminine traits had low esteem. Such findings suggest that masculine traits continue to relate to esteem in both sexes, while feminine traits in males do not.

The extent to which matters such as race and socioeconomic status affect the adolescent's self-esteem depends a great deal on the meaning and value individuals attach to such matters. For instance, blacks and other minorities reared by parents and other role models who do not accept the negative stereotypes of their racial group may develop positive self-concepts. Although they may experience discrimination from time to time, they do not necessarily have low esteem (Louden, 1980). As a matter of fact, in the study by Rust and McCraw (1984) cited above, black adolescents exhibited significantly higher self-esteem than whites.

Change and Stability in Self-Esteem

Although adolescents' feelings about themselves vary somewhat, their overall self-esteem tends to remain fairly stable throughout adolescence. The period of greatest fluctuation occurs during early adolescence, with 12 and 13 year olds exhibiting the highest self-consciousness and lowest self-esteem. The rapid, unpredictable changes in the body at this stage of development may have a negative impact on body image. But adolescents' self-esteem appears to suffer even more because of changing schools. This was demonstrated in a series of longitudinal studies by Blyth, Simmons, & Carlton-Ford (1983) comparing changes in adolescents' self-esteem among students attending schools using the 8–4 plan and those using the 6–3–3 plan. The results showed that school transitions generally have a temporary negative effect on adolescents' self-esteem. Girls in the 6–3–3 arrangement showed two dramatic drops in self-esteem during each of the moves to a new school, that is, between the sixth and seventh grades and between the ninth and tenth grades. Although such changes were less pronounced among the boys, the boys in the 6–3–3 arrangement experienced slight drops in self-esteem between the ninth and tenth grades. In contrast, boys and girls attending schools using the 8–4 plan experienced a gradual rise in self-esteem throughout adolescence. For girls, changing schools following the sixth grade may be upsetting because it coin-

SELF-EFFICACY

Self-efficacy, or the sense of personal mastery, is a more specific concept than self-esteem. According to Albert Bandura (1977), who originated the concept, *self-efficacy* refers to our belief or expectation that we can perform successfully in a particular situation. Unlike the notion of self-esteem, especially global self-esteem, self-efficacy depends primarily on the specific situation. For instance, you may have a great deal of self-confidence in your ability to drive your car in city traffic this afternoon while harboring considerable doubts about your ability to do well on your history test in school tomorrow. Not surprisingly, measures of perceived self-efficacy (asking people to predict their levels of self-confidence in a particular situation) actually predict effective performance with impressive accuracy.

Bandura suggests that we may enhance our sense of self-efficacy through any one or a combination of the following ways.

1. Increase your experiences of successful performance accomplishments in specific areas, such as improving your test grades in a particular course in school.

2. Choose a challenging task, put forth your best effort, and perform it yourself.

3. Observe competent models in the areas of performance of interest to you.

4. Elicit feedback from teachers, counselors, and mentors, especially specific suggestions on how you can improve your performance.

5. Learn to manage emotional arousal more effectively, especially the use of relaxation techniques, so that you can perform well under stress.

Once established, positive expectations about your abilities may generalize somewhat to new situations. Yet, our sense of self-efficacy or self-mastery is best thought of as a collection of *specific* self-evaluations.

Albert Bandura. "Self-efficacy: Toward a Unifying Theory of Behavior Change," *Psychological Review*, 1977, 84, 191–215.

cides with the onset of puberty and the beginning of dating. Changing schools in the ninth grade appears to have negative effects among both sexes primarily because of the presence of older youth. As a result, the younger students who have just changed schools become less active in school activities, are more prone to use alcohol and drugs, and are more likely to feel anonymous—all of which has a deleterious effect on their self-esteem.

Educators have long realized that self-esteem is a critical factor in learning, with high-esteem students forging ahead academically and low-esteem students falling behind. Consequently, as Barbara Lerner (1985) points out, since the 1960s many educators have advocated giving priority to raising students' self-esteem. To do this, teachers have provided students with constant praise and encouragement, insuring that they experience a feeling of success in school as immediately and as often as possible. At all costs, students' self-esteem must be protected from injury through criticism or failure.

Hard-won achievements generally boost young people's self-esteem. (Ken Karp)

An unintended result, however, has been the awarding of good grades for mediocre accomplishments, sometimes accompanied by a lack of basic competence. Lerner contends that an undue emphasis on "feel-good-now" esteem has encouraged an orientation toward instant success and a constant hunger to get more for less. By contrast, the emphasis on excellence in the 1980s has given rise to a different approach, namely, that of "earned self-esteem" through the development of competence. Self-esteem is based on one's success in measuring up to standards at school, work, and home. Such esteem is hard won and develops slowly. Students must learn to persist in the face of momentary unhappiness, self-criticism, and occasional failure in order to acquire this sort of self-esteem. For example, when the minimum competence testing movement was introduced, educators of the "self-esteem-now" approach were appalled. In Florida, more than three-fourths of the black students who took that state's minimum competence test failed on the first attempt. A federal judge declared the program unconstitutional and issued an injunction against it in 1979. But by 1983, the injunction was lifted, and for good reason. Although students who initially failed the test may have suffered a temporary loss of esteem, they redoubled their efforts. And by the fifth try, more than 90 percent of them passed the test. Along with their diplomas, they also received a healthy dose of earned self-esteem.

THE SEARCH FOR IDENTITY

The issue of personal identity—the sense of who one is—is at the heart of the task of self-change. The particular combination of changes occurring at this stage of life evoke a variety of personal concerns, such as, "Who am I?" "Where am I headed?" "How well do I get along with males, with females?" The manner in which such issues are resolved becomes a crucial factor in the individual's personality development.

Erikson's Psychosocial Theory

The idea that late adolescents are especially concerned with the pursuit of their personal identity has long been associated with Erik Erikson's name. Although Erikson was trained in the psychoanalytical tradition, he later modified many of the Freudian ideas in the light of his own thought and experience. Essentially, Erikson has widened the potential application of psychoanalytic theory by transforming Freud's psychosexual theory of development into a more inclusive psychosocial theory of development. Whereas Freud focused on the child's psychosexual development within the family, Erikson takes into account the individual's psychosocial relationships within the larger society. Furthermore, whereas Freud's stages covered only the years between birth and puberty, Erikson's stages extend throughout adulthood into old age.

developmental stages: Periods of development when individuals are especially ready to learn some things more than others.

One of Erikson's (1980) major ideas is that development is cumulative, in the sense that each part of development has its special time of ascendancy until all the parts of development are integrated into a fundamental whole. More specifically, the individual's genetic potential unfolds throughout a lifelong sequence of eight developmental stages. Each stage of development is set by the biological readiness for a particular kind of growth, at least throughout the first four stages. In addition, each stage consists of a positive ability to be achieved along with a related threat or vulnerability. (See Table 7–1.) Optimal growth consists of the successful resolution of the developmental crises at each stage, with one's overall personality composed of the cumulative strengths and weaknesses acquired in each stage. The greater the relative strengths acquired, the stronger one's personality.

adolescent identity crisis: The characteristic struggle accompanying the adolescent's search for a revised, personal identity.

The Adolescent Identity Crisis

Erikson holds that the characteristic life crisis of adolescence is the establishing of a self-chosen identity. Actually, identity formation is a lifelong task that begins in infancy and continues into old age. What makes the search for identity so critical at adolescence is the combination of changes that occur at this age, especially the dramatic bodily changes, sexual maturation, cognitive changes, and the growing involvement with peers and emancipation from one's family.

developmental tasks: Capabilities, skills, and attitudes that must be acquired in growing up, especially in the appropriate critical stages of development.

The developmental task of this period is the achievement of a positive

identity confusion: Sense of confusion over one's personal identity or who one is.

identity based on one's emerging adulthood, with the attendant danger of identity confusion. Although this is a period of great growth potential, it is also a time when individuals suffer more deeply than ever before or again from the confusion of roles. Yet, Erikson holds that the disturbances of this period are more aptly regarded as an aggravated life crisis rather than a psychological disorder. A certain amount of conflict, experimentation, and self-doubt are essential for achieving a firm self-chosen identity. Otherwise, those who take up their parents' goals of success too easily may find themselves with only a weak, untested character, leaving them vulnerable to subsequent developmental crises.

Erikson deliberately uses the rather broad term *identity* to capture the overall richness of the adolescent's experience. At the same time, Erikson distinguishes at least three related meanings of the term. First, identity refers to the sense of sameness or continuity between one's past and present selves. Ordinarily, adolescents retain a fairly strong sense of continuity with their past, affirming many aspects of themselves while rejecting others. Second, identity also pertains to the integration of one's private and public selves. At this age individuals must strive to integrate the selves that they know themselves to be with the more role-oriented selves that others know them by. Otherwise, gross misperceptions and illusions about themselves may hinder adolescents' efforts to achieve realistic goals and find an acceptable place in society. Third, identity also refers to the relationship between one's present self and one's future or potential self. More often than not, it's the concern for the "self I *can* be"

Erik Erikson, pioneer in studying the adolescent identity crisis. (Rick Stafford)

Table 7–1   ERIKSON'S DEVELOPMENTAL STAGES

	1	2	3	4	5	6	7	8
VIII								INTEGRITY versus DESPAIR
VII							GENERATIVITY versus STAGNATION	
VI						INTIMACY versus ISOLATION		
V	Temporal Perspective versus Time Confusion	Self-Certainty versus Self-Consciousness	Role Experimentation versus Role Fixation	Apprenticeship versus Work Paralysis	IDENTITY versus IDENTITY CONFUSION	Sexual Polarization versus Bisexual Confusion	Leadership and Followership versus Authority Confusion	Ideological Commitment versus Confusion of Values
IV				INDUSTRY versus INFERIORITY	Task Identification versus Sense of Futility			
III			INITIATIVE versus GUILT		Anticipation of Roles versus Role Inhibition			
II		AUTONOMY versus SHAME-DOUBT			Will to Be Oneself versus Self-Doubt			
I	TRUST versus MISTRUST				Mutual Recognition versus Autistic Isolation			

Horizontal row V shows the overall range of issues involved in the adolescent identity crisis; vertical column 5 represents the contributions of the first four developmental stages to the identity crisis.

Source: E. H. Erikson, *Identity, Youth and Crisis* (New York: W. W. Norton & Company, Inc., 1968), p. 94. Copyright © 1968 by W. W. Norton & Company, Inc. Reprinted by permission.

that evokes the most anxiety in late adolescents. After all, who fully knows his or her potential self? This is something which must be discovered through an active process of trial and error, and involves considerable anxiety. In the process adolescents must reject some existing aspects of themselves, which

---

ERIKSON'S YOUTH

Erikson's own development is itself a dramatic example of the adolescent identity crisis. "No doubt," he says, "my best friends will insist that I needed to name this crisis and to see it in everybody else in order to really come to terms with it in myself." Erikson readily admits, "They could, indeed, quote a whole roster of problems related to my *personal* identity" (1975, p. 25).

Erik Erikson was born to a Danish Jewish mother, and a Danish father who abandoned Erikson's mother before he was born. His father died at about the time of Erikson's birth. During his first several years, Erikson lived alone with his mother, an artist who traveled frequently. When he was 3 years old, his mother married a German-Jewish pediatrician and settled in southern Germany. His stepfather was a highly respected man who served as the leader of the local synagogue. Although the marriage provided Erikson with security, it also posed thorny problems about his identity. For example, his parents attempted to "forget" his earlier years with his mother, though Erikson always knew better. Later, as a blonde, blue-eyed youth, Erikson was called "goy" (non-Jewish outsider) at the synagogue but "Jew" by his classmates at school (Roazen, 1976).

During his teens, like other youth with artistic or literary aspirations, Erikson felt alienated from everything his family stood for. "At that point," he said, "I *set out* to be different" (1975, p. 28). After graduating from the *gymnasia* (high school), Erikson went to art school, then took to wandering. In those days art was considered more a way of life than a specific career with an acceptable niche in society. As a result, even though he had enough artistic talent to consider it a vocation, he saw himself as a traveling "Bohemian." He recalls those years as an important part of his training, however, acknowledging that highly creative people tend to resolve their identity crisis somewhat later than other youth.

By his midtwenties, Erikson was teaching art at a school near Vienna, where he met his wife-to-be. She was a Canadian-born American (Gentile) who was taking her training and personal work in psychoanalysis. Through her, Erikson became interested in psychoanalysis and was admitted into the circle of Freud's followers. He underwent a personal analysis by Freud's daughter, Anna Freud, who strengthened his interest in children and adolescents. By his late twenties, Erikson had become a psychoanalyst, but the lack of formal degrees left him somewhat insecure about his professional identity. After an unsuccessful attempt to practice in Denmark, he came to the United States. Upon becoming a naturalized citizen, he took the name of "Erikson," retaining his stepfather's name Homburger as his middle name. Although some authorities claim that this was Erikson's attempt to acknowledge his real father, others contend he was simply naming himself after the early Danish discoverer of America (Roazen, 1976).

Erikson's life illustrates, perhaps to an extreme, his view that adolescence is better understood as an aggravated life crisis than as a psychiatric disturbance. As he himself discovered, adolescents must often test the extremes before settling on a considered course.

are known, while affirming potential aspects of themselves which are not fully known. Understandably, this is a risky process filled with promise and self-doubt. It's no wonder some young people do not fully resolve their adolescent identity crisis.

Although the search for identity may become totally preoccupying and assume dramatic forms, it tends to occur gradually and unconsciously in the lives of most adolescents. As James Marcia (1980) reminds us, the achievement of a new identity "gets done by bits and pieces." Sometimes there are major decisions to be made dealing with whether to take a job after high school or attend college. At other times, the decisions may seem trivial, such as which courses to take and whom to date. In most cases, decisions are not made once and for all but have to be made again and again. One of the hallmarks of a successful identity is flexibility. That is, having resolved one's identity crisis at adolescence doesn't preclude future crises and decisions. Rather, the successful resolution of one's identity at adolescence provides the necessary stability with which to explore subsequent revisions. As Erikson (1968) says, an optimal sense of identity is experienced as a "sense of psychological well-being, . . . a feeling [of being] at home in one's body," "knowing where one is going," and the assurance of "anticipated recognition from those who count."

Problems in Identity Formation

negative identity: The adoption of socially undesirable roles or traits.

A major pitfall in the struggle for a self-chosen identity is the adoption of a negative identity. According to Erikson (1980), everyone harbors a negative identity, in the sense of identity fragments which individuals have to submerge within themselves as undesirable or irreconcilable. Normally, these aspects of one's self remain unconscious, as the dark, unruly side of one's overall identity. They are expressed, if at all, in fantasies and occasional uncharacteristic behaviors by the individual. But some individuals are more apt to act out their negative identities, especialy those consumed with defiance, failure, or fear. For example, a mother became so worried about her brother's alcoholism that she responded selectively to those traits in her son's behavior associated with her brother's fate. As a result, this negative identity became more real to her son than all his natural attempts at being good.

The adolescent's choice of a negative identity frequently represents a desperate attempt at becoming "somebody" or achieving mastery when positive alternatives are blocked or seem remote. Adolescents who have repeatedly disappointed or failed their parents, teachers, or friends may prefer being somebody "totally bad" or a "nobody" to being a "not-quite-somebody." A negative identity often plays a prominent role in the lives of young people who are constantly in trouble, such as the school dropout, habitual delinquent, or drug abuser. Marginal youths in the ghettos of large cities often assume a negative identity because it is the most available and reinforced in their impoverished environment. Even youth in the affluent suburbs who are having difficulty coping with problems at home or at school may temporarily become defiant and rebellious, preferring the company of like-minded friends rather than those with more positive goals. A major hazard of this age is that such negative identities will be accepted as permanent by parents, teachers, courtroom judges, and friends—and thus by youths themselves. In these instances, adult frustration and disgust may unwittingly push troubled youth into acting out their negative identities as a kind of self-fulfilling prophecy. As one girl told her parents, "Since you feel I'm a tramp, I might as well be one." Wise adults, on the other hand, will look beyond the negative identity to the underlying needs it fulfills, and encourage youth to find more positive and rewarding ways of achieving a sense of identity.

When the search for self proves too difficult or overwhelming, adolescents may experience identity confusion—sometimes called identity diffusion. Such a state is characterized by a fragmented, incoherent sense of self. Individuals who are confused about themselves tend to exhibit high levels of anxiety and a strong sense of inadequacy along with a marked tendency toward rigid behavior (Rothman, 1984). The sense of confusion is not restricted simply to one's self-identity but is manifested throughout the individual's life. Such adolescents may have trouble concentrating and making decisions. They may have trouble holding down a job, managing their time, and making friends. The state of identity confusion may be expressed in opposite ways. On the one hand, such individuals may lose themselves in the crowd, becoming extreme conformists among their peers and fixated with fads. They may also identify with strong leaders and become likely candidates for cults. On the other hand, individuals who are confused about themselves may go to the opposite extreme, withdrawing from their family and peers. They may drop out of school and have few, if any friends and little social life. Such individuals are often seen as loners and drifters.

Since some degree of role confusion is an inherent part of the adolescent identity crisis, almost every adolescent experiences occasional moments of bewilderment and self-doubt. But those who are persistently plagued by a confusion of their personal identities are at odds with themselves and the world. Even then, such individuals can be viewed along a continuum. At one end are those with the mildest degree of confusion—the normal "lost" adolescents—who outwardly may function fairly well but inwardly have little sense of who they are or where they are headed. At the other end are those who are severely disturbed—the schizophrenic adolescent who needs psychological treatment and perhaps hospitalization.

VARIATIONS IN IDENTITY STATUS

By now, it may be apparent that adolescents cannot simply be divided into two groups—those who have achieved a positive identity and those who

Table 7—2 THE FOUR STATUSES OF IDENTITY

		CRISIS	
		Present	Absent
COMMITMENT	Present	Identity achievement	Identity foreclosure
	Absent	Identity moratorium	Identity confusion

identity status: The various ways adolescents attempt to resolve the adolescent identity crisis, such as identity achieved, foreclosure, moratorium, and confusion.

remain confused about themselves. Instead, it is more realistic to view the adolescent's struggle with identity along a continuum, with identity achievement at one end and identity confusion at the other end. As a result, in order to empirically test Erikson's theoretical notions about identity, James Marcia (1980) and others have used a fourfold concept of identity status—identity achieved, moratorium, foreclosure, and identity confusion. These identity statuses are best seen as four different modes of dealing with the adolescent identity crisis and provide a greater variety of styles of identity formation than does Erikson's simple dichotomy of identity versus identity confusion. Individuals are usually classified into the various identity statuses on the basis of in-depth interviews and questionnaires. The major criteria used are the presence or absence of a critical period of exploration or decision making (crisis) and the extent of the adolescent's personal involvement (commitment) in two key areas—career and ideology, or beliefs and values. (See Table 7—2.)

We'll describe some of the characteristics of individuals classified in each identity status, the pattern of changes in identity status that occurs throughout adolescence, some of the factors which facilitate identity achievement, as well as the sex differences involved.

Types of Identity Status

identity achieved: The successful attainment of personal identity by those who have faced and resolved their adolescent identity crisis.

Individuals who are classified as identity achieved have faced the identity crisis and have already resolved many of the issues involved. Typically they have chosen a career goal and have a good sense of their own values. At the same time, they are flexible and able to change their life goals depending on their experience. Also, as Harold Bernard (1981) found, these individuals are able to appraise their parents realistically so that they tend to form "part identifications" with them, that is, identifying with some aspects of their parental models they find attractive and rejecting parts they find unappealing. They also tend to choose friends and romantic partners who will help them to become

Young people in the identity moratorium status are still exploring various interests and career goals. (Laimute Druskis)

identity moratorium: The identity status that accompanies a prolonged adolescent identity crisis in which few, if any, commitments are made.

less dependent on their parents. Such individuals tend to set realistic goals for themselves and perform well under stress. An example would be a young man who is majoring in engineering because he does well in subjects like math and physics and enjoys working with mechanical things. Although he disagrees with his parents on some issues, like sex and politics, he identifies with many of their middle-class values and gets along with them reasonably well.

Those classified in the *identity moratorium* status are generally experiencing a delayed or drawn-out exploration of their identity. They have made few, if any, firm commitments to a careeer or personal values. Understandably, they have high levels of anxiety, suggesting the continued awareness of an unresolved life crisis. They are also less certain of their values than identity achievers and are more apt to change their minds. They may be less confident in their choice of a college major, and are more likely to change it or drop out before making a firm choice. Students in this group are more likely to be experimental, rebellious, and critical of the establishment. They also have a great deal of guilt about disappointing their parents or about their potential for doing so (Bernard, 1981). An example would be a student who changes her college major several times before dropping out. Even when she returns to college, she remains undecided whether to combine marriage with a career or not.

identity foreclosure: The identity status associated with the avoidance of the adolescent identity crisis, resulting in premature commitments.

Individuals classified in the *identity foreclosure* status have largely avoided any substantial identity exploration usually through premature choices endorsed by their parents. On the surface, at least, they appear much like the identity achievers. But the appearance of maturity comes at a price. Inwardly, they experience a high level of anxiety, depression, and defeatist thoughts and feelings, suggesting that even though they have specific goals they are greatly concerned about fulfilling them and others' expectations of them (Rothman, 1984). By and large, individuals in identity foreclosure avoid experimentation and conflict. Instead, they show great respect for authority and tend to choose friends and romantic partners who are substitute objects of dependency (Bernard, 1981). An example would be the student who marries his high school girlfriend, takes a position in his father's company, settles down in his parents' community, and only later in life comes to regret such choices.

People classified in the *identity confusion* status have largely avoided the adolescent identity crisis. Not surprisingly, these individuals exhibit very high levels of anxiety, rigid or stereotyped behavior, and felt inadequacy (Rothman, 1984) As a result, they put off making life choices, feeling they aren't ready for them. Some lose themselves in an endless absorption with social life, sex, and drugs. Others become loners and drifters, shifting from one interest or job to another. Still others may experience a profound emotional disturbance and need hospitalization at one time or another. An example would be the young woman who attends a college with minimal academic standards, grows bored, drops out, and gets married. She soon becomes restless and unhappy and attempts suicide, eventually turning to psychotherapy.

Changes during Adolescence

Since Erikson regards the resolution of the adolescent identity crisis as an ongoing process, we would expect individuals to shift from one identity status to another as they progress in their personal development. Numerous investigations have shown this to be the case, with much of the significant development of identity occurring in late adolescence or the post-high school years.

The early adolescent years generally are not conducive to the development of a firm identity. Perhaps most homes and schools do not provide the atmosphere and range of alternatives that encourage adolescents to experiment and explore their lives at this stage. If anything, most adolescents are just beginning the move into moratorium status. In one study, Sally Archer (1982) examined identity development in regard to career goals, religious beliefs, political views, and sex-role preferences among students in the sixth, eighth, tenth, and twelfth grades. She found that the identity-achievement status

WHAT ABOUT YOUR OWN IDENTITY DEVELOPMENT?

Looking back, how would you characterize your own experience of the adolescent identity crisis? That is, to what extent did you explore meaningful alternatives in regard to your identity and, if so, make some sort of commitment in the various aspects of your identity development, such as career goals, sexual attitudes and practices, values, and religious beliefs? For instance, did you arrive at a satisfying career goal (identity achieved) with little or no trouble? Or did you experience considerable indecision and change (moratorium or identity confusion) before achieving a firm career goal?

In reflecting on her marriage at an early age, one woman said, "At the time I felt like an achiever. But now I realize my first marriage respresented more of a foreclosure pattern." One young man who changed his degree major twice during college felt confused about his career goals at that period of his life. Yet, looking back, he said, "Actually, I think it was a moratorium experience. Now I realize it's normal to change your career goals at that age."

Which aspect of your development has been the least difficult in achieving a satisfying identity? Which area has given you the greatest trouble? How would you account for this?

increased in frequency with each successive grade level. But the confusion and foreclosure statuses remained the most common categories at all grade levels for both sexes.

The most significant differences in identity status tend to occur in late adolescence, between 18 and 21 years of age. For instance, in a study by Adams and Fitch (1982) college students in their freshman, sophomore, and junior years were observed for a 1-year period. During this time, about half of the students remained stable in their identity status; the other half either advanced or regressed in their identity status. Over half of the students who were classified in the confusion and moratorium statuses at the outset had moved out of these categories by the end of this period. As expected, many of the students in the moratorium status had progressed into the identity-achievement status.

By the time students reach their senior year in college, they are more likely to have achieved a firm sense of identity. After surveying the literature on the subject, Alan Waterman (1982) reports that college seniors of both sexes are more apt to have resolved their identity crisis and hold a stronger sense of personal identity than those just entering college. This is especially true in regard to career goals. But it is less apparent in other areas such as religious belief. If anything, the college atmosphere tends to undermine students' traditional religious beliefs, meaning that students who would be classified in a foreclosure status in their religious beliefs tend to move into the moratorium status. Students were most likely to remain in the confused status in regard to their political views, especially those who were unsure of their political commitments or apathetic in this area.

Factors Facilitating Identity Achievement

The greater proportion of identity achievers among college seniors compared to those beginning college raises the question of the role college plays in the achievement of identity. Since college provides a socially acceptable atmosphere for the exploration of ideas, social roles, values, and career goals, it may promote a temporary moratorium in the search for identity as well as the eventual attainment of a self-chosen identity. At the same time, *all* youth this age may experience significant growth in their identities, regardless of whether they attend college or not. Then again, perhaps it is the individuals who are progressing well toward a clear identity who are more likely to enroll in college and remain there, so that identity development encourages attendance at college rather than the other way around. Most likely the role college plays in identity development varies considerably from one individual to another.

A study by Morash (1980) comparing working-class youth with college students, many of whom came from a middle-class background, suggests that working-class youth experience the identity crisis differently. Working-class youth were more apt to be in the identity-achievement and identity-confusion statuses, whereas college students were more frequently in the moratorium or foreclosure status. A related study by Munro and Adams (1977) found that

---

IDENTITY AND SOCIAL CHANGE

Today's youth are likely to have a more personalized identity than their counterparts of a generation ago. At least, this is what Louis Zurcher (1977) found in his study using the Kuhn and McPartland Twenty Statements Test. In this relatively unstructured, open-ended procedure, each student is asked to answer quickly the question "Who am I?" twenty times. The answers are then scored according to several basic categories.

During the 1950s, Zurcher found that students characteristically gave answers that identified them with a social role or institutional status. For example, students said, "I am a student," "I am an American," and "I am a female." Such responses indicated that their personal identities were largely social, reflecting the relatively stable and widely accepted social order of that time.

Since the 1970s, however, Zurcher found that students give a different type of response. The responses reflect a self-concept less closely identified with social roles or institutions. For example, students are more likely to describe themselves with more personal statements like "I am happy," "I am searching," and "I am a serious person." Such statements reflect a self-identity based more on personal characteristics than social roles as in the 1950s. The prevalence of such personalized identities has been interpreted as a sign that contemporary students are more at home among constant change. In fact, such a personalized, changeable self may be a more functional type of identity for coping with accelerated social change. It also reflects the high value young people place on personal life styles and intimate relationships in an impersonal society.

L. A. Zurcher, *The Mutable Self* (Beverly Hills: Sage Publications, Inc., 1977).

working-class youth 18 to 21 were more likely than college students to have established a sense of identity in the area of ideology, or beliefs and values. The most likely explanation for such findings is that working-class youth go through the adolescent identity crisis in a shorter period of time. Apparently, many working-class youth move directly to the state of commitment in regard to career and values without the benefit of the prolonged moratorium available to (middle-class) college students. Even though the identity achievement of working-class youth resembles the foreclosure status in many respects, it may be adaptive for individuals who enter work and family roles directly from high school.

The attainment of formal operational thinking also aids in the resolution of the identity crisis. Individuals who have reached this stage of thought are better able to deal with problems reflecting identity-related issues. Positive role models in the home plus democratic parenting generally encourage decision making and the attainment of a satisfying, self-chosen identity. Then too, the presence of a variety of alternatives in the environment along with the pressure for making some sort of commitment promotes the development of a positive identity. Individuals from a restricted or homogeneous background may be more likely to adopt a foreclosure status (Waterman, 1982).

Although common sense suggests that the emotional stress of a parental divorce might thwart the achievement of identity, this is not necessarily the case. One study of high school girls found that two-thirds of those in the identity-achieved status came from homes disrupted by death or divorce, compared to only one-fifth of those in the other identity statuses who came from broken homes (St. Clair & Day, 1979). As mentioned earlier, having to cope with such crises may temporarily hurt the adolescent's self-esteem while at the same time challenging them to struggle all the harder to resolve their own identity crisis.

Sex Differences in Identity

Since males and females are socialized somewhat differently from birth on, it shouldn't be surprising that they approach the adolescent identity crisis with different emphases. Much of the reason has to do with the growing importance of adult sex roles that normally accompanies physical and sexual maturation, that is, the increased pressure to behave in sex-appropriate ways. As a result males tend to become more career oriented, more emotionally independent of their families, and to move toward identity achievement faster than females (Marcia, 1980). On the other hand, females tend to become more relationship oriented, investing themselves more deeply in intimate friendships and love relationships than males. Such differences may in part reflect the practical considerations associated with the traditional sex roles, namely, the male's role as breadwinner and the female's role as wife and mother. However, because of changing sex roles young women tend to experience

intimacy: Interpersonal closeness between two or more persons, such as friends or lovers, which may or may not involve sexual intimacy

More young women are exploring nontraditional roles. (U.S. Coast Guard Official Photo)

greater conflict between the traditional and nontraditional expectations of women, especially between family and career goals (Morgan & Farber, 1982). Young women who identify strongly with the traditional feminine role characteristically exhibit low levels of anxiety, indicating that the foreclosure status continues to be stable and adaptive for women. At the same time, women in the foreclosure status also report high levels of depression, defeatest thinking, and a sense of inadequacy, suggesting that they are intuitively aware of the limitations and vulnerability of the traditional feminine role (Rothman, 1984).

A major difference between the sexes has been the relationship of identity and intimacy. Erikson (1980) holds that individuals who have achieved a firm sense of identity in late adolescence are better prepared to establish intimate relationships in young adulthood. Since healthy intimate relationships, such as friendship, love, and marriage, involve closeness without the surrender of one's personal identity, individuals who are sure of themselves are freer to enter into the give and take of close relationships. By contrast, those with a weak or unclear identity may become so fearful of losing themselves that they can only enter into superficial, unstable, or addictive relationships. Sure enough, studies of college students have shown that men and women with a firm sense of identity enjoy deeper and more committed relationships than those who have not yet developed a clear identity (Kacerguis & Adams, 1980). But what about the young women who define themselves primarily in terms of their close relationships, with little or no consideration of a career goal? Actually, it appears that women follow a greater variety of patterns than men in integrating their identity and intimacy needs. Some young women adopt a pattern similar to males, focusing on their educational and career goals before resolving their intimacy needs. A larger proportion of young women follow the traditional pattern, resolving the crises in close relationships of love, sex, and marriage before clarifying their career goals. Still others follow an androgynous pattern, experiencing crises in both realms simultaneously (Hodgson & Fischer, 1979). Today, such factors as the increased flexibilty of sex roles, the growing range of career options for women, the diminished obligation to have children, and the growing importance of family relationships for men all promote greater overlap between the sexes. Indeed, much of the research on identity development in the 1980s has pointed to similarities rather than differences in the ways men and women cope with the identity crisis in late adolescence and young adulthood (Grotevant & Thorbecke, 1982).

SUMMARY

The Self-concept

1. We began the chapter by pointing out that the adolescent's personal development depends partly on his or her self-concept—all those perceptions of "I" or "me" together with the feelings, beliefs, and values associated with them.

2. Largely because of the cognitive changes that occur at this stage of development, adolescents begin to perceive themselves in a more abstract and individualized way.

3. A critical aspect of the adolescent's self-concept is self-esteem—the individual's personal judgment of his or her own worth.

4. After an initial period of heightened self-consciousness and low esteem during early adolescence, self-esteem tends to rise steadily, with late adolescents generally reporting higher levels of self-esteem.

The Search for Identity

5. The particular combination of changes that occur during adolescence awakens the individual's search for identity—the sense of who one is—a major concern in Erik Erikson's psychosocial view of development.

6. Erikson holds that the developmental task during late adolescence is the achievement of a positive, self-chosen identity, with the attendant risk of identity confusion.

7. When the alternatives to a positive identity are thwarted or seem remote, adolescents may adopt a negative or socially undesirable identity.

8. In extreme instances when the search for self proves too difficult or overwhelming, adolescents experience identity confusion, which may range from a mild state of not knowing who one is to a severe disturbance.

Variations in Identity Status

9. Much of the research in this area has utilized four identity statuses—identity achievement, moratorium, foreclosure, and identity confusion—providing for a greater variety of styles in dealing with the adolescent identity crisis than described by Erikson.

10. Individuals tend to move from one identity status to another as they struggle with the adolescent identity crisis, with the identity-achievement status the most stable category and moratorium the least stable.

11. How well adolescents handle the identity crisis depends on many factors, especially whether they have attained formal operational thinking and have access to opportunities for exploration and decision making (such as those normally present in a college education).

12. Traditionally, males and females have approached the adolescent identity crisis with different emphases, with males being more career oriented and females more relationship oriented. However, recent studies point to greater similarities than differences between the sexes in their handling of the adolescent identity crisis.

REVIEW QUESTIONS

1. To what extent do you think the adolescent's self-esteem varies from one situation to another?

2. Why is early adolescence characterized by heightened self-consciousness and low self-esteem?

3. Explain Erikson's concept of the adolescent identity crisis.

4. To what extent does Erikson's personal life illustrate his view of the adolescent identity crisis?

5. Explain the concept of negative identity, giving an example of this.

6. What does it mean for an adolescent to be in a state of identity confusion?

7. Explain the four types of identity status used in the research on identity development.

8. Which identity status—moratorium or foreclosure—represents the greater risk of adolescents failing to find themselves?

9. How important is the attainment of formal operational thinking for the successful resolution of the adolescent identity crisis?

10. In what ways have males and females approached the adolescent identity crisis differently?

8   Peers

■ **PEER RELATIONS**
Parent and Peer Influence
Changing Patterns of Parent/Peer Influence
Function of Peers
Social Acceptance

■ **PEER GROUPS**
Cliques and Crowds
Changes in Peer Groups
Peer Conformity

■ **FRIENDSHIP**
Intimacy with Parents and Peers
Changing Patterns of Friendship
Same-Sex and Opposite-Sex Friendships

■ **DATING AND EARLY MARRIAGE**
Dating
Dating Patterns
Adolescent Marriage

■ **SUMMARY**

■ **REVIEW QUESTIONS**

In order to see what adolescents do in a typical week, 75 high school students were given electronic pagers that beeped at different times of the day from early morning to late at night (Csikszentmihalyi & Larson, 1984). Each time the students were beeped, they were asked *where* they were, *what* they were doing, and *who* they were with at the time.

Almost half the time, adolescents were at home, usually in their bedrooms. One-third of the time they were at school, primarily in the classroom. And about one-fourth of their time was spent in public places, such as at work, at a friend's house, or in an automobile. What were they doing? Slightly over one-fourth of the time adolescents were engaged in productive activities, mainly related to school work. Almost one-third of the time, they were involved in maintenance activities, such as doing chores, eating, or personal care. And almost half the time they were engaged in leisure activities, such as socializing with others or watching television.

An analysis of the people they were with showed that adolescents spend considerably less time with adults than they do with their peers. Adolescents spend about one-fifth of their waking time with their families, whether parents or siblings. Another one-fourth of the time is spent in solitude. But fully half of the time is spent with peers, about equally divided between classmates and friends. In fact, as adolescents move from the ninth to the twelfth grade, they spend an increasing proportion of their time with their peers. When talking with peers, adolescents feel friendly, sociable, happy, and free. But when talking with adults, adolescents report feeling passive and constrained in the relationship.

PEER RELATIONS

Studies such as the "beeper" survey confirm the familiar observation that as children reach adolescence they spend more time with their peers. Although parents sometimes regard this as a sign that they are being displaced by their son's or daughter's friends, this is not necessarily the case. Instead, most adolescents continue to identify strongly with their parents; thus, it would be more accurate to say that adolescents regard their parents and peers as different reference groups for different aspects of their lives.

Parent and Peer Influence

peers: People who are about the same age.

reference groups: Groups to which adolescents look for approval and support.

antisocial behavior: Behavior that is against the principles of society; unsociable.

The relative influence parents and peers exert on the lives of adolescents depends on several factors, including the adolescent's maturity relative to his or her age, the parent–adolescent relationship, and the particular issue involved.

The adolescent's orientation toward parents and peers generally changes with age. In one study (Berndt, 1979) students in grades 3 through 12 were asked to make choices on a number of behaviors, indicating their relative orientation toward parents or peers. In the third grade, parent and peer influences were directly contradictory, with young people still strongly dependent on their parents' wishes. By the sixth grade, however, parent and peer influences were less polarized and functioned in different areas, so that parents were more influential in some areas and peers in others. By the ninth grade, there was greater opposition between parent and peer influences, largely because of adolescents' growing independence from their parents and increasing conformity to peer-oriented behaviors and values. By the eleventh and twelfth grades, peer influence had begun to decrease somewhat as adolescents exhibited less conformity to peer-endorsed antisocial behaviors and acquired greater independence in decision making.

The degree of influence parents have over their teenagers also depends on how well they get along with each other. Although parental influence gener-

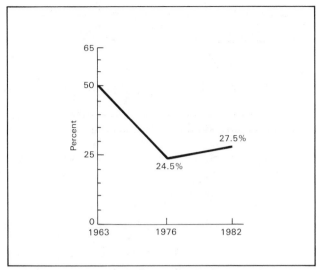

FIGURE 8–1 CHANGES IN PARENT ORIENTATION BY BOYS AND GIRLS COMBINED IN 1963, 1976, AND 1982.

Source: Hans Sebald, "Adolescents' Shifting Orientation toward Parents and Peers: A Curvilinear Trend over Recent Decades," *Journal of Marriage and the Family* (February 1986): 9. ©1986 by the National Council on Family Relations. Reprinted by permission.

ally decreases as students advance from the sixth to the twelfth grade, mostly because of the growing independence of adolescents themselves, the decrease tends to be much less in some families than others. Parents who retain a great deal of influence in the lives of their adolescents generally show a higher level of interest, understanding, and helpfulness toward their teenagers than do other parents (Larson, 1980). When teenagers are unduly influenced by their peers, it is more likely because of something lacking in the parent–adolescent relationship than because of the greater attractiveness of peers.

self-fulfilling prophecy: Expectations about behavior or events that act to increase the likelihood that such behaviors will occur.

Parents who do not know what to expect of their adolescents or who fear being displaced by their adolescent's peers are engaging in a self-fulfilling prophecy, helping to make it so. Parents who neglect or mistreat their adolescents also leave them more vulnerable to undue peer influence. It is also important to realize that adolescents differ greatly among themselves. Some teenagers habitually conform to their parents' wishes, whereas others are more oriented toward their peers.

Changing Patterns of Parent/Peer Influence

The extent to which adolescents seek advice from their parents or peers has shifted somewhat over the past few decades, as seen in a comparative study. Hans Sebald (1986) asked identical questions regarding 18 items to a comparable number of teenagers in 1963, 1976, and 1982, and then compared

curvilinear pattern: Pattern consisting of a curved line or lines.

their responses. Sebald found that the combined data for boys and girls forms a curvilinear pattern over time, reflecting a decline in parent orientation between the 1960s and 1970s and its partial recovery during the 1980s. (See Figure 8–1.) Yet, this profile applies more to boys than girls, primarily because the changes in parent and peer orientation have been different for girls than boys. Whereas girls were more parent-oriented and boys more peer-oriented in the 1960s, this pattern became more balanced, if not reversed to some extent, by the 1980s. In 1976, for instance, 37 percent of the boys and 35 percent of the girls indicated they would look to their parents for "advice on personal problems." By 1982, 49 percent of the boys sought their parents' advice for personal problems while only 18 percent of the girls did so.

A second major finding was that the extent to which adolescents seek advice from their parents or peers depends largely on the particular issues involved, a finding similar to earlier studies. For the most part, concerns about education, career goals, and spending money tend to be more parent oriented. By contrast, social activities—such as which club to join and whom to date—are strongly peer-oriented. (See Tables 8–1 and 8–2.)

Sebald also found an increase in the "myself" orientation—the extent to

Table 8–1   TEENAGE BOYS SEEKING PARENTAL ADVICE IN 1976 AND 1982 (in percents)

Issues	1976 (n = 100)	1982 (n = 110)	Change
1. On what to spend money	31	54	+23
2. Whom to date	4	14	+10
3. Which clubs to join	6	8	+ 2
4. Advice on personal problems	37	49	+12
5. How to dress	6	11	+ 5
6. Which courses to take at school	33	59	+26
7. Which hobbies to take up	6	14	+ 8
8. In choosing the future occupation	43	65	+22
9. Which social events to attend	4	8	+ 4
10. Whether to go or not go to college	51	65	+14
11. What books to read	8	14	+ 6
12. What magazines to buy	4	8	+ 4
13. How often to date	12	14	+ 2
14. Participating in drinking parties	18	19	+ 1
15. In choosing future spouse	27	35	+ 8
16. Whether to go steady or not	14	15	+ 1
17. How intimate to be on a date	8	11	+ 3
18. Information about sex	24	38	+14
Average percentages	18	30	+12

Source: Hans Sebald, "Adolescents' Shifting Orientation toward Parents and Peers: A Curvilinear Trend over Recent Decades," *Journal of Marriage and the Family* (February 1986): 7. © 1986 by the National Council on Family Relations. Reprinted by permission.

which teenagers rely primarily on their own decisions. The questionnaire submitted to the teenagers provided three categories for each of the 18 issues—parents, peers, or undecided. Whereas all responses of the 1963 sample were made within these categories, a number of the 1976 and 1982 respondents spontaneously added comments such as "myself," "personal opinion," and "figure it out myself." In 1982, there was a marked increase of such unsolicited responses, especially among girls. Such responses support Larson's (1980) earlier findings that neither parent nor peer orientation preclude adolescents' ability to make their own decisions. It appears that the majority of adolescents decide many things for themselves, judging mostly by the situation itself. Whether they seek help from their parents or peers depends largely on which they regard as the most competent guide for the matter at hand.

Function of Peers

Although peer influence tends to supplement rather than usurp parental influence, peers play a vital role in the lives of adolescents in several ways.

Table 8–2   TEENAGE GIRLS SEEKING PARENTAL ADVICE IN 1976 AND 1982 (in percents)

Issues	1976 (n = 100)	1982 (n = 110)	Change
1. On what to spend money	55	60	+ 5
2. Whom to date	17	13	− 4
3. Which clubs to join	7	7	0
4. Advice on personal problems	35	18	−17
5. How to dress	9	2	− 7
6. Which courses to take at school	44	64	+20
7. Which hobbies to take up	10	9	− 1
8. In choosing the future occupation	64	56	− 8
9. Which social events to attend	10	11	+ 1
10. Whether to go or not go to college	70	64	− 6
11. What books to read	8	7	− 1
12. What magazines to buy	6	2	− 4
13. How often to date	57	27	−30
14. Participating in drinking parties	26	18	− 8
15. In choosing future spouse	42	24	−18
16. Whether to go steady or not	32	16	−16
17. How intimate to be on a date	17	9	− 8
18. Information about sex	43	40	− 3
Average percentages	31	25	− 6

Source: Hans Sebald, "Adolescents' Shifting Orientation toward Parents and Peers: A Curvilinear Trend over Recent Decades," *Journal of Marriage and the Family* (February 1986): 8. © 1986 by the National Council on Family Relations. Reprinted by permission.

First, peers may help adolescents make the transition from a family orientation to a peer orientation at a critical time in the individual's development. After years of living in the close, dependent relationships of the family, adolescents' attempts at increased autonomy are accompanied by intense emotions, often marked by ambivalence and conflict. The presence of friends may alleviate some of the turmoil involved, by providing sympathetic feedback, a sense of emotional security, and a status independent of the family.

Second, peers also provide teenagers with opportunities for interpersonal relationships that are more characteristic of future adult relationships. Adolescents learn to invest themselves more fully in their friendships, gradually and often painfully learning what they can reasonably expect and not expect from their friends. With the attainment of sexual maturity, adolescents must also learn how to get along with opposite-sex friends in a new way, integrating love and sex. Participation in all types of peer groups may help adolescents to learn alternate models of leadership and sex-role expectations, thus supplementing those learned in the family. This may be especially valuable to adolescents from disadvantaged backgrounds.

Third, as we have already discussed, peers also serve as reference groups, providing adolescents with alternative standards for judging their experience and behavior. Although individuals of all ages need reference groups, such groups are especially important during this period of rapid change. And since people tend to compare themselves with those most like themselves, it is natural for adolescents to prefer the company of other teenagers. Many of the awkward changes of this period, whether they be facial pimples or social rejection, become somewhat less unsettling when shared with understanding friends who may be experiencing the same thing.

Finally, peers help adolescents define their own personal identities. This is an especially important function when you consider that the increased orientation to peers comes at a period when personal identities are fluid and open to change. Peer interaction may encourage positive self-discovery through exploring new roles, such as the shy girl who discovers her capacities for self-expression through acting in school plays. Of course, it can work the other way too, as in the case of the adolescent who adopts a negative identity partly because of the influence of a "bad" crowd. Then too, there is always the danger of undue peer influence on adolescents, which is more apt to occur among teenagers from emotionally unsupportive families.

Social Acceptance

Like adults, adolescents have a strong desire to be recognized and accepted by their peers. Although the importance of social acceptance has declined somewhat in the past two decades, more than 8 out of 10 teenagers feel that being liked and accepted by one's peers is "very important." Generally, the desire for social acceptance is most pronounced among teenagers of the upper middle class, whose parents are in managerial, business, or professional positions. Working-class teenagers are much less concerned about peer popularity (Sebald, 1981). Also, the emphasis on ethnic pride and cultural traditions in recent years has made acceptance within one's own ethnic group even more important than in the larger society.

The adolescent's popularity is partly dependent on his or her personality. Generally, adolescents who are well-liked are at ease with themselves and make others feel accepted and involved in social activities. Thus, well-liked adolescents tend to be friendly and outgoing. They're also apt to have a moderately high level of self-acceptance and self-esteem, so that they can interact with a variety of adolescents, not just those like themselves. Having a good sense of humor, being cheerful, and being fun to be with usually increases one's popularity among peers. Physical attractiveness, intelligence, and social skills also count for a lot (Hartup, 1983; Sebald, 1981). Teenagers who are not well-liked tend to exhibit the opposite personal traits from their

Being in the right crowd is especially important for young women's popularity. (Laimute Druskis)

more popular counterparts. That is, they tend to be shy, and in some cases aloof. Having a low level of self-acceptance, they're apt to be self-centered, anxious, and defensive, all of which tends to alienate others. In some instances, such individuals may attempt to compensate for their marginal status by acting sarcastic, aggressive, or conceited, thereby compounding their problems of rejection.

The adolescent's willingness to join in social activities and dress and talk in ways that are acceptable to his or her peers also plays an important role in social acceptance. Hans Sebald (1981) found that the leading reason for peer popularity was social conformity, and the main reason for social rejection was nonconformity among one's peers. At the same time, the third leading cause of peer popularity was "being yourself." Sebald suggests that this apparent contradiction may express an all-American value dichotomy that affirms both individualism and conformity. Thus, adolescents may be liked and admired for being themselves, but only up to a point. Those who are too different from their peers, especially in socially undesirable ways, tend to be rejected by them. Similarly, while habitual "loners" are generally unpopular, teenagers who spend up to a third of their waking time alone tend to get along better with their peers than those who spend either too much time alone or too much time socializing with others (Csikszentmihalyi & Larson, 1984).

value dichotomy: Division into two opposing groups of values.

The adolescent's achievements and activities continue to be a major consideration in peer popularity. Although the relative importance of one's accomplishments and activities varies somewhat depending on the type of school, community, sex-role expectations, and dominant values, the criteria for peer popularity have remained remarkably consistent in different eras. Joel Thirer and Stephen Wright (1985) surveyed 600 high school students in six states. They found that men and women rated being an athlete as the leading criterion for men's popularity and being in the "leading crowd" as the main criterion for women's popularity. But on most other items, men and women were ranked somewhat differently by their same-sex and opposite-sex peers. (See Table 8–3.) For instance, both men and women ranked being a "leader in activities" higher for their own sex than for the opposite sex. Similarly, both sexes ranked "coming from the right family" dead last for their own sex, but not for the opposite sex. This suggests that teenagers are especially concerned to be judged by their own accomplishments, which they have more control over than "coming from the right family." Yet, they are not always so understanding when judging their opposite sex peers.

The students were also asked how they wished to be remembered, whether as a "brilliant student," "athletic star," or the "most popular" person. Among women, 43 percent selected "brilliant student," 40 percent selected "most popular," and 17 percent selected "athletic star." In contrast, 37 percent of the men selected "athletic star," 34 selected "brilliant student," and 29 percent selected "most popular," confirming some of the sex-role differences cited earlier.

Table 8–3 PEER POPULARITY BY SAME-SEX AND OPPOSITE-SEX

MEN	
Rated by women	**Rated by men**
Be an athlete	Be an athlete
Be in leading crowd	Be in leading crowd
Come from the right family	Leader in activities
Have a nice car	High grades, honor roll
High grades, honor roll	Come from the right family

WOMEN	
Rated by men	**Rated by women**
Be in leading crowd	Be in leading crowd
High grades, honor roll	Leader in activities
Come from the right family	Be a cheerleader
Be an athlete	High grades, honor roll
Have a nice car	Be an athlete
	Come from the right family

Source: Adapted from Joel Thirer and Stephen D. Wright, "Sport and Social Status for Adolescent Males and Females," *Sociology of Sport Journal* (1985): 167 and 169.

PEER GROUPS

Adolescents spend much of their time in groups. Some groups are character-ized by an official organization, with rules for how members may be admitted and function within the group. Examples would be the various types of social and interest groups—drama club, yearbook staff, honor society, athletic teams, and religious groups. But adolescents probably spend an even greater amount of time in informal groups, which are characterized by less formal organization and stability. For instance, two or more teenagers may share lunch daily during the week or telephone each other regularly.

Cliques and Crowds

In a well-known study, Dunphy (1980) found that the majority of adolescents belong to two types of groups—cliques and crowds—which differ both in their size and function.

group: Two or more persons who share a common interest or goal.

Cliques are essentially small groups (less than 10 members in Dunphy's study) that meet mostly for personal communication and sharing. Activity in cliques tends to occur spontaneously, such as meeting in the hallway at school. Members are attracted to each other on the basis of similar interests, personalities, schools, neighborhoods, or religious affiliations. Although ado-lescents tend to form same-sex cliques in early adolescence, they participate more in opposite-sex cliques as they grow into middle and late adolescence.

Crowds are larger size groups (more than 10 persons in Dunphy's study) that meet primarily for organized social activities, such as parties or dances. Crowd activities occur mostly on the weekends, in contrast to clique activities, two-thirds of which occur during the week. Most crowd activity includes both sexes. Dunphy also found that the crowd was essentially a collection of cliques, with membership in the latter required for belonging to the crowd.

About 30 percent of the boys and 20 percent of the girls in Dunphy's study did not belong to either type of group. These adolescents are often referred to as the "outsiders" or "loners," though the particular individuals so labeled may change from year to year. Some loners do not join groups because they have been rejected by their peers. Others deliberately choose not to join group activities, even when offered opportunities to do so; an example is the habitually shy adolescent who feels more comfortable with ideas, things, or nature than with people. Sometimes, marginal individuals may get together in groups of their own. Interestingly, the more creative and gifted adolescents are less likely to join groups, or even if they do, they are less comformist in them during high school and college. All this reminds us that although membership in peer groups is desirable, it is not essential for normal development in every instance.

Changes in Peer Groups

Dunphy found that the relationship between adolescent cliques and crowds changes throughout adolescence, and that the sequence of changes is somewhat predictable. Furthermore, it appears that these changes are an integral part of the adolescent's socialization, especially in the transition from preadolescent, same-sex roles to the opposite-sex roles characteristic of adults. (See Figure 8–2.)

In stage 1, boys and girls are still active in the same-sex peer groups characteristic of preadolescence. Boys tend to join each other mostly to do things together, and they form somewhat larger and more stable groups than girls. The latter get together more for personal sharing, and form smaller and more intimate groups than boys.

Stage 2, characterized by more interaction between same-sex cliques, often takes place during the junior high-school years. Teenagers now spend more time away from their homes and neighborhoods and have a wider selection of friends to choose from. Because girls are often taller and more mature than boys at this stage, most opposite-sex relationships occur in group activities. Individual dating tends to be the exception at this stage. Group parties and dances offer boys and girls more security, mostly because of the awkwardness and superficial antagonism between the sexes at this period.

In Stage 3, the upper-status members of the same-sex cliques begin interacting in more personal, boy-girl relationships. Adolescents begin dating at

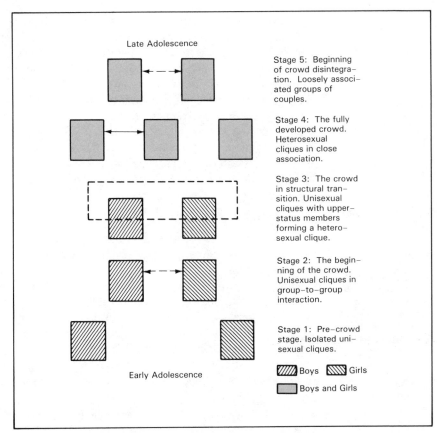

Late Adolescence

Stage 5: Beginning of crowd disintegra- tion. Loosely associ- ated groups of couples.

Stage 4: The fully developed crowd. Heterosexual cliques in close association.

Stage 3: The crowd in structural tran- sition. Unisexual cliques with upper- status members forming a hetero- sexual clique.

Stage 2: The begin- ning of the crowd. Unisexual cliques in group-to-group interaction.

Stage 1: Pre-crowd stage. Isolated uni- sexual cliques.

Early Adolescence

Boys Girls
Boys and Girls

FIGURE 8–2 STAGES OF PEER-GROUP DEVELOPMENT IN ADOLESCENCE. Dunphy found that the sequence of changes in adolescent peer-group formation re- mains remarkably persistent despite variations of time and place. But he deliberately avoided specifying an average age for the onset of each stage because of the wide variation in ages among different groups of adolescents

Source: D. C. Dunphy, "The Social Structure of Urban Adolescent Peer Groups," *Sociometry, 26* (1963): 236. Used by permission of the publisher and author.

this stage, with the earlier-maturing boys and girls usually taking the lead. Teenagers now form opposite-sex cliques, which become the core of the adolescent crowd. The latter is denoted by the dotted line rectangle in Figure 8–2. Notice that these adolescents retain membership in both opposite-sex and same-sex cliques.

By stage 4, the majority of adolescents have begun dating, which leads to a transformation in the clique system. Now the crowd becomes composed mostly of opposite-sex cliques, with a great deal of close association between them. Although Dunphy refers to this stage as the fully developed crowd, he also notes that the crowd structure tends to last only long enough to ensure that its members have acquired the basic sex-role characteristics necessary for adult heterosexual relationships.

Stage 5 is designated as the beginning of crowd disintegration. Upon graduation from high school, individuals tend to take different paths; some leave school to work and others to attend college. More couples are also going steady, becoming engaged, or getting married. All of this makes for less need for the support of the old adolescent cliques and crowd.

Dunphy (1980) found that this sequence of changes in adolescent peer groups tends to persist despite the variations of time and place. He also found that teenagers enter the respective stages at increasingly earlier ages, mostly because of earlier maturation and the beginning of adolescence.

Peer Conformity

conformity: Changes in one's attitudes or behaviors because of real or imagined pressure from others.

Adolescents frequently dress and act the way they do partly out of conformity to their peers. Sometimes peer conformity may be rather obvious, such as the boy who engages in a delinquent act mostly because of pressure from his buddies. But teenagers are often influenced by their peers in more subtle ways, especially in the need for approval. A common example is the adolescent who begins smoking tobacco, using alcohol, or experimenting with drugs, mostly to win the approval of his or her friends. Then there are instances in which individuals modify their behavior because they have changed their attitudes and values, such as the teenager who joins a religious group because of a change in faith. More often than not, conformity to peers reflects a combination of these influences.

Conformity to peers generally increases throughout childhood, peaking in early to midadolescence and declining slowly afterwards. The developmental trend in peer conformity can be seen in a study by Thomas Berndt (1979) which included students from the third through the twelfth grades. Students were administered a questionnaire containing 30 hypothetical situations which measured three types of conformity—prosocial, neutral, and antisocial. One prosocial item consisted of agreeing to a peer group's request to help a classmate with a project when one really wanted to do something else. A neutral item consisted of agreeing with peers to engage in an activity the student was not really interested in. An antisocial item consisted of the adolescent's response to a peer's request to engage in a delinquent act. The results showed that the mean conformity score for prosocial and neutral items rose to a peak in early adolescence (grade six for prosocial; grade nine for neutral items) with a gradual decline thereafter. But the mean score for antisocial behavior rose markedly to a peak in the ninth grade and then declined to the twelfth grade. This later age for the peak in conformity to antisocial peer behavior corresponds to the increased socialization in the peer culture and incidence of delinquency generally seen at this age.

At the same time, peer conformity varies widely among adolescents and within a particular individual's behavior according to a number of influences. A major factor is the age and maturity of the adolescent, especially because the cognitive capacity to internalize the rules of society increases with age throughout grade school. The attainment of formal thought or adult-type reasoning in late adolescence usually brings a greater understanding of the complexity of social behavior, along with a corresponding decrease in automatic compliance with the rules (Polovy, 1980). Adolescents with high self-esteem and a wide range of competencies are much less likely to conform to their peers than those with low self-esteem and few abilities (Hartup, 1983). By contrast, teenagers with low social status are more apt to conform to their peers (Coleman, 1980). The particular issue and situation at hand, as discussed earlier, are very important considerations. Accordingly, adolescents are more likely to conform to their peers in certain matters, such as dress, music preferences, which clubs to join, and whom to date (Sebald, 1986). Then too, there are a number of characteristics about peer groups which influence the extent of the adolescent's conformity. Generally, the more cohesive the group, the greater the proportion of conformers in the group. But the presence of individuals who oppose the group tends to reduce the proportion of conformers. The newer and less experienced an individual is in the group, the more likely he or she is to conform to the group. Also, the more difficult the task the group is engaged in and the less competent the adolescent feels in this area, the greater the tendency to conform to one's peers (Feldman, 1985).

Girls have been portrayed as more prone to peer conformity than boys, partly because of girls' greater concern about peer acceptance. Yet it is becoming increasingly evident that such differences have been exaggerated. After a careful analysis of many studies in this area, Alice Eagly (1983) suggests that many of the findings are suspect because of a potential biasing factor, namely the sex of the researcher. Eagly found that 80 percent of the studies were conducted by men, who tended to obtain larger sex differences than did women researchers. Generally, experiments conducted by women have found little or no sex difference in conformity behavior. Although Eagly (1981) and her colleague Linda Carli did not find a statistically significant relationship between conformity in different situations and sex, they did find a trend in that direction, suggesting the need for continuing research in this area.

statistical significance: The extent to which an obseved difference between two samples reflects a real difference rather than chance.

FRIENDSHIP

Among the most important aspects of adolescents' lives are the close relationships with their friends. Although friendship is important at all ages, it is not until mid- or late adolescence that friendships take the form of intimate rela-

From adolescence on, friendships involve intimate sharing and loyalty. (Laimute Druskis)

tionships, that is, those characterized by trust, self-disclosure, and loyalty. One major factor at work at this stage of development is the adolescent's emerging autonomy, or sense of being a separate person. Another important factor is the cognitive changes that enable adolescents to see situations from another person's point of view, thereby increasing their capacity for empathy and helpfulness. As a result of these developments, from this stage on, individuals experience a greater need for intimacy and an increased capacity to enter into close relationships.

Intimacy with Parents and Peers

friendship: An informal, affectional relationship between two or more individuals.

"As I entered adolescence," wrote one girl, "suddenly friends became very important to me. There were two or three of us who called each other every night and shared everything." This is the essence of friendship, being able to share yourself without fear of being betrayed. Yet, contrary to popular opinion, the increasing intimacy between adolescents and their friends is not necessarily accompanied by a decrease in intimacy with their parents. This was brought out in a study by Hunter and Youniss (1982), which measured intimacy between adolescents and their mothers, their fathers, and their best friends. Adolescents ranging in age from 9 to 19 years of age were asked to rate the three relationships according to four measures of intimacy: their willingness to discuss personal problems with the person, the degree of intimate knowledge shared, their ability to talk through disagreements, and feelings of companionship. The results showed that intimacy between teenagers and their best friends increased steadily with age, with the highest level of intimacy occurring among the 19 year olds. From 12 years of age on, teenagers reported the relationships with their best friends as more intimate than those with their parents. However, the difference was due mostly to the increased intimacy with peers rather than decreased intimacy with parents. The levels of intimacy between teenagers and their mothers remained steady throughout adolescence, and at a higher level than for fathers. Intimacy between adolescents and their fathers fluctuated across the age groups, increasing slightly up to 12 years of age, then decreasing to 15 and remaining steady afterwards.

Such findings are similar to those regarding the relative importance of parent and peer influence discussed earlier. That is, the growth of intimacy between teenagers and their best friends supplements rather than displaces intimacy between teenagers and their parents. It is possible for adolescents to be friends with their parents, though for most people this tends to occur at a later stage of development after young people feel they have achieved their independence.

Changing Patterns of Friendship

The meaning and quality of friendships also change throughout adolescence, mostly because of the developmental and social changes that are occurring in their lives. Friendship gradually evolves from superficial, activity-centered relationships to more emotionally intense and intimate relationships.

When boys or girls in early adolescence, around 10 to 14 years of age, speak of a friend, it is likely to be someone they enjoy doing things with. And since boys and girls this age have different interests and engage in different sorts of peer activities, most of their friends are of the same sex. The actual relationship between friends tends to be rather superficial, with qualities such as self-disclosure and loyalty just beginning to be evident at this stage of adolescence, especially among girls (Berndt, 1982).

By midadolescence, around 15 to 17 years of age, friendships become more emotionally intense and relationship oriented, and also more unstable. For one thing, adolescents are becoming more independent of their parents and more emotionally involved with their peers. Then too, feeling anxious and confused about their bodily changes and intensified sexual feelings, teenagers are looking for someone they can confide in, who will be trustworthy, and who can offer emotional support. As a result, friendships become more mutual, emotionally intense, and are characterized by greater intimacy with

LONELINESS

Throughout adolescence and the transition to adulthood, individuals are especially susceptible to loneliness—the state of unhappiness at being alone accompanied by a longing for companionship. According to a survey among people of all ages, 18 to 25 year olds suffer the most from loneliness. The intensity of loneliness decreases steadily by age, with people over 70 being the least lonely. A major reason is that loneliness is largely a state of mind resulting from the gap between one's desires for closeness and the failure to find it. Having loosened the ties with their parents, young people are actively seeking intimacy with their peers. Yet, many of them have not developed the social skills and relationships to satisfy their intimacy needs (Rubenstein, Shaver, & Peplau, 1982).

Youths who blame their loneliness on their personal inadequacies ("I'm so lonely because I'm unattractive") make themselves even lonelier. They are more apt to try to overcome their loneliness by focusing on their own efforts ("I'll stop studying so much and get out and meet more people"). Many young people might benefit from social-skills training, learning how to initiate a relationship and conduct a meaningful conversation with others. Carolyn Cutrona (1982) points out that people who are habitually bothered by loneliness may also need to change their attitudes. She found that college students who overcame their initial loneliness at school were no more likely to initiate a conversation with strangers, join groups, or attend parties than students who remained lonely. The biggest difference between the two groups was in their attitudes. Students who eventually overcame their loneliness persisted in their efforts to socialize with others despite some initial loneliness at the beginning of the school year. In contrast, those who suffered the most from loneliness lowered their goals and rationalized their loneliness. Yet, their efforts to convince themselves they didn't need friends or that they were too busy with school to pursue a more active social life made them even lonelier. Consequently, it may not be enough to encourage young people to join groups and acquire social skills. We must also help them deal with their pessimistic attitudes about finding friends.

peers than with parents (Hunter & Youniss, 1982). The greater degree of emotional dependency and identification in relationships at this age brings more satisfaction in friendship but also greater vulnerability than in earlier stages. Consequently, midadolescent friendships tend to blow hot and cold, with sudden, dramatic changes and bitter feelings when friends break up. Even when adolescents dream of their friends, such dreams are often marked by negative fears of separation and abandonment, attesting to the intense emotional involvement of adolescent friends (Roll & Millen, 1979).

By late adolescence, from 18 years of age on, the passionate quality of friendship gives way to a calmer, more stable relationship. On the one hand, individuals are capable of forming close, meaningful relationships characterized by trust, self-disclosure, and loyalty. For instance, one study found that between the fifth and eleventh grades adolescents were increasingly apt to say they knew what their friends "feel" about things even when not told, and also, to "feel freer" to talk to their friends about anything (Sharabany, Gershoni, &

Hofman, 1981). At the same time, late adolescents have greater intellectual and emotional maturity, bringing them greater understanding of the complexity of human relationships. Having had more experience in making friends by this age, they tend to be more realistic about what to expect from a friendship. They are also more tolerant of individual differences and can appreciate what others bring to a relationship without the need to have their friends be "just like themselves." As a result, friendships between late adolescents tend to be emotionally close and stable relationships resembling adult friendships.

Same-Sex and Opposite-Sex Friendships

Although opposite-sex friendships increase throughout adolescence, they don't become common until late adolescence, toward the end of high school. During early adolescence, boys and girls keep pretty much to friends of their own sex, mostly because they have different interests and activities at this age. Boys and girls tend to feel uncomfortable in each other's presence, as reflected in the joking and teasing so characteristic of this stage. For instance, if a boy is seen talking with a girl in the school cafeteria, each of them is apt to be teased about it by his or her same-sex friends. However, by midadolescence, a greater proportion of 15 and 16 year olds are choosing friends of the opposite sex, with girls generally choosing a larger number of them than boys. The transition from same-sex to opposite-sex friends occurs about the same time boys and girls are beginning to socialize in mixed crowds. Thus opposite-sex friends are more common in dating situations, parties, and the crowd activities at school.

Opposite-sex friendships generally supplement rather than displace same-sex friendships. For instance, in one study the likelihood of opposite-sex friendships increased throughout high school, while the number of same-sex friendships remained constant: of opposite-sex friends girls reported an average of two while boys reported an average of only one; of same-sex friends, girls reported an average of five while boys reported an average of four (Sharabany, Gershoni, & Hofman, 1981). Girls are also more apt to report close friendships with boys who are older, often at another school, whereas boys generally list girls the same age or younger. All of this suggests that friendships among late adolescents flourish in the context of dating (Blyth, Hill, & Thiel, 1982).

Although friendship is not necessarily more important for one sex than the other, it tends to have a different meaning for girls than for boys. More specifically, intimacy appears to play a more central role in girls' friendships. Girls report higher levels of intimacy both in their same-sex and opposite-sex friendships, especially from midadolescence on. Girls also express greater anxiety over intimate relationships at all age levels. Tensions, jealousies, and conflicts between close friends are more common among girls than boys. A

Most older adolescents have friends of both sexes. (Laimute Druskis)

greater proportion of the negative themes among girls involves rejection or exclusion from friendship, whereas those of boys involve outright disputes over property, leisure activities, and girlfriends (Sharabany, Gershoni, & Hofman, 1981; Coleman, 1980). All of this does not mean that intimacy is absent from boys' friendships. Rather, boys tend to express their intimacy needs in more subtle ways, through shared activities rather than the direct satisfaction of emotional needs.

DATING AND EARLY MARRIAGE

Many of the friendships with the opposite sex occur in the context of dating, which as you may recall is the practice by which a boy or girl agree to meet at a specific time and place, either alone or in a group, for a social engagement. Dating begins earlier and plays a more important role among young people in the United States than in many other parts of the world. This may be accounted for in part by the prolonged adolescence of modern youth plus the characteristic emphasis on choosing one's own friends and marriage partners in Western society. In any event, youth now reach physical and sexual maturity at an earlier age and tend to marry at a later age than in the past, giving them considerably greater experience in dating than their counterparts in earlier eras.

Dating

Dating serves several purposes in the lives of young people. First, the practice of dating provides youth with opportunities for sharing leisure activities, such as attending a concert or a party. Dating is also an important means of learning the social and interpersonal skills needed to get along with members of the other sex, such as how to initiate a relationship and how to end an unwanted relationship. One of the most important functions of dating is establishing and maintaining mutually satisfying close relationships. Dating also provides an opportunity for sexual experimentation within mutually acceptable limits. Although a certain amount of testing behavior occurs in dating, most sexual intercourse among young people occurs within close, steady relationships rather than casual dating. Finally, dating also provides valuable experience for mate selection. By the time young people reach their early to midtwenties, most have had at least one serious relationship with the other sex, with the majority having had several or more such relationships (Kephart, 1977).

How soon teenagers begin dating depends largely on the particular adolescent, his or her parents, and peer influence. Some observers (Dornbush et al., 1981) have found that adolescents generally begin dating when members of their particular clique have begun to date, regardless of the adolescent's age or maturity. Although this varies somewhat from one student population to another, one study (Hansen, 1977) showed that one-fourth of the high school students began dating by the age of 13; over half started dating by 15 or 16; and less than a fourth had their first date at 16 or later. Teenagers who begin dating relatively early or late are more apt to exhibit certain problem behaviors. Wright (1982) found that teenagers who received early parental permission to date—15 years of age in this study—not only exhibited greater self-confidence and autonomy than others their age, but also were more likely to report using alcohol and marijuana, and drinking and drug-abuse problems. Those who received late parental permission felt less sure of themselves, felt more dependent on their parents, complained more about their parents' strictness, and were also more apt to be bothered by suicidal ideas.

Dating Patterns

According to a national survey of high school seniors, 4 out of 10 teenagers date occasionally (two or three times a month or less). About 5 out of 10 teenagers date at least once a week or more. And 1 out of 10 seniors never dates. Among those who date the most frequently, three times a week or

WHAT ABOUT THE BASHFUL, SHY ADOLESCENT?

Chances are he or she has a more difficult time making friends, especially with the opposite sex.

Shyness is the tendency to avoid contact or familiarity with others. It is especially acute at adolescence, mostly because of the rapid body changes as well as the increasing capacity for self-consciousness and abstract thought which appears at this stage.

In a study of shyness among high-school and college students Philip Zimbardo and his colleagues (1974) found that 82 percent of the students regarded themselves as shy at some point in their lives. Although about half this number felt they had outgrown shyness, over 40 percent of them labeled themselves as presently shy. Most of them did not like being shy.

Shy adolescents have more trouble making friends because they are often misperceived by others in a negative way. Shy persons tend to be regarded as aloof, bored, disinterested, condescending, cold, and hostile. When treated accordingly, they may feel even more isolated, lonely, and depressed. As a result, they overindulge in the normal process of self-monitoring, thus increasing their self-criticalness, their concern for the impression they make on others, and their shyness.

Zimbardo found that shyness covers a wide range of behaviors. At one end of the spectrum are those who are not especially apprehensive about being with people when necessary, but who prefer being alone most of the time. These are the adolescents who feel more comfortable with ideas, nature, or working with things. In the middle range are those who are easily embarrassed, reflecting their lack of self-confidence and social skills. An example would be the awkward, socially inept adolescent who hesitates to ask for a favor or a date. At the other extreme are adolescents whose shyness serves as a kind of neurotic self-imprisonment. These are the individuals who judge themselves with impossible rules, leading them to avoid unfamiliar situations and people and the possibility of rejection.

Our society aggravates shyness by the emphasis on competition, individual success, and personal responsibility for failure. Parents unintentionally encourage shyness by stressing individual achievement and social approval as the primary measures of an adolescent's self-worth.

Shyness can be overcome partly through getting teenagers involved in something outside themselves, like total absorption in a task, role playing, or dramatics. It may help to identify specific situations that elicit shyness, and to provide opportunities for practicing the social skills needed in them, as in assertiveness training. It also helps to realize that shyness is entirely "normal" for adolescents and will be outgrown in most instances.

Did you ever have any trouble with shyness in adolescence? If so, have you mostly outgrown it?

P. Zimbardo, P. Pilkonis, and R. Norwood, *The Silent Prison of Shyness* (Glenview, IL: Scott, Foresman and Company, 1974). Copyright © 1974 by Scott, Foresman and Company. Reprinted by permission.

more, are a greater proportion of girls than boys and youth not bound for college (Bachman, Johnston, & O'Malley, 1980).

When young people 16 to 21 were asked where they preferred to go on their first date, going to the movies or a party received the highest number of votes; dinner dates, dancing, study dates, and "parking" dates ran far behind. On subsequent dates, parties ranked the highest, though nearly a fifth of the youths preferred a quiet evening at home as a more conducive setting for getting to know someone (Gaylin, 1978).

The proportion of teenagers going steady with someone has remained remarkably stable over time. A nationwide Gallup Youth Survey (Gallup, 1979) showed that 28 percent of the teenagers were dating only one person at the time, with a larger percentage of girls (32) than boys (23) going steady. The proportion of individuals going steady increased with age, from 16 percent of the 13 to 15 year olds to 35 percent of the age 16 to 18 year olds currently going steady with someone. More girls than boys reported going steady at all ages, with the discrepancy probably reflecting several factors. First, many girls were probably going with older boys not included in the study. Second, when asked if they were in love, a larger percentage of girls (61) than boys (42) said they were in love with their partners, indicating girls were more likely to be in one-sided relationships. Such discrepancies suggest that girls and boys may not always perceive their relationships in the same way.

It is widely assumed that boys and girls differ in their orientation toward dating, with boys putting more emphasis on sex and girls placing greater emphasis on the affectional aspects of dating, such as affection, trust, companionship, and romance. But a study by McCabe and Collins (1979) reveals that such matters are more complex. Participants' attitudes were measured on psychobiological behaviors (physical intimacy and sex) and psychoaffectional behaviors (emotional involvement). Then based on their sex-role scores as well as their biological sex, the participants were divided into six categories: masculine, androgynous, and feminine males; and masculine, androgynous, and feminine females. The results showed that the desire for physical intimacy and sex depended heavily on sex-role orientations. Generally, masculine and androgynous adolescents of both sexes were more likely to desire physical and sexual intimacy than feminine adolescents. Also, the differences between the sexes in such matters diminished with increasing age, with females increasingly more inclined toward physical and sexual involvement as they got older. Furthermore, as the dating relationship deepened, there was greater compatibility between the sexes regarding their desire for physical and sexual intimacy, with practically no difference in the desire for sex among older youth going steady. More than three-fourths of the older youths going steady desired heavy petting, oral sex, and sexual intercourse. Finally, with the deepening of the dating relationships, the desire for mutual affection increased among all groups, suggesting that the desire for affection is less polarized between the sexes than popularly believed.

Adolescent Marriage

Although the importance of marriage and raising a family has declined somewhat in the past few years, the great majority of people eventually marry—most of them in their twenties. At the same time, the median age of first marriage in the United States has been rising and is now 25 for men and 23 for women (U.S. Bureau of the Census, 1986). When high school seniors were asked their views about marriage, more than three-fourths of them indicated they plan to marry. Only a small proportion of them, about 5 percent, felt sure they did not want to marry, with the the rest being uncertain. Most young people would prefer to wait at least 2 years after marriage before having children (Bachman, Johnston, & O'Malley, 1980).

Despite such sentiments, a significant proportion of young people marry during adolescence, with about one-third of all marriages having at least one partner in his or her teens (U.S. Bureau of the Census, 1984). The single biggest reason for adolescent marriage is that the girl is pregnant. In fact, the younger the bride, the more likely she is to be pregnant. About half the time

It's like being grounded for eighteen years.

Having a baby when you're a teenager takes away more than your freedom, it takes away your dreams.
The Children's Defense Fund.

(Children's Defense Fund)

teenage girls with a premaritally conceived pregnancy marry the father of their child. A disproportionate number of teenage pregnancies occur among those in the lower socioeconomic groups, in many of which the girl already has another child. Even when there is no pregnancy involved, teenage marriages occur more frequently among those in the lower socioeconomic groups— accompanied by lower education, higher rate of school dropouts, and lower incomes—all of which lessens a couple's chances for a successful marriage.

Unfortunately, marrying to legitimize the birth of the child reduces the likelihood that the mother will return to school after pregnancy. Teenage mothers who remain unmarried are more apt to continue living with their parents and to receive financial support and encouragement to continue their schooling. For some reason black teenage mothers, whether they marry or not, are more likely than whites to attend school following the birth of their first child. Yet, teenage mothers who marry *before* the birth of their child are less likely to separate from their husbands in later years. By contrast, those who delay their marriage until after the birth of the child are much more apt to separate and divorce in later years (McLaughlin, 1986). In one study, 60 percent of the premaritally pregnant mothers 17 years of age and younger were separated or divorced within 6 years of their marriage. About 20 percent had split up their marriages within the first year (Guttmacher, 1977).

Couples who marry in their teens, whether the woman is pregnant or not, are much more likely to report problems in their marriages and divorce in subsequent years than those who marry in their twenties. Brides 17 years of age and younger are three times as likely to separate or divorce—and husbands the same age twice as likely—as those who marry in their twenties. Individuals who marry in the late teens fare slightly better, but they also divorce in greater numbers than those who delay marriage until their twenties (Guttmacher, 1977).

Alan Booth and John Edwards (1985) found that a major factor in the greater marital instability of teenage marriages is "inadequate role performance." That is, those who marry in their teens tend to have less than an adequate role model in the home as well as insufficient time and exposure to family life to fulfill the spouse role. One woman who grew up in a home seeing her father beat her mother was encouraged by her mother to marry early. The message was, "Find a husband, get married, get out of here." Many of these young people lack the personal maturity for marriage, especially at that age. Boys are unready to assume the responsibility of marriage and parenthood. And teenage girls often resent the loss of freedom. Moreover, the ability to enter into satisfying emotional intimacy depends on the attainment of a strong sense of personal identity. Individuals must know who they are before they can share themselves meaningfully in a marriage relationship. In most cases, those who marry in their teens have not had enough experience in finding themselves to share deeply with someone else. In some instances individuals who marry in their teens may grow together and achieve an even more meaningful marriage because of the extra struggles they have shared. But more often, the responsibilities of marriage are perceived by young partners as restraints, prematurely limiting their freedom to grow. All too often the people who seek a divorce in their thirties and forties to "become free" are those who probably married too early in the first place.

SUMMARY

Peer Relations

1. As children reach adolescence they spend an increasing proportion of time with their peers—those of their own age.

2. Adolescents' orientation toward their peers tends to supplement rather than displace the orientation toward their parents, so that parents and peers serve as alternate reference groups for different matters.

3. The relative importance of parent influence, compared to that of peers, declined between the 1960s and 1970s but has reversed somewhat in the 1980s, especially for boys.

4. Peer relations help to ease the adolescent's transition from a family to peer orientation, provide opportunities for adultlike relationships, create alternate reference groups, and serve as a means of clarifying their personal identities.

5. How well teenagers are accepted by their peers depends primarily on such factors as the adolescent's personality, achievements, and participation in social activities.

Peer Groups

6. The majority of adolescents belong to two types of peer groups: *cliques* (small groups that meet mostly for personal sharing) and *crowds* (larger-sized, groups that meet primarily for organized social activities).

7. Throughout adolescence, the pattern of participation in peer groups changes from isolated same-sex cliques to crowds composed of mixed-sex cliques, thereby helping adolescents to make the transition from unisexual to heterosexual roles.

8. Conformity to peers generally increases throughout childhood, peaking in early to midadolescence and declining slowly afterwards, mostly because of the attainment of formal thought.

Friendship

9. Adolescence brings a greater need and capacity to enter close relationships, with the increasing intimacy between peers supplementing rather than displacing closeness between adolescents and their parents.

10. The meaning and quality of friendships changes throughout adolescence, with friendship evolving from superficial, activity-centered relationships to more emotionally intense and intimate relationships resembling adultlike friendships.

11. Although opposite-sex friendships increase throughout adolescence, they are not common until late adolescence, toward the end of high school.

Dating and Early Marriage

12. Much of the friendship between the sexes occurs in the context of dating—the practice by which a boy and a girl agree to meet at a specific time and place, either alone or in a group, for a social engagement.

I3. Half the adolescents have begun to date by 15 or 16 years of age, and they date, on the average, once a week or more.

14. There is a tendency for girls to emphasize the affectional aspects of dating and for boys to place greater emphasis on sex, but such differences diminish with age and the deepening of the dating relationship.

15. Couples who marry during adolescence tend to separate and divorce in greater numbers in later years than those who wait until their twenties to get married.

REVIEW QUESTIONS

1. To what extent does the increased peer orientation at adolescence supplement rather than displace the orientation toward parents?

2. What are some of the functions of peers in the lives of adolescents?

3. Explain the criteria for social acceptance of adolescents.

4. What is the difference between cliques and crowds? Give an example of each type of group.

5. Describe the characteristic sequence of changes in peer group participation during adolescence as studied by Dunphy.

6. What are some factors affecting adolescents' conformity with their peers?

7. Explain the characteristic patterns of friendship that occur during early, mid-, and late adolescence.

8. Explain some of the sex differences in friendship between boys and girls.

10. What are some of the hazards of adolescent marriage?

9 Adolescents at School

- **SECONDARY SCHOOLS**
  Compulsory Education
  The Curriculum
  Schools for Early Adolescents

- **THE SCHOOL ENVIRONMENT**
  School and Class Size
  Teachers
  School and Classroom Climate

- **SELECTED FACTORS AFFECTING SCHOOL ACHIEVEMENT**
  Learning Ability
  Exceptional Students
  Family and Socioeconomic Background
  Dropping Out

- **IMPROVING THE SCHOOLS**
  What People Think of Public Schools
  New Directions
  The School Adolescents Would Like

- **SUMMARY**

- **REVIEW QUESTIONS**

Kathy is a senior and an honor student at Harrisburg High School. She gets along well with her peers and is active in extracurricular activities. Kathy has been the editor-in-chief of the yearbook, has been vice-president of the student council, has taken part in several school plays, and has played on the girls' lacrosse team. She has applied to several Ivy League colleges, with an eye toward eventually studying law. Mark, also a senior, attends the same high school, but associates with a different group of students. He is taking a business program and makes average grades. Mark spends the mornings in classes at school and the afternoons and weekends working at a fast-food restaurant. Although Mark's job has enabled him to buy a late-model sports car, he often feels too tired to study and does not participate in extracurricular activities. Upon graduation, Mark plans to enter the hospitality and management program at the local community college.

SECONDARY SCHOOLS

secondary schools:
Those coming next in se-
quence after elementary
schools, e. g., middle
schools and high
schools.

Kathy's and Mark's orientations toward school illustrate, perhaps to the extreme, some of the different types of students who attend comprehensive
public high schools. Such schools are the result of a dramatic change that has
taken place in public education since the 1920s. Prior to that time, the high
school was essentially a voluntary institution for a minority of adolescents,
mostly those from affluent families who were interested in preparing for college. Since that time, the high school has become a mandatory institution for
virtually all adolescents, including many students who have little or no interest
in the traditional academic subjects. Consequently public high schools are
expected to provide a wide variety of programs for a diverse group of students, making it difficult for educators and the public alike to agree on the
goals and priorities of secondary education.

Compulsory Education

The biggest single reason for such a change, of course, has been the passage
of compulsory education laws by the various states. By the late nineteenth
century, half the states had passed compulsory attendance laws. Today, such
laws are in effect in every state, with the average cutoff age being 16. However, in recent years some legislators have proposed lowering the required
age to 14 so that students who have little interest in school can begin work.
Others want to raise the required age to 18. Still others would require 2 years
of college. Meanwhile, the existing laws have dramatically increased the proportion of adolescents enrolled in school.

An elderly woman was telling about the high school reunion of her class of
1910 in South Philadelphia. "Out of 400 students who entered the ninth
grade," she said, "only 99 graduated from high school." She was surprised to
discover that this was about average for that era. Nationwide in 1910, only a
third of all 14 to 17 year olds were enrolled in school, with about half of them
graduating from high school. Each decade since then, however, has seen a
larger percentage of teenagers enrolled in school as well as graduating from
high school. By 1984, over 94 percent of all 14 to 17 year olds were enrolled
in school, with over three-fourths of them graduating from high school (U.S.
Bureau of the Census, 1985). The exact figures, of course, vary considerably
from one school to another, depending largely on the socioeconomic characteristics of the students and community. (See Figure 9–1.)

Compulsory education has both advantages and disadvantages. On the
plus side, our system of compulsory education has raised the level of literacy
in the general population, essential for participation in a democratic government and society. Access to free public education is also based on the ideal of
providing everyone in our society with the opportunities to make the most of
his or her potential, regardless of one's origins, race, sex, or religion. Since the

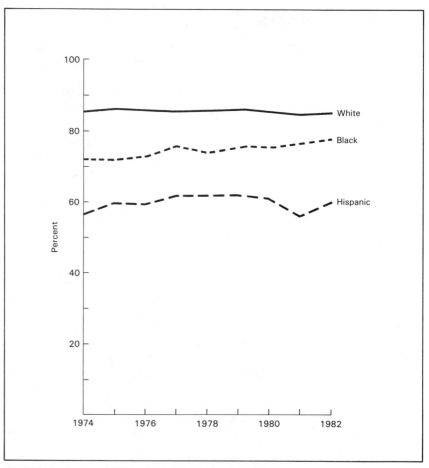

FIGURE 9–1. PERCENT OF 20 TO 24 YEAR OLDS WHO HAVE COMPLETED 4 YEARS OF HIGH SCHOOL.

Source: Valena White Plisko and Joyce D. Stern, eds., *The Condition of Education,* 1985 edition (Washington, DC: National Center for Educational Statistics, Government Printing Service, 1985), p. 74.

median: A measurement that falls in the exact middle of a distribution of scores; half the scores fall above this number and half fall below.

median educational level is now over 12 years of schooling, graduation from high school also makes it possible to aspire to the better paying jobs, with high school dropouts having the highest rate of unemployment. On the minus side, compulsory education has not been without its problems. First and foremost, if everyone is required to attend school, then schools should be equally accountable to all types of students—the gifted, the handicapped, the socially disadvantaged, and so on,—something virtually impossible to do in practice. Yet, when students feel disinterested or uninvolved in their studies, they become bored and inclined toward problem behavior, including costly vandalism. Student discipline continues to be a leading problem in many schools (Gallup, 1985). Although many states have either adopted or are considering

Schools have become a more significant part of adolescents' lives than in the past. (Ken Karp)

adopting stricter standards of discipline, they must continue to deal with attendance problems and school dropouts. Moreover, compulsory attendance laws have been difficult to enforce. In some instances they may violate students' rights and be prohibitively expensive to defend legally.

The Curriculum

curriculum: A fixed program of studies required for graduation from high school or college.

The comprehensive high school offers a broad range of educational programs. These generally include an intellectually rigorous program for the most able students who plan to attend college, a career-oriented program for those who anticipate working immediately after high school, and a general education for other students.

The *college preparation program* aims to prepare students for success in college. Although about half of all high school students nationally are enrolled in college preparation curricula, some affluent, prestigious high schools send more than their share of graduates to college, including the best colleges. In addition, high schools have enriched their courses to ease the transition to college-level work. More than three-fourths of all high schools now offer specialized academic courses in such subjects as anthropology, psychology, and sociology, and more than half give credit for college courses on college campuses (Dearman & Plisko, 1979).

The *vocational program,* though traditionally enrolling fewer students than the college preparatory curriculum, has grown in popularity in recent years.

Students in business/commercial programs may prepare for jobs in such fields as accounting, computer sciences, and secretarial services. Those in vocational/industrial programs may prepare for jobs in fields such as food services and automobile repair. Students in these programs generally spend about half their time in general education courses and the rest in special courses and on-the-job training. In addition, there are high schools devoted entirely to vocational or technical education.

The *general education program* provides a basic education for students who are not presently committed to a college- or career-oriented program. Traditionally, this program has attracted a larger proportion of low-achieving students, with a higher rate of dropouts. But the increased number and range of elective courses throughout the 1970s have attracted a larger proportion and a more diverse group of students to the general curriculum. Students have received academic credit for work experience outside the schools as well as a variety of courses aimed at improving life skills and personal growth (National Commission on Excellence in Education, 1983).

One of the perennial problems facing high schools is how much emphasis should be given to academic rigor and how much should be devoted to preparing students for their larger role in society, including the more socially relevant courses. Since adolescents spend much of their waking day at school, the schools have been called upon to shoulder a disproportionate share of socializing the young. In addition to the basic academic skills, schools have added courses on an endless variety of subjects such as driver education, sex education, drug education, and the like. With limited budgets, however, the schools cannot accomplish everything equally well. As a result, public sentiment and educational trends tend to swing back and forth between these two emphases. For instance, since the late 1960s the demand for more relevant and practical courses has resulted in a wider range of elective courses, thereby increasing the proportion of students in the general curriculum. However, in recent years there has been a move back to the "basics," partly in response to lower scores on student achievement tests and the threat that America is losing its competitive edge in the world market. This has resulted in an emphasis on academic excellence, including the traditional academic subjects of English, mathematics, the natural and social sciences, with the addition of a new basic skill in computers. Yet, it remains to be seen if this recent trend will produce the intended result, and/or whether it will further alienate students from their own interests and needs.

Schools for Early Adolescents

Another major issue in secondary education is tailoring the curriculum to the different developmental needs of early and mid- to late adolescents. Traditionally, the recognition that puberty begins about 12 years of age prompted school systems to separate seventh and eighth graders from elementary school students and place them in a junior high school. Yet, all too often the junior high school became a scaled-down version of the high school, with similar curricula and extracurricular activities without necessarily meeting the needs of younger adolescents. Furthermore, studies have shown that changing from the sixth grade in an elementary school to the seventh grade in a junior high school tends to be especially stressful for students because of the combined effects of the school change and the onset of puberty. Consequently, since the 1960s there has been a move toward different age groupings which more nearly reflect the physical, cognitive, and social needs of younger adolescents—as seen in the development of middle schools. The characteristic groupings of the middle school tend to avoid school changes at the critical stage of puberty. At the same time, there is a tremendous variation in such schools, with some arrangements composed of students from the sixth through eighth grades, others including the fifth through eighth grades, and still others including the ninth grade (Lipsitz, 1983).

middle schools: Schools organized around the characteristic needs of early adolescents.

In one study (Blyth, Simmons, & Carlton-Ford, 1983), students in schools with the 6–3–3 arrangement (in which students changed schools twice) were compared with students in schools using an 8–4 arrangement (in which students changed schools only once between the eighth and ninth grades.) The students were observed throughout a period of 5 years and compared on measures of self-esteem, participation in extracurricular activities, grade-point averages, and feelings of anonymity as they moved through the transitions in their respective school systems. The results showed that adolescents' self-esteem is affected differently in each school system. In the 8–4 arrangement, both boys and girls experienced a steady rise in self-esteem throughout the five years. Boys also showed a rise in self-esteem in the 6–3–3 arrangement, except in the transition between the ninth and tenth grades when self-esteem remain unchanged. But girls in the 6–3–3 arrangement experienced two significant drops in self-esteem during each of the transitions, that is, between the sixth and seventh grades and again between the ninth and tenth grades. Participation in extracurricular activities also increased in the 8–4 arrangement, especially among girls. Eighth-grade girls in the 8–4 system participated in about twice as many activities as eighth grade girls in the other system did. Tenth-grade girls in the 8–4 system participated in about three times as many activities as did their counterparts in the 6–3–3 system. Overall, the results showed that students felt more anonymous and made lower grades each time they had to change schools, regardless of when the transitions occurred.

6–3–3 arrangement: An educational sequence consisting of elementary schools, middle or junior high schools, and high schools.

8–4 arrangement: An educational sequence consisting of elementary or lower schools and high schools.

Another study (Blyth, Hill, & Smyth, 1981) evaluated the impact of older students on younger students in two arrangements: when ninth graders were grouped with seventh and eighth graders and when ninth graders were combined with tenth graders in a separate school. The investigators found that the presence of older students tends to have negative effects on younger students at all ages. Surprisingly, ninth graders suffered more from being combined with tenth graders than seventh and eighth graders did when combined with ninth graders. When combined with the tenth graders, ninth graders were less likely to participate in school activities and to assume leadership positions, and

More attention is being given to the developmental needs of early adolescents. (Ken Karp)

more likely to use alcohol and drugs and to suffer from feelings of anonymity. Similarly, seventh and eighth graders suffered when combined with ninth graders, usually by precociously engaging in behaviors not ordinarily seen until high school. In contrast, the removal of ninth graders generally had positive effects on the younger adolescents. For instance, when ninth graders were not present, seventh-grade girls became more active in school activities, dated more often but had sexual intercourse less frequently, and exhibited higher self-esteem.

Administrators and teachers realize that continuing research is needed in order to evaluate the effects of middle schools, especially the extent to which they meet the needs of early adolescents. Simply changing the age groupings or designating them by different labels such as "middle school" rather than junior high school is not enough. But so far, the more effective middle schools have shown that it is possible for students to be happy and productive at an age that is ordinarily associated with negativism or being "on hold."

THE SCHOOL ENVIRONMENT

Students entering a new school, whether a middle school or a high school, initially may be impressed by the appearance of the buildings and campus, and how spacious or crowded the hallways are. But once they have begun

school environment: All those conditions and surroundings that affect the student's experiences in school.

attending classes, they become more aware of their *human* environment—what their classes are like, how interesting and approachable their teachers are, and how friendly the other students are. These are just a few of the many factors in the school environment affecting the students' academic achievement and satisfaction in school.

School and Class Size

The number of students enrolled in public high schools has more than doubled in the past 40 years, mostly because of the baby boom following the Second World War. At the same time the number of schools has not increased proportionately, meaning that large public high schools have become a familiar part of many communities. Now that the baby boomers are growing older, high school enrollment has begun to decline somewhat in many parts of the country, helping to alleviate the problem. But with much of the population living in urban communities, a significant proportion of students continues to attend large public high schools. Critics contend that large high schools are the spawning ground for many of the ills of public education, such as student passivity and boredom, violence, and dropping out of school. Smaller schools, it is claimed, would be beneficial to many students, with the optimal-sized high school being anywhere from 500 to 1000 students. Although research studies have generally shown that school size does not have a major effect on academic achievement, there are other effects—such as greater prevalence of antisocial behavior in large schools—which may influence the learning climate of schools (Anderson, 1982).

Mark Grabe (1981) investigated the effects of school size among 1500 adolescents of both sexes in 20 high schools of varying sizes. He found that both large and small high schools have their advantages and disadvantages, similar to earlier findings. Large schools tend to provide greater breadth and depth of course offerings, better prepared teachers, and a greater diversity of extracurricular activities as well as auxiliary services like school counselors. Even though small schools usually have less impressive academic offerings and faculty, they tend to provide better student-teacher communication, more interaction between teachers, and greater student participation in extracurricular activities. With a greater ratio of potential activities to students in small schools, such students are generally more involved in school. Many of the positive results of small schools, such as greater improvement of the students' abilities and feelings of satisfaction, come from the students being active performers rather than spectators. Although the more active students in large schools show many of these same benefits, a larger proportion of students do so in small schools. The difference is especially striking for marginal students, who are less likely to drop out in smaller high schools, even when they have the same intelligence, grades, and family backgrounds as those who drop out of larger schools. At the same time Grabe found that a few students in small schools feel more alienated than those in large schools, mostly because of the pressure to participate. When students lack the ability, motivation, or interest to meet the greater demand for participation in small schools, they may suffer from negative self-images and a loss of self-esteem.

Participation in high school activities also has important implications for one's later involvement in the adult community. A followup study of a national sample of adolescents done when the subjects were 30 years old showed that, independent of social origins, ability, or academic performance, greater involvement in high school extracurricular activities led to greater involvement in the adult community, including voting behavior (Hanks & Eckland, 1978).

The issue of class size has also received a lot of attention, though a review of the literature in this field has shown no consistent support for small classes (Anderson, 1982). After conducting an intensive investigation of students in 12 inner-city schools, Michael Rutter (1983) maintains that variations of class size within the usual range—30 students plus or minus 10— have no significant effect on academic achievement. Students in a class of 40 tend to learn as much as those in classes with 20 students. At the same time, there are certain situations in which small classes are especially effective, for example remedial classes with low-ability students (Anderson, 1982). Some schools already have a policy of expanding many of their regular classes of 25 to 30 students by several students each, thereby allowing other teachers to teach small classes for students who need specialized instruction.

Teachers

Teachers play a major role in determining whether the adolescent's school experiences will be rewarding or frustrating. When parents of public school students are asked to rate their adolescent's teachers, two-thirds of them give the teacher an A or B rating. (See Table 9 - 1.) Whereas administrators tend to be concerned about the teacher's professional qualifications, mastery of subject matter, and teaching methods, students care more about the teacher's attitudes, personality, and how well a teacher relates to his or her students.

When students are asked, "What do you like about a teacher?" their answers are instructive. In one national survey (Norman & Harris, 1981), this question was posed to 160,000 students aged 13 to 18 with the following results. The most commonly mentioned evaluations included: the teacher "is fair, grades fairly, and doesn't pick on students." Yet, nearly three-fourths of the students felt that teachers favor bright students. After fairness, other desirable characteristics of the teacher were ranked in this order: knows his or her subject; enthusiastic about his or her subject; helpful with kids' homework; likes kids; gives few homework assignments; and can maintain adequate discipline in the classroom. According to other studies, students generally prefer warm, friendly teachers who possess sufficient self-confidence and

Table 9–1 **HOW PARENTS RATE THE TEACHERS IN THEIR CHILDREN'S SCHOOLS, 1985**

Using the A, B, C, D, FAIL scale, what grade would you give the teachers in the school your oldest child attends?

Public School Parents	A %	B %	C %	D %	FAIL %	Don't Know %
TOTAL	22	48	21	5	2	4
Education						
College	24	48	17	4	1	6
High school	18	46	25	5	3	3
Occupation						
White collar	21	53	17	2	1	6
Blue collar	24	40	25	6	3	2
Oldest child attends						
High school	16	46	26	7	3	2
Elementary school	25	48	20	3	2	2
Oldest child's class standing						
Above average	29	51	16	3	*	1
Average or below	14	42	29	7	5	3

*Less than one-half of 1%.

Source: The 17th Annual Gallup Poll of the Public Attitudes toward the Public Schools, Alec M. Gallup, in Phi Delta Kappan (September 1985):38.

poise to deal with students' suggestions and criticisms without making students feel inferior. Students also like teachers who are trustworthy and can set reasonable limits without being harsh. Adolescents respond best to teachers who exercise natural authority based on their greater experience and wisdom, rather than exerting arbitrary authority or abdicating their authority and trying to become "pals" with their students (Feeney, 1980).

No matter how many desirable characteristics a teacher may have, there are few characteristics that are *always* effective. A great deal depends on the interaction between the teacher and students in the classroom situation, including such factors as age and grade of students, their ability and performance, and compatibility between teacher and students. As a result, recent research has focused more on the teacher's *behavior* and interactions with students.

There is greater realization that teachers tend to treat various students differently and that such treatment generally leads to different results. For instance, Rutter (1983) found that teachers who adopt a teaching style which actively involves students in the learning process tend to get more effective results than those who treat students as passive objects of learning. Teachers who are hostile and domineering tend to affect student performance and behavior adversely. Teachers often adopt more positive expectations and give more generous praise to high achievers, which in turn encourages them to give their best and continue their high achievement. However, differential treatment is sometimes based on stereotypes of race and socioeconomic and sex differences, with negative results. For instance, some studies of young adolescents have shown that girls receive less criticism than boys but also less praise. And when girls are treated differently than boys in this way, they adopt lower expectations of their own achievements. There is also a tendency for teachers to prefer students who don't "make waves." Several studies have shown that education majors and student teachers—compared to psychology majors and teacher-corps interns—give more favorable ratings to students who are perceived as conforming, orderly, and rigid, and give unfavorable ratings to those who are seen as independent, active, and assertive (Minuchin & Shapiro, 1983).

The teacher's behavior is also influenced by the student's behavior. In a study by Natriello and Dornbusch (1983), one of the more consistent findings was that the teacher's classroom behavior was shaped more by the student's *behavior* than by the student's characteristics. In contrast to many earlier studies, the student's race and sex had no significant effect on the teacher's standards, warmth, or helpfulness. Instead, the student's achievement record

The learning climate in the classroom depends a lot on the interaction between teacher and students. (Laima Druskis)

and classroom behavior had the greatest impact on the teacher's behavior, though not always as predicted. For instance, teachers sometimes gave more favorable attention to low-achieving students who were needy and receptive to help.

School and Classroom Climate

learning climate: Refers to the combined influence of a variety of factors that affect the student's learning experiences in school.

What teachers do in the classroom is usually a major factor in the learning process. But there is an increasing realization that it is the interaction or combined effect of various factors that affect the student's overall learning experience. Research in this area aims at discovering the learning climate of the school or classroom, and tends to focus on process variables such as teacher-student interaction rather than on structural variables such as school size. After reviewing the literature in this area, Carolyn Anderson (1982) points out that while many studies have yielded inconclusive or conflicting results, some have discovered significant factors in the learning process that are not ordinarily found in studies using the results of standardized achievement tests.

Classroom Environment Scale: Measures the ways schools and classrooms differ and how such differences affect students' academic performance and behavior.

In one approach, Edison Trickett (1978) and his colleague Rudolph Moos have used a questionnaire called the Classroom Environment Scale (CES). Students complete a questionnaire measuring nine variables: involvement, affiliation, order and organization, rule clarity, task orientation, competition, teacher control, teacher support, and innovation. The results of this line of research have shown that schools and classrooms differ markedly in their overall climate of learning.

Students generally are more satisfied in classes which combine moderate structure with high student involvement and high teacher support. Students in these classes are encouraged to participate and are given ample opportunities for innovation and responsibility. Students tend to be least satisfied in classes that are too task-oriented and tightly controlled by the teacher. Students in these classes are more likely to feel anxious, disinterested, and angry. At the same time, the desirable classroom climate varies somewhat by subject matter, with rule clarity and teacher control being considered more important in classes in business and vocational courses than in English and social studies.

In their study of inner-city schools, Michael Rutter (1983) and his colleagues found several factors similar to those discussed above that were significantly related to the learning climate. That is, students did best in those classes in which teachers held positive expectations of their students and reasonable, well-defined standards of academic performance and behavior. The teacher's commitment to improving student performance and time spent on lessons were important factors. Student involvement was encouraged, with students being active participants in the learning process. Although classroom-management skills were essential in creating a conducive atmosphere for learning, effective teachers emphasized a system of incentives and rewards rather than punish-

---

TIME SPENT ON HOMEWORK

How much time did you spend on homework each week when you were in high school? Was it at least 6 or 7 hours?

According to a survey by the U.S. Census Bureau (1984), the median amount of time spent on homework by students in public high schools is 6.5 hours a week, or about 1 hour and 20 minutes per day. Girls generally report doing more homework than boys. But the sharpest difference in time spent on homework is between types of schools, with students in private high schools doing 14.2 hours of homework weekly, more than twice the national average.

The National Commission on Excellence in Education (1983) has recommended that public high school students be assigned far more homework. The parents of such students agree by a ratio of five to three, with 49 percent favoring increased homework and 37 percent opposed. Parents whose teenagers have average or below-average grades, nonwhites, residents of inner cities, and people living in the western United States are somewhat more likely to favor increased homework (Gallup, 1985).

At the same time, parents seem less helpful than they might be in assisting students with their homework. On the average, parents spend only 1 1/2 hours per *week* helping their adolescents with homework, barely as much as these students spend on their homework each *day*. One-third (34 percent) of all parents do not spend any time assisting young people with their homework (Gallup, 1986).

What do you think?

---

ment. Students in classrooms exhibiting these characteristics demonstrated better academic work, attended class more regularly, and were less likely to engage in delinquency or antisocial behavior than students in other classes.

Such studies suggest that it is the *combined* effect of different variables that most influence the student's learning experience. One of the encouraging results of this approach is the discovery that many of the significant factors in the learning process are under the control of the school staff and thus can be improved.

SELECTED FACTORS AFFECTING SCHOOL ACHIEVEMENT

School achievement is also affected by many other factors not directly under the control of the school. The student's intelligence or learning ability is a very important though controversial ingredient in academic achievement. So is the educational level in the home and the encouragement and support students receive, or don't receive, from their parents. Family and socioeconomic influences shape students' attitudes toward learning before they enter school, and these become more pronounced with each passing year (Coleman, Hoffer, & Kilgore, 1981). Consequently, whether students do well in school and go on to college or do poorly in high school and eventually drop out frequently depends on the combined effect of these background factors and the school environment, rather than the school alone.

Learning Ability

learning ability: Intelligence or capacity for acquiring and applying knowledge.

standardized achievement tests: Achievement tests based on large samples and established norms.

intelligence tests: Standardized tests designed to assess one's intelligence or capacity for learning.

IQ (intelligence quotient): A widely used measure of intelligence, calculated as a ratio of one's mental age to his or her chronological age (in children) or by comparing one's test scores to those of a standardized sample (in adults).

Traditionally, learning ability has been measured by intelligence tests. Much of the reason for this has to do with the high correlations between measures of intelligence and standardized achievement test scores, which are usually consistently higher than the association between teachers' reports and achievement test scores. In recent years, however, critics of intelligence testing have pointed out that students from disadvantaged backgrounds tend to score lower on intelligence tests partly because such tests reflect the highly emphasized, traditional academic skills of white middle-class culture. Accordingly, the use of intelligence tests to place students from minority groups in slow classes or vocational programs is thought to further restrict their opportunities for learning (Olmedo, 1981). Largely because of the controversy surrounding intelligence testing, educators tend now to use intelligence tests much more selectively than in the past. In many schools intelligence tests are used mostly to determine if a student needs remedial help—preferably along with other measures such as teacher reports—and even then such tests should be subject to periodic review.

Much of the current interest in learning ability is based on a broader view of intelligence than implied in the traditional IQ tests. For instance, Robert Sternberg in his book *Beyond IQ* (1985) suggests that all of us are governed by three aspects of intelligence: componential, experiential, and contextual. Each aspect of intelligence is explained in a subtheory. In his *componential* subtheory, Sternberg distinguishes between "performance" components that are used in solving a problem and "metacomponents," or the executive learning strategies, used in selecting which performance components are needed to solve a given problem, how to monitor ourselves while solving it, and how to evaluate it after we're done. For example, suppose you were traveling in England and wanted to know how many English pounds you could get for 500 dollars. Metacomponents are the processes used to determine how the problem is to be solved, such as determining what each dollar is worth in pounds at the time. The performance components are the processes which are used in the actual solution, such as multiplying the amount a dollar is worth in pounds by 500. Although these metacomponents cannot be changed as easily or directly by instruction alone as the performance components, in the long run they are very important to one's overall intelligence and learning.

The *experiential* subtheory emphasizes the importance of insight and creativity in intelligence. It includes several abilities such as the selective encoding of knowledge, and the selective combining and selective comparing of the same. As an example of selective encoding of knowledge, Sternberg cites Sir Alexander Fleming's discovery of penicillin. One of Fleming's experiments was spoiled when a sample of bacteria was contaminated by mold. Most people would have become disgusted and thrown it out. But not Fleming.

THREE GRADUATE STUDENTS

Robert Sternberg illustrates his triarchic theory of intelligence with stories of three hypothetical graduate students—Alice, Barbara, and Celia.

Alice seemed very smart according to conventional views of intelligence. She had nearly a 4.0 grade average in college, an extremely high score on the Graduate Record Exam (GRE), and excellent letters of recommendation. In the first year of graduate school, Alice did very well on multiple-choice tests and was good at analyzing arguments and criticizing other people's work. Yet, as she advanced in the program, it became apparent that Alice lacked the ability to come up with good ideas of her own, thus limiting her contribution.

Barbara, the second student, had a different kind of record. Her college grades were not exceptional and her GRE scores were low by Yale standards. However, her letters of recommendation said that Barbara was extremely creative and did superb research. Realizing that creativity is a precious quality, Sternberg wanted to accept her into the program. When Sternberg was outvoted, he hired Barbara as a research assistant. Sure enough, Barbara's work and ideas proved to be just as good as her former professors said, and in time she was admitted into the program. Furthermore, some of Sternberg's most important work was done in collaboration with her.

Celia, the third student, had grades, GRE scores, and letters of recommendation that were good but not great. Accepted into the program, she did all right. But her work was not outstanding. Later, however, she turned out to be the easiest student to place in a good job. Celia had learned how to play the game. She did the type of research that was valued in the journals and submitted her papers to the right journals. Although Celia lacked Alice's superb analytical ability and Barbara's creative ability, she possessed a high level of "street-smarts"—a valuable quality that doesn't show up in traditional IQ tests.

From Robert J. Trotter, "Profile—Robert J. Sternberg: Three Heads Are Better than One," *Psychology Today*, August 1986, p. 60. Reprinted from *Psychology Today* Magazine, Copyright © 1986 American Psychological Association.

Instead, he realized that the mold which killed the bacteria was more important than the bacteria—an example of selective encoding—an insight that eventually led him to the discovery of a substance in the mold he called "penicillin."

The *contextual* subtheory emphasizes adaptation, or the practical applications of intelligence in dealing with the environment. In order to better understand this aspect of intelligence, Sternberg and Richard Wagner studied people in two careers, business and psychology. People who had achieved success and prominence in these two careers were asked what qualities were needed to be practically intelligent in their fields. One quality business executives and psychologists agreed on was something Sternberg called "tacit knowledge." Sternberg and Wagner constructed a test for such knowledge and administered it to junior and senior personnel in business and psychology. Their results suggest that tacit knowledge comes through learning from experience. Business executives who scored high on this test exhibited better performance ratings, higher salaries, and more merit raises than those who scored low. Similarly, psychologists who did well on the test, compared to those who did not, had published more research, presented more papers at conventions, and were at the better universities.

Most people combine these various aspects of intelligence in different proportions. In some cases, individuals may excel in a given aspect of intelligence but not necessarily in others. See the examples cited in the box "Three Graduate Students." Currently Sternberg is developing a multidimensional abilities test which will measure intelligence in a much broader way than the traditional IQ tests. Instead of giving individuals a number etched in stone—a common misunderstanding of IQ tests—this test can be used as a basis for assessing one's intellectual strengths and weaknesses in the everyday world as well as in school.

Exceptional Students

exceptional students: Students whose special talents or limitations affect their academic functioning significantly more than the average student.

A perennial challenge facing high schools is how to deal with significant differences in learning ability. Since the comprehensive high school is oriented toward the typical or average student, those who deviate from this norm are less likely to receive an education that matches their needs and interests. Of special concern are the exceptional students, those whose abilities tend to be extremely high or low in relation to their peers. At one extreme are the adolescents who are gifted in some way, whether the intellectually gifted—

Students with a variety of disabilities now attend regular classes. (Ken Karp)

learning disability: An impairment in learning ability because of a neurologically-based limitation such as perceptual handicaps or minimal brain dysfunction.

dyslexia: Impairment of the ability to read, often because of a genetic defect or brain injury.

mainstreaming: Educational policy of integrating students with special needs into classrooms and social interaction with regular students.

often determined by intelligence tests—or those who have a special talent or creativeness that sets them apart from others. At the other extreme are students who suffer from some type of learning handicap, whether intellectual, emotional, or behavioral. For instance, some adolescents with learning disabilities have normal intelligence but show a marked discrepancy between their estimated academic potential and their performance. Although a great deal of attention has been focused on the failure to learn to read, known as dyslexia, students may also suffer from verbal, perceptual, attention, and memory disabilities, frequently accompanied by behavior problems. It is not always clear whether the problem behaviors manifested by such students are the result of academic failure or a more generalized frustration. But many of these students have emotional problems independent of their learning abilities, which must also be taken into account (Kronick, 1978).

One approach has been to develop programs for students with special needs, such as gifted adolescents and those with learning handicaps. See Figure 9–2. Many experts feel that students who differ significantly from their peers in learning ability or talent benefit from educational programs oriented toward their special needs. Yet funding for these programs is frequently limited, and the public may have mixed feelings about spending money for the gifted when so many students need training in remedial skills (Gallup, 1985). Furthermore, with the passage of legislation in the mid-1970s, the emphasis has turned toward mainstreaming exceptional students into regular classes whenever possible. The success of mainstreaming varies widely from one school to another and depends on such matters as the ratio of special students to the school's resources, the attitude and skill of administrators and teachers, and the understanding and acceptance of other students. Some students with mild disabilities may readily enter regular academic programs. Students with more marked learning handicaps may divide their time between special and regular classes, while still others may require special educational programs or schools. At the same time, there has been a movement toward focusing on the individual student's strengths and weaknesses rather than categorizing students into global categories. In many places, each student who is designated exceptional must have an individual educational program, in which the student, parents, and teachers agree on what the student will study. In the case of gifted adolescents, this might involve the opportunity to accelerate through the regular program at a faster pace, possibly taking college-level courses sooner than others. Or it might take the form of enrichment, with students having special activities and classes in addition to their regular classes.

There are advantages and disadvantages of each approach. On the one hand, separate classes with specially trained instructors may more readily meet the needs of exceptional students than the regular academic programs. But separating students solely on the basis of their learning ability may also

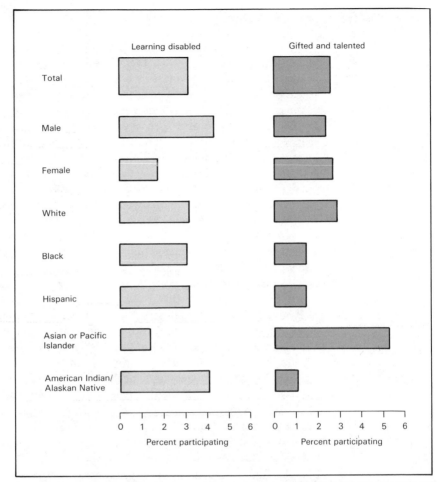

FIGURE 9–2 PARTICIPATION IN PUBLIC ELEMENTARY/SECONDARY LEARNING DISABLED PROGRAMS AND GIFTED AND TALENTED PROGRAMS, 1980.

Source: Valena White Plisko and Joyce D. Stern, eds., *The Condition of Education,* 1985 edition (Washington, DC: National Center for Educational Statistics, Government Printing Service, 1985), p. 193.

have negative psychological and social effects, such as the loss of self-esteem, lowered expectations about their abilities, and social stigma among students in the lower-ability programs (Rutter, 1983). Furthermore, there is some evidence that in many situations mixed-ability classes may offer academic and social advantages for low-ability students without significantly limiting the progress of high-ability students. The impact of mixed-ability classes seems to be even greater on the low-ability students, who appear to be more dependent on the class norm than higher-ability students (Veldman & Sanford, 1984).

Family and Socioeconomic Background

How well adolescents do in school has a great deal to do with the kinds of homes they come from—their parents' education, income, family size, and values. First of all, there is a strong positive association between the parents' education and the student's achievement in school. Students whose parents never completed high school generally score below the national median on most areas of achievement tests, whereas those whose parents completed high school score at or above the median. Students with college-educated parents score even higher (Bachman & O'Malley, 1980). To a large extent the mother's education is more significant for girls, while the father's education plays a greater role for boys (Rumberger, 1983). The values and childrearing practices in the home also play an important role in the student's achievement in school. High achievers in school tend to come from homes in which the parents emphasize personal characteristics and values that are conducive to learning. Such parents value independence, competence, curiosity, and knowledge. They also tend to give their adolescents a lot of attention and approval, with students in these families growing up to become self-confident, self-directed learners (Bachmann & O'Malley, 1980)

Parental expectations also play a significant role in adolescents' educational achievements. In one study, parents' educational aspirations for their adolescents appeared to be as significant as the student's own ability in predicting the latter's educational aspirations and achievements. Furthermore, the influence of parental aspirations for their young was even greater than their teenagers' perceptions of them, implying that parents influence their adolescents' achievements in subtle as well as direct ways, without adolescents necessarily being aware of this (Davies & Kandel, 1981).

At the same time, parental influence on young people's academic achievement depends partly on the extent to which adolescents identify with their parents. Adolescents growing up in a middle-class home with warm, caring parents are especially apt to identify with their parents' educational aspirations and values. Yet, every teacher and counselor knows students who have difficulty in school because they are rebelling against parental pressure for them to succeed in school, whereas other students are highly motivated to achieve in school despite negative parental models and lack of support at home.

The importance of the family is closely meshed with socioeconomic factors. The effects of family background are especially strong for students in the lower socioeconomic groups, and this fact accounts for much of the racial differences in educational achievement (Rumberger, 1983). Thus, adolescents from the lower socioeconomic groups generally come from homes with lower incomes, less-educated parents, and more brothers and sisters, and receive less attention and support than those in higher socioeconomic groups. In contrast, adolescents from the middle and higher socioeconomic levels tend to come from families with higher levels of education and income, from smaller families (which is especially significant among whites), and to receive greater recognition and support for their educational achievements. As a result, middle-class adolescents generally score higher on achievement tests, make higher grades in school, and complete more years of school than their counterparts from the lower socioeconomic groups (Garbarino & Asp, 1981). At the same time, it is well to remember that these are the general patterns of academic achievement and socioeconomic background. There are many exceptions to the rule throughout all socioeconomic groups.

Dropping Out

Because dropping out of high school is less prevalent today than it has been in the past, many people think the problem no longer exists. But the facts say otherwise. The dropout rates for whites have remained at the same levels or increased somewhat, especially among middle-class whites (Rumberger, 1983). On the other hand, while the proportion of minority youth who drop out of school remains high, the *rate* of dropouts among blacks has actually decreased in recent years (Burton & Jones, 1982). If anything, dropping out of school has even greater consequences than it did in the past: Dropouts can expect to hold the lower-level jobs and have the highest unemployment rates.

Figures on high school dropouts tend to be somewhat inconsistent partly because of the different methods used to calculate the rate of dropouts. The National Center for Educational Statistics calculates the dropout rate as the difference between the total number of public high school graduates and the number of public school students enrolled in the ninth grade four years previously. According to this approach, the average rate of high school dropouts for the nation in 1982 is 27 percent. On the other hand, the U.S. Census Bureau uses household interviews in which respondents are asked the highest grade of school completed by themselves and other members of the household. According to this approach, the dropout rate for the same period would be 16 percent (Cooke, Ginsburg, & Smith, 1985). Furthermore, the actual dropout rate varies widely according to many factors, such as age, sex, race, and other socioeconomic variables. (See Table 9–2.) Dropout rates also vary widely among different types of schools, with many inner-city schools in poor neighborhoods having dropout rates in the 50-percent bracket.

Students leave school for a variety of reasons, many of them school-related (Rumberger, 1983). Most young people who drop out of school dislike school, are doing poor academic work, or have been suspended or expelled. But students also drop out of high school for other reasons. Males tend to drop out because they would rather work, or feel they must work for economic reasons. Females are more likely to drop out because of pregnancy and marriage. (See Table 9–3.) Family background strongly influences the adolescent's tendency to drop out of school and accounts for much of the racial differences in dropout rates. Youths with the highest dropout rates

Table 9—2   HIGH SCHOOL DROPOUTS 14 TO 24 YEARS OLD, BY AGE AND RACE, 1970 TO 1984

Age and Race	Number of Dropouts (1,000)				Percent of Population			
	1970	1975	1980	1984	1970	1975	1980	1984
Total dropouts*†	4670	4974	5212	4784	12.2	11.5	12.0	11.2
16-17 years	617	715	709	485	8.0	8.6	8.8	6.8
18-21 years	2138	2557	2578	2297	16.4	16.3	15.8	14.9
22-24 years	1770	1553	1798	1845	18.7	14.5	15.2	14.6
White†	3577	3861	4169	3831	10.8	10.5	11.3	10.8
16-17 years	485	594	619	421	7.3	8.4	9.2	7.1
18-21 years	1618	1980	2032	1852	14.3	14.7	14.7	14.5
22-24 years	1356	1169	1416	1429	16.3	12.6	14.0	13.5
Black†	1047	1024	934	789	22.2	18.5	16.0	13.2
16-17 years	125	116	80	55	12.8	10.2	6.9	5.2
18-21 years	500	540	486	373	30.5	27.0	23.0	16.9
22-24 years	397	337	346	339	37.8	27.8	24.0	20.5

*Includes other race groups not shown separately.

†Includes persons 14-15 years, not shown separately.

Source: U.S. Bureau of the Census. *Statistical Abstract of the United States, 1985,* 106th ed., (Washington, DC: U.S. Government Printing Office, 1985), p. 148.

generally come from large families (more so among whites than blacks), have parents of low educational attainment and aspirations, are from lower socio-economic groups in the inner city or are youths who themselves were born outside the United States. The inability to speak and read English is a major disadvantage, as seen in the high dropout rate among Hispanic youth who are not native born.

The act of dropping out of school appears to be the culmination of a long process involving repeated poor performance in school, frustration and fail-ure, and dislike for school. Youths with low intelligence tend to drop out of school in greater numbers than those with average or higher intelligence, especially in the lower grades. Yet intelligence itself is not always a decisive factor, especially since the majority of high school dropouts have at least average intelligence. In these cases, youths have the ability to complete school but fail to do so for a variety of reasons, as explained earlier. At the same time, we have pointed out how the student's attitude and achievement in school are shaped in large measure by influences from the home and socioeconomic background. Consequently, some investigators (Lloyd, 1978) have been able to identify future high school dropouts by the third grade with 75 percent accuracy, using such variables as family and socioeconomic factors, elemen-tary school grades, intelligence scores, and achievement test scores. About half of the individuals who drop out of school do so as ninth or tenth graders

Table 9–3 PRIMARY REASON HIGH SCHOOL STUDENTS DROP OUT, AGES 14 TO 21, BY RACE AND SEX (in percents)

Reason for leaving school	Female				Male				Overall
	Black	Hispanic	White	Total	Black	Hispanic	White	Total	Total
School Related	**29**	**21**	**36**	**32**	**56**	**36**	**55**	**53**	**44**
Poor performance	5	4	5	5	9	4	9	9	7
Disliked school	18	15	27	24	29	26	36	33	29
Expelled or suspended	5	1	2	2	18	6	9	10	7
School too dangerous	1	1	2	1	0	0	1	1	1
Economic	**15**	**24**	**14**	**15**	**23**	**38**	**22**	**24**	**20**
Desired to work	4	7	5	5	12	16	15	14	10
Financial difficulties	3	9	3	4	7	9	3	5	4
Home responsibilities	8	8	6	6	4	13	4	5	6
Personal	**45**	**30**	**31**	**33**	**0**	**3**	**3**	**2**	**17**
Pregnancy	41	15	14	19	0	0	0	0	9
Marriage	4	15	17	14	0	3	3	2	8
Other	**11**	**25**	**19**	**20**	**21**	**23**	**20**	**21**	**19**
Total percent	**100**	**100**	**100**	**100**	**100**	**100**	**100**	**100**	**100**

Source: National Longitudinal Survey of Youth Labor Market Experience, In Russel W. Rumberger, "Dropping Out of High School: The Influence of Race, Sex, and Family Background," *American Educational Research Journal* (Summer 1983): 201.

(Grant & Snyder, 1984). By this time, students' poor records, school absences, and behavior problems have become painfully evident. Peer influence also weighs heavily at this age; when friends are not doing well in school or are dropouts already, potential dropouts are even more at risk. Then too, in many states, age 16 is the cutoff for enforcing compulsory education laws.

However, since the majority of high school dropouts have the ability to complete school, most school districts have developed a variety of programs to keep students from dropping out. A major aim of these programs is to make school a more rewarding experience for students, and the earlier in the student's school experience such programs can be implemented, the better. In one program (Ruby & Law, 1982), ninth graders identified as potential dropouts were given special assistance in addition to their regular classes to improve their attitude, attendance, and academic achievement. As a result class attendance and behavior improved, with more than two-thirds of these students eventually graduating from high school compared to none of the controls who had not received such help.

IMPROVING THE SCHOOLS

Now that high schools have become such an important part of adolescents' lives, they are expected to take on a broad range of programs and issues that have social, moral, economic, and political dimensions. As such, schools have become easy targets for criticism by practically everyone, including parents, educators, and legislators. We could easily devote an entire chapter to what people think of public high schools and their suggestions for improving them. However, in this final section we'll confine our attention to (1) the results of a national public opinion poll, (2) a look at some of the recommendations for educational reform by a national commission, and (3) the kind of school students would like.

What People Think of Public Schools

One of the most reliable sources of public opinion is the Annual Gallup Poll of the Public's Attitudes toward the Public Schools (Gallup, 1986). This poll is based on personal interviews with a representative national sample of 1552 adults in all areas of the nation and in all types of communities. And it includes parents of students who attend public, parochial, and private or independent schools.

When asked to rate the public schools, 41 percent of the public give the public schools an A or B. (See Table 9–4.) As has always been true, parents who have students in the school system express the most favorable views. When asked to grade the particular school their oldest child attends, almost two-thirds of parents give the schools an A or B rating. A similar proportion of students and teachers give the schools an above-average grade. Furthermore, parents whose oldest child has better than average grades express the most favorable attitudes of all. (See Figure 9–3.)

Parents generally support efforts to improve the quality of public education. About 9 out of 10 public-school parents endorse the idea of competency testing for teachers when they are hired, though only 6 out of 10 support the

Table 9—4   RATING OF THE PUBLIC SCHOOLS, 1986
             (in percents)

Students are often given the grades A, B, C, D, and FAIL to denote the quality of their work. Suppose the public schools themselves, in this community, were graded in the same way. What grade would you give the public schools here—A, B, C, D, or FAIL?

GRADE	National totals	No children in school	Public school parents	Nonpublic school parents
A	11	8	18	11
B	30	28	37	29
C	28	27	29	29
D	11	11	11	16
FAIL	5	5	4	11
Don't Know	15	21	1	4

Source: The 18th Annual Gallup Poll of the Public Attitudes toward the Public Schools, Alec M. Gallup, in Phi Delta Kappan (September 1986): 47.

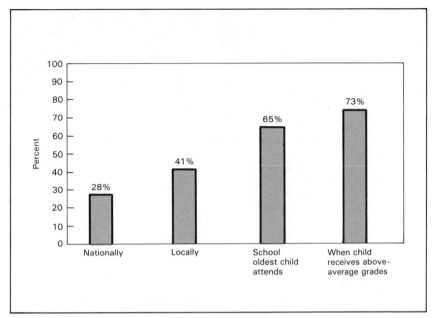

FIGURE 9–3 AMERICANS WHO GIVE THE PUBLIC SCHOOLS A GRADE OF *A* OR *B*.

Source: The 18th Annual Gallup Poll of the Public Attitudes toward the Public Schools, Alec M. Gallup, in *Phi Delta Kappan* (September 1986): 46–48.

more controversial idea of merit pay for teachers. The great majority of parents support a rigorous academic program of studies, including courses in math, English, history/U.S. government, science, and computer training for college-bound students and to a lesser extent for other students. At the same time, about half of parents want their children to have the option of taking courses in business, career education, health education, physical education, and sex education. And three-fourths of parents believe vocational training should be offered for students not planning to attend college (Gallup, 1986).

When parents of public school students were asked about the biggest problems facing local schools, more than half mentioned either drugs (27 percent) or discipline (23 percent). However, as in the past, parents tend to see such problems as being less prevalent in the school their oldest child attends than in the local schools generally. (See Table 9–5.) Since we'll discuss the problems of drugs in the school in a later chapter, it may be more helpful to take a look at the perennial problem of lack of discipline—rated as the most important problem in 16 of the 17 Gallup polls. When asked how the problem of discipline should be handled in the schools, the public and parents alike prefer the least harsh method. Over two-thirds of parents endorse the idea of giving special training to teachers in how to deal with problem students. And over half of parents support the idea of required

**Table 9—5   BIGGEST PROBLEMS FACING THE
PUBLIC SCHOOLS, 1986 (in percents)**

Here are the responses of parents of public school students when asked: "What do you think are the biggest problems with which the school your oldest child attends must deal?"

	School own child attends	Local schools
Use of drugs	16	27
Lack of discipline	15	23
Lack of proper financial support	9	15
Large schools/overcrowding	9	6
Poor curriculum/poor standards	7	10
Parents' lack of interest	4	5
Teachers' lack of interest	4	6
Moral standards/dress code	4	5
Difficulty in getting good teachers	3	6
Pupils' lack of interest/truancy	3	2
Drinking/alcoholism	3	5
Lack of needed teachers	3	1
Lack of respect for teachers/other students	3	4
Fighting	3	2
Parents' involvement in school activities	3	1
There are no problems	8	4
Miscellaneous	24	24
Don't know	12	4

(Figures add to more than 100% because of multiple answers.)

Source: The 18th Annual Gallup Poll of the Public Attitudes toward the Public Schools, Alec M. Gallup, in *Phi Delta Kappan* (September 1986): 45.

classes for parents of problem students along with suspension of these students (Gallup, 1985).

New Directions

Educators, school boards, and legislators also have plenty of ideas about the kinds of improvements needed in schools. At the national level, many of the criticisms and recommendations for educational reform have been expressed in the report of the National Commission on Excellence in Education (1983), with a major emphasis on improving the quality of public education. It appears that Americans are now among the most educated people in the world in terms of years of school completed. More than three-fourths of those over age 25 have completed high school, with about one-third of them having some college. Yet, judging by a variety of measures, including SAT and achievement test scores, the quality of education has declined since the 1970s. Furthermore, grades have risen while homework has decreased, making the rise in grades suspect. As a result a large proportion of public school students do not have the basic study skills needed for a quality education. Colleges, business, and the military are spending millions on remedial education.

Many of the commission's recommendations are aimed at upgrading the standards of education, making better use of time in school, and improving the quality of teaching.

1. *Upgrading standards.* An increasing proportion of students are in the general curriculum, taking fewer academic courses and more electives in the areas of "life education" and "personal growth." As a result, the commission recommends upgrading graduation requirements to include more courses in language, math, and the sciences and improving academic rigor throughout the entire curriculum.

2. *Better use of time.* Noting that American students spend less time in school than students in many other nations, the commission proposes that students spend more time in school. This includes a longer school day, a longer school year, and a more efficient use of time in school.

3. *Improving teaching.* Some of the perennial problems cited by the commission have been the difficulty of attracting talented individuals to teaching, teacher's low salaries—especially in the fields of math, sciences, languages, and special education—and teachers' lack of involvement in decision making. Among the recommendations are improving the standards of the teaching profession, better salaries, use of personnel from outside education, and greater involvement of teachers in the decision-making process.

Unfortunately, the commission didn't propose *how* these measures might be implemented. And since the public has generally opposed raising local taxes for financing public schools (Gallup, 1986), it remains to be seen how many of these recommendations will be carried out.

The School Adolescents Would Like

Adolescents themselves are rarely consulted on the kind of school they would like. And even if they were, their ideas would probably vary considerably according to the particular groups of students asked. But the student responses in one such survey are suggestive. Millicent Poole (1984) surveyd 1596 adolescents in 32 schools, with the student sample stratified on the basis of sex and socioeconomic differences. She found that the majority of students were satisfied with their schools the way they were, confirming the results of the Gallup poll. Where changes are desired, they are mostly in terms of attitudes and procedures rather than structural changes.

The most frequent and emotional responses concern the issue of authority and discipline in school. Most students realize the need for rules and discipline, but they question the use of a blind, authoritarian approach. The students want a more democratic policy in which the individual's freedom would be encouraged as long as this did not infringe on the freedom of others. In terms of the curriculum, students want a greater diversity of subjects and more attention paid to practical experience and vocational relevance. Teachers are regarded as a significant source of student satisfaction—or dissatisfaction—in school. But students feel that two major changes are needed. One is the need for teachers who are more competent in regard to qualifications, knowledge, experience, and classroom-management skills. The other need is for better teacher communication and teacher-student relationships. Students want teachers who are friendly, understanding, approachable, and helpful to students. They also want similar advisers available for personal and career counseling. Issues such as class size and better buildings are of less concern.

The concerns for more competent and caring teachers—as well as more democratically oriented policy on student discipline—are similar to those expressed by the National Commission and the Gallup poll, cited earlier. But the desire for greater attention to practical experience and vocational relevance is at odds with the current emphasis on academic excellence. Such a discrepancy is a reminder of a perennial issue facing schools, that is, whether to emphasize academic rigor or relevance to life. As mentioned earlier in the chapter, the pendulum of educational priorities has swung back toward an emphasis on academic rigor in recent years. But these concerns need not be so polarized. Courses that help prepare students for life beyond high school need not be lacking in academic rigor. Nor should the concern for basic skills and academic excellence be pursued in a competitive way without encouraging the love for learning itself and the value of learning in preparing students for life in a constantly changing world. Instead, the diversity of students requires that schools continue to be responsive to a wide variety of needs.

SUMMARY

Secondary Schools

1. Largely because of compulsory-education laws, secondary schools now enroll the vast majority of adolescents of school age, making for a diverse group of students.

2. One of the continuing challenges facing the schools is providing programs that serve the needs of all students equally well, with student apathy and discipline being a problem in many schools.

3. The comprehensive high school offers a broad range of educational programs, which usually includes a college preparation, vocational training, and general curriculum.

4. The widespread development of middle schools in recent years is an attempt to design schools which more nearly reflect the physical, cognitive, and social needs of early adolescents than has the traditional organization of secondary schools.

The School Environment

5. Although large and small schools have their advantages and disadvantages, students tend to become more involved in smaller schools, with marginal students less likely to drop out.

6. Teachers who are fair-minded and actively involve students in the learning process tend to be more effective than those who treat students as passive objects of learning.

7. While the learning climate of classrooms is a function of many factors, students generally do better in classes that combine moderate structure with positive teacher expectations and support.

Selected Factors Affecting School Achievement

8. Learning ability, once measured exclusively in intelligence tests, is now approached more in terms of cognitive development along with a variety of other measures.

9. Exceptional students, ranging from gifted adolescents to those with learning handicaps, are dealt with in a variety of ways that usually combine special programs with regular classes.

10. How well students do in school depends to a great extent on their family and socioeconomic backgrounds, with students who come from middle-class homes and who have college-educated parents generally doing better than those from lower socioeconomic groups.

11. Although students drop out of school for many reasons, a major factor is the dislike of school, with males inclined to go to work and females leaving school because of pregnancy or marriage.

12. The act of dropping out of school is generally the culmination of a long process of frustration and failure at school; today, many schools offer a variety of programs aimed at making school a more positive experience for those at risk.

Improving the Schools

13. The public is generally supportive of high schools. Almost three-fourths of public-school parents give their adolescents' schools an above-average grade.

14. A major theme of education reform in the 1980s has been the concern to improve the quality of secondary education, with the emphasis on academic rigor.

15. At the same time, many adolescents want to retain a diversity of subjects in school. They want greater attention paid to practical experience and vocational relevance.

16. Since the comprehensive high school has a wide variety of students, care must be taken that the improvement of academic skills is not made at the expense of preparing students for their larger role in society.

REVIEW QUESTIONS

1. What are some of the advantages and disadvantages of compulsory education?

2. Explain the concept of the comprehensive high school.

3. In theory, how is the middle school like and unlike the junior high school?

4. What are the advantages and disadvantages of large and small high schools?

5. What are some of the characteristics of effective teachers?

6. How important is the adolescent's family and socioeconomic background for his or her academic achievement?

7. List some of the common reasons students drop out of school.

8. Explain how dropping out of high school is generally the culmination of a long process beginning earlier in the student's academic development.

9. According to the annual Gallup poll, how do public-school parents view their adolescents' high schools?

10. What are some of the recommendations by the National Commission on Excellence in Education for improving the quality of education?

10 Sexuality

- ■ **SEXUAL ATTITUDES**
  Changing Attitudes toward Sex
  How Consistent Are Attitudes and Behavior?
  Sex and Love

- ■ **SEXUAL BEHAVIOR**
  Masturbation
  Physical Intimacy and Sexual Intercourse
  Homosexuality

- ■ **PROBLEMS IN SEXUAL BEHAVIOR**
  Use of Contraceptives
  Premarital Pregnancy
  Sexually Transmitted Diseases

- ■ **SEX EDUCATION**
  Sources of Sex Information
  Sex Education in the Home
  Sex Education in the School

- ■ **SUMMARY**

- ■ **REVIEW QUESTIONS**

Laura, an attractive 17 year old, is a top student in her class. She is also popular among her peers. For the past six months, Laura has been sexually active with her boyfriend Brad, an 18 year old. Yet, for the first couple of months, neither of them was concerned about the possibility of pregnancy. Laura put off getting any type of contraceptive partly because of her belief that someone her age was not as likely to become pregnant as an older person—a belief she now admits is "dumb." One reason Laura began using contraceptives was the emotional anguish she shared with one of her girlfriends who had just recently become pregnant unexpectedly. Similarly, Brad felt that the risk of contracting a sexually transmitted disease pertained mostly to adults. Recently, however, he has begun to use condoms partly out of the realization that AIDS is a potential problem for all sexually active people— not just homosexuals.

SEXUAL ATTITUDES

Laura and Brad's sexual attitudes and practices are typical of their peers in several ways, as we'll see throughout this chapter. First, they built up an affectionate relationship with each other before engaging in sex. Even then, their initial act of sex was a spur-of-the-moment thing without any discussion of birth control; only later did they begin using contraceptives. But, once they discussed the matter, both of them felt equally responsible for practicing birth control. Then too, Laura and Brad admit that they became sexually active at an earlier age than they think is desirable. And yet finally, for many reasons, Laura and Brad have become more cautious and responsible in their sexual behavior than many of their counterparts of a decade ago.

Teenagers' attitudes toward sex often reflect the times in which they live. For instance, in 1960 only one in ten teenagers felt "sexually behind" others his or her age. But by 1980, one in five teenagers felt this way, largely because of the extensive changes in sexual attitudes and practices that occurred in the previous decade or so (Offer, Ostrov, & Howard, 1981). Also, American teenagers seem to be more preoccupied with sex than their peers in other parts of the world. As you may recall from the worldwide survey cited in the opening chapter, almost all of the American teenagers surveyed regarded premarital sex as a major problem facing youth today. Yet, this contrasts sharply with youth in other countries, who rank premarital sex far behind other issues such as drug abuse (*U.S. News and World Report,* 1986).

Changing Attitudes toward Sex

Adolescents' attitudes toward sex continue to change, partly because the youth in each generation are exposed to different social changes and trends and grow up somewhat differently. The attitudes of today's youth are probably best understood in light of the pervasive changes in social attitudes that occurred in the late 1960s and 1970s and the countertrends in the 1980s.

Adolescents' attitudes toward sex generally became more liberal in the period from the late 1960s throughout the 1970s as an integral part of the social changes that brought about an emphasis on civil rights, racial equality, and women's rights. The impact of these changes has had several major implications for youth's attitude toward sex.

First, there has been a greater openness and honesty in sexual matters. You can hardly pick up a newspaper or watch television without being aware that issues such as rape, sexual abuse of children, and abortion are treated more candidly than a generation ago. Young people brought up in this atmosphere naturally talk about sex more openly and honestly with their peers, though to a lesser extent with their parents and teachers. They are also more inclined to discuss sensitive subjects, such as homosexuality, and they want better sex education.

Second, youth are more accepting of premarital sex compared to their counterparts of a generation ago. Legal marriage is no longer required as a sanction for sex. For the large majority of young people, sex is now acceptable within an "affectionate" or "meaningful" relationship. For instance, in 1965 70 percent of the women and 33 percent of the men in a large southern university felt premarital sex was wrong. But by the mid-1970s, only 21 percent of the women and 20 percent of the men held this view (Robinson & Jedlicka, 1982).

Third, youth and adults alike enjoy greater personal choice in regard to sexual matters than in the past. By the 1970s, Sorenson (1973) found that two-thirds of adolescents 13 to 19 years of age admitted that when it comes to sex "I do what I want to do" regardless of what society thinks. Such a change reflects the growing suspicion of established institutions like the school and church—if not adult authority in general—as well as the trend toward individualism, personal growth, and self-expression. At the same time, this approach has made greater demands on the personal judgment and maturity of the individual than the past approach of conformity with a set standard; the result of the less conformist approach is that many youth feel that it is getting harder to know right from wrong in sexual matters.

Young people growing up in the 1980s have become somewhat more conservative in their attitudes toward sex compared to their counterparts in the 1970s. But, as Daniel Yankelovich (1981/1986) points out, the trend toward conservatism in sex is part of the overall change in social attitudes because of such forces as an uncertain economy and a tight job market. Nevertheless, when young people in their teens and twenties were asked how they felt about the changes in attitudes toward sexual conduct that occurred in the 1960s and 1970s, about as many felt such changes had been "mostly bad" (44 percent) as thought they had been "mostly good" (50 percent): Such survey results reflect a more conservative stance than in earlier eras. After talking with many leaders at schools and campuses around the country, John Leo (1984) concluded that the sexual revolution has peaked and that sexual attitudes and behavior have become more conservative for a number of reasons. For one thing, the baby boomers are growing up and settling down into family life, thereby adopting a less adventurous attitude toward sex. Then too, there is a growing fear of sexually transmitted diseases, such as herpes and especially AIDS. Consequently, Robinson and Jedlicka (1982) found that while attitudes toward sex had grown progressively more liberal between 1965 and 1975, they became more conservative by 1980. For instance, when college students were asked if they felt that a man or a woman who had sex with a great many partners was immoral, more people of both sexes agreed with this statement in 1980 than in 1975. The swing back toward traditional values may be partly a corrective to the changes of the sixties and seventies. It isn't that the sexual revolution has been rebuffed, but rather that such changes have been absorbed into society, with sex being put in better perspective. As a result, while

sexual revolution: The social trend toward more liberal sexual attitudes and behaviors, especially during the 1960s and 1970s.

sexually transmitted diseases: Infections transmitted primarily by sexual intercourse.

Adolescents are more sexually sophisticated than they were in the past but also more cautious about casual sex. (Camerique)

today's students are more sexually sophisticated than their counterparts in the past, they tend to be more cautious about casual sex.

How Consistent Are Attitudes and Behavior?

Not surprisingly, adolescents, like adults, do not always act in a manner consistent with their atttiudes. In an effort to find the relationship between adolescents' sexual attitudes and behavior, Laurie Zabin and her colleagues (1984) collected data from about 3500 junior and senior high school students attending four inner-city schools. They found that the majority of young people already hold attitudes and values that are consistent with responsible sexual conduct, but not all of them translate these attitudes into their personal behavior.

sexual intercourse: Sexual activity involving the penetration of the female vagina by the male penis, characteristically accompanied by pelvic thrusting and orgasm for one or both partners.

First off all, it appears that many adolescents are initiating sexual intercourse at an earlier age than they themselves regard as desirable. When students were asked "What is the best age for a woman (or man) to have sex for the first time?" 83 percent of the sexually experienced adolescents reported they had experienced their first intercourse at a younger age than they considered desirable. (See Table 10–1.) Almost half of the sexually experi-

Table 10–1 PERCENTAGE OF SEXUALLY ACTIVE ADOLESCENTS WHOSE AGE AT FIRST INTERCOURSE WAS YOUNGER THAN, THE SAME AS, OR OLDER THAN THE AGE THEY GAVE AS THE BEST AGE FOR FIRST INTERCOURSE, ACCORDING TO SEX, RACE, AND CURRENT AGE

Sex, race, and current age	Younger	Same	Older	Total
Female	**86.3**	**10.6**	**3.1**	**100.0**
White	88.3	10.5	1.2	100.0
≤15	96.2	2.5	1.3	100.0
≥16	84.6	14.2	1.2	100.0
Black	85.7	10.6	3.7	100.0
≤15	87.5	9.6	2.9	100.0
≥16	84.3	11.4	4.3	100.0
Male	**80.7**	**13.6**	**5.8**	**100.0**
White	75.9	16.8	7.3	100.0
≤15	84.7	11.9	3.4	100.0
≥16	71.7	19.1	9.2	100.0
Black	83.2	11.9	4.9	100.0
≤15	90.0	6.2	3.8	100.0
≥16	75.5	18.3	6.2	100.0

Female N = 1032'   ' Male N = 1060

Source: Laurie S. Zabin, Marilyn B. Hirsch, Edward A. Smith, and Janet B. Hardy, "Adolescent Sexual Attitudes and Behavior: Are They Consistent?" *Family Planning Perspectives* (July/August 1984): 182.

virgin: A person who has not experienced sexual intercourse.

enced adolescents reported a best age for first intercourse older than their current age. As you might expect, virgins of both sexes expressed a preferred age for first intercourse 2 or 3 years older than did sexually experienced adolescents: about 18 years of age compared to 16 (for women) and 15 (for men) years respectively.

When asked if they believed premarital sex was wrong, the proportion of young women who felt this way was only slightly higher than the proportion of men: 39 percent and 32 percent respectively. Overall, attitudes were more closely related to one's sexual experience than gender. Over half the virgins felt that premarital sex was wrong, compared to only one-fourth of the sexually experienced adolescents. But for the 25 percent who had had intercourse, despite their belief to the contrary, attitudes and behavior were widely divergent.

Young women are much more likely than men to desire a strong relationship or marriage before engaging in sex. At the same time, about half of both sexes require no stronger ties than the dating relationship to initiate sex. Virgins of both sexes demand stronger relationships before engaging in sex than nonvirgins do, with about 30 percent of the men and 40 percent of the women virgins holding marriage as a requirement for sex. Women are also twice as likely as men to report having had a strong relationship with their last partner. In contrast, about four times as many men as women report having had casual sex, that is, sex with someone they had just met or didn't know well. Such results confirm the familiar finding that women are more concerned than men about having sex within an affectionate relationship with commitment. Yet it is noteworthy that 18 percent of the young women and 25 percent of the men saw the relationship in which they last had intercourse as **birth control:** Control of weaker than they desire for engaging in sex, indicating an inconsistency between their attitudes and behavior.

Although there is considerable overlap between adolescents' beliefs and practices in regard to birth control, there are inconsistencies as well. For instance, adolescents were asked for a true or false answer to the question "I would only have sex if one of us was using some kind of birth control." Among those who answered "true" to this statement, 21 percent of the women and 26 percent of the men, or about one-fourth of the adolescents, admitted they hadn't used a contraceptive device during their last intercourse. And the failure to use a contraceptive device was even higher among those who answered "false"—more than 50 percent for men and women alike. As expected, adolescents who felt that one or both partners is responsible for birth control are much more likely to use contraceptives than those who feel neither partner is responsible. Furthermore, contraceptive use is strongly related to the strength of the couple's relationship before engaging in sex, with about 7 out of 10 adolescents of both sexes with a strong relationship using some contraceptive device compared to only 4 out of 10 of those with a weak relationship.

These findings show that while there is frequently a significant relationship between adolescents' sexual attitudes and reported behavior, there is a large minority in each area for whom attitudes and behavior are at odds. Although some adolescents may recognize that there is an inconsistency between their ideals and behavior, for the majority this is probably not the case. Some of the attitude-behavior inconsistency may be due to the influence of peers, family, community, and mass media. That is, questions about premarital sex or the best age for first intercourse may trigger a response that is influenced by these pressures. In other cases, inconsistencies may result from changes in attitude over time. But the inconsistency between general attitudes and personal behavior suggests that sex education should include helping adolescents translate their attitudes into practice, through decision making and the like, rather than assuming that changes in attitudes will automatically influence behavior.

birth control: Control of how many children a woman will have and when she has them, through the use of contraceptives or fertility awareness techniques such as the rhythm method.

contraceptives: Birth-control devices, such as birth-control pills.

Sex and Love

The link between sex and love—both in attitude and behavior—is an important concern among youth, though this varies considerably according to such factors as the age and sex of the individuals. Generally, the older the youth and the more sexually experienced they are, the more liberal their attitudes. For instance, while 32 percent of the high school students in Zabin's (1984) study favor saving sex for steady relationships or marriage, only 21 percent of college students feel this way. A majority of college students in one survey—70 percent of women and 79 percent men—feel that "love enriches sexual relations but is not necessary for enjoyment" (Crooks & Bauer, 1983). Couples living together tend to hold even more liberal views, with only 12 percent of them disapproving of sex without love (Blumstein & Schwartz, 1983).

Most studies show that a larger proportion of males than females favor casual sex or sex with someone they don't know well or love (Zabin, 1984; Crooks & Bauer, 1983). Yet, as we've seen in the last chapter, such differences between the sexes tend to diminish with increasing age and the deepening of the dating relationship. There is little or no difference in regard to the degree of sexual intimacy or affection desired among older youth going steady or those living together (McCabe & Collins, 1979; Blumstein & Schwartz, 1983). At the same time, Robert Crooks and Karla Bauer (1983) have observed that many youths who embrace the notion that sex and love belong together may nevertheless have sex with someone they do not love. Such encounters often result in guilt feelings, which may weigh more heavily on some individuals than others. One young woman, who had grown up with the idea that you don't have sex until you're in love, had her first sexual experience with a man to whom she was only physically attracted. "The sex was OK, but the guilt afterwards was terrible," she said. "I thought myself to

Most sexual intercourse among adolescents occurs between partners who are going steady. (Camerique)

be emancipated from all that childhood indoctrination. But I wasn't" (Crooks & Bauer, 1983). Although guilt feelings over sex are less prevalent than in the past, about one out of four young women and one out of five young men experience guilt feelings over sex. And with the resurgence of the view that sex and love belong together, youth who engage in sex without love may become all the more susceptible to guilt (Rubenstein, 1983).

Ultimately, the relationship between sex and love depends largely on the values of the people involved. This was brought out in a study of 231 couples in college who had been dating steadily for about 8 months (Peplau, Rubin, & Hill, 1977). The couples were surveyed in depth about a number of things pertaining to sex and love, including their sexual attitudes and values, emotional intimacy, frequency of sex, and whether they had initiated sexual intercourse early or late in their relationship. On the basis of their data, the researchers divided the couples into three groups according to the ways each couple viewed sex and love. The sexually traditional couples, a minority of couples, believed that abstaining from sex until marriage indicated their love and respect for each other. The sexually moderate couples felt that sex is an expression of love that only comes with time. Accordingly, they didn't begin having sex until they had been dating about 6 months. Couples with sexually liberal values viewed sex as an activity that is enjoyable in itself, and which may help the partners to get to know each other better. Not surprisingly, they generally began having sexual intercourse soon after they met.

The researchers then grouped the couples according to whether they had begun sex early or late in the relationship: The link between sex and love was much stronger among the later-sex couples. Those who began having sex later in their relationship were more often in love, felt more intimate ties, and were more inclined to marry their partners than those who had begun sex earlier in the relationship. They also engaged in sex less frequently, averaging two to three times a week compared to four or five times a week in the early-sex couples. Yet, a followup study conducted 2 years after the original survey showed that there was no significant relationship between the couple's sexual activity and their overall satisfaction with their relationship or its permanence. Couples who abstained from sex, or who engaged in it early and often, stayed together or broke up with equal frequency. Such findings suggest that there is no single pattern of sexual values or behavior that consistently increases the development of intimacy or love. A major factor is the compatibility between the individuals' attitudes and values, including their views regarding the relationship between sex and love.

SEXUAL BEHAVIOR

In some ways, it's even more difficult to generalize about adolescents' sexual behavior than about their sexual attitudes. For one thing, sexual behavior is an intensely private matter; researchers have to rely on self-reports. Also, as we've already mentioned, adolescents' sexual behaviors are not always consistent with their expressed attitudes. Then too, sexual practices vary widely among different groups, with the incidence of premarital intercourse having increased more among middle-class college students than among working-class youth. Finally, sexual practices vary widely among individuals. Keeping these points in mind, let's look at several major areas of sexual behavior.

Masturbation

Masturbation, which may begin in childhood for some, has become a common practice by adolescence, especially among boys. By 16 years of age, about three-fourths of the boys and half the girls report having masturbated. By 18 years of age, the incidence of masturbation is slightly higher among both sexes (Haas, 1979). Among college-aged youth, practically all (92 percent) of the men and two-thirds of the women engage in masturbation (Story, 1982). Boys tend to masturbate more regularly and frequently than girls. But masturbation practices vary more widely among girls. Some girls masturbate regularly, others sporadically, and still others not at all. Youth of both sexes often use fantasy and, to a lesser extent, erotic pictures and stories to heighten arousal during masturbation. Although such fantasies may sometimes involve improbable feats and partners, in most instances sexual fantasies involve having sex with a loved partner (Miller & Simon, 1980).

Masturbation may serve a variety of purposes among adolescents. Since sexual arousal and orgasm are among the most intensely pleasurable experiences available to people of all ages, it should not be surprising that adolescents frequently masturbate mostly for the pleasure of it. Masturbation also affords temporary release from anxiety and tension. College-aged women report that their primary motives for masturbating are to experience the pleasurable sensations of sex and to achieve physical release from sexual tension. About one in five college women learn to masturbate by copying their partner's petting technique (Clifford, 1978). Masturbation is also a valuable means of learning about the body and experimenting with one's sexual response. Men may experiment with self-stimulation to achieve better control over ejaculation. Women may use masturbation as a way of learning which forms of stimulation best help them to achieve orgasm. Masturbation is also commonly used as a substitute when a sexual partner is not available. Yet, masturbation is also practiced among dating youth. About three-fourths of late adolescents of both sexes who are going steady express a desire for mutual masturbation as one among other forms of achieving sexual pleasure (McCabe & Collins, 1979).

Probably the greatest change in masturbation practice is the reduced anxiety and guilt associated with it. Yet, some youths worry about whether they masturbate too much. According to one survey, one-fourth of college students fear that they masturbate too frequently (Winer et al., 1977). Interestingly, it also appears that many youth are reluctant to discuss masturbation, largely because they have misgivings about achieving pleasure through self-stimulation. Perhaps this is because the sense of adventure and the closeness and stimulation of a love relationship that accompanies sexual intercourse are missing in masturbation.

Physical Intimacy and Sexual Intercourse

Adolescents tend to engage in more intimate forms of physical, erotic contact and at younger ages than in the past. By the midteens, 14 to 15 years of age, the great majority of adolescents of both sexes have held hands, kissed and hugged their partners. By 16 or 17 years of age, most of them have engaged in deep kissing, touching the girl's breast, and fondling the genitals of the other sex. The more intensely erotic behaviors generally occur about the same time teenagers begin having sexual intercourse. Oral-genital sex is more apt to occur in the late teens, after the initiation of sexual intercourse (DeLamater & MacCorquodale, 1981).

oral-genital sex: Mouth-to-genital contact to create sexual pleasure.

The average age at which young women have their first sexual intercourse is 16.2 years, compared to 15.7 years for young men. Young women tend to have their first intercourse with a partner nearly 3 years older than themselves. Men generally have their first intercourse with a partner less that 1 year older. Black teenagers of both sexes generally initiate sex at a younger age than do whites. Overall, it appears that 50 percent of the women 15 to 19 years of age and 70 percent of the men aged 17 to 21 have experienced sexual intercourse. Comparative studies indicate that the proportion of young people who have ever had intercourse has increased about 20 percent since the early 1970s, with much of the increase occurring among white women (Dreyer, 1982: Zelnik & Shah, 1983).

Young women's first sexual intercourse generally occurs with someone toward whom they feel an emotional commitment. More than 6 out of 10 young women in Zelnik and Shah's study (1983) said they had been going steady or were engaged to their first sexual partner. By contrast, fewer than 4 out of 10 young men said they had been going steady or were engaged to their first sexual partner. Young men were much more likely to have had their first intercourse with a friend or someone they had just met. (See Table 10–2.) The influence of peers on adolescents' sexual activity varies somewhat with the individual. For instance, one study showed that friends of either sex had no significant influence on the sexual activity of black males, black females, or white males. In contrast, white females were more likely to be influenced by the sexual behavior of their best female friend or best male friend. A white female virgin whose best friends of both sexes were sexually active usually became sexually active sooner than a virgin whose best friends were not sexually active (Billy & Udry, 1985).

Table 10—2 **PERCENTAGE OF WOMEN AND MEN RANKED BY RELATIONSHIP WITH THEIR FIRST SEXUAL PARTNER AND AGE AT FIRST INTERCOURSE**

Relationship with first partner	Women			Men		
	<15 (N=273)	15–17 (N=555)	≥18 (N=103)	<15 (N=305)	15–17 (N=294)	≥18 (N=64)
Engaged	3.9	8.8	18.7	0.4	0.8	0.0
Going steady	44.4	61.9	46.1	20.0	46.2	47.9
Dating	28.9	21.6	29.0	18.6	22.4	12.6
Friends	13.2	4.3	5.4	54.4	20.0	26.7
Recently met	9.6	3.4	0.8	6.6	10.6	12.8
Total	100.0	100.0	100.0	100.0	100.0	100.0

Source: Melvin Zelnik and Farida K. Shah, "First Intercourse among Young Americans," *Family Planning Perspectives* (March/April 1983): 66.

The initiation of sex seems to be a spur-of-the-moment decision for most young people. Only 17 percent of the young women and 25 percent of the young men in the national survey cited earlier said they had planned their first intercourse. Women who were going steady with their first partner were most likely to have planned intercourse, while young men who had just met their partner shortly before intercourse were the most likely to have planned the sexual act. Fewer than half of either sex used contraceptives, either by themselves or by their partner. Generally, those who had planned intercourse were more likely to have used contraceptives.

The proportion of young people who have ever engaged in sexual intercourse rises steadily with age. The pattern continues into the college years, with decreasing differences in the percentage of men and women who have ever engaged in sexual intercourse. The exact percentage of students who have engaged in sex varies widely from one campus to another. But according to the National Center for Health Statistics, four out of five women try sex before marriage (Schmid, 1985). It appears that sexual activity has become an integral part of the overall adolescent experience and is engaged in by all types of youth, not just the agressive, rebellious male or dependent, security-minded female. Young people of both sexes are more likely to engage in sexual intercourse to satisfy their erotic, sexual needs or because sexual intimacy has become an acceptable part of close relationships with a caring partner (Dreyer, 1982).

Homosexuality

Homosexuality is a topic that arouses considerable anxiety among adolescents. At earlier ages, children are exposed almost exclusively to heterosexual models, at least visibly. This, plus the assumption in our society that everyone

A Gay Pride march in New York City. (Ken Karp)

is expected to be heterosexual unless "proven" otherwise, means that most adolescents come to think of themselves as heterosexuals. Yet, the process of affirming one's sexual identity that occurs at puberty may well evoke a certain degree of self-doubt in the adolescent's mind about his or her own sexual identity. The awkwardness and frustrations in their sexual relationships with the opposite sex may also arouse anxiety and self-doubt. Consequently, it is not uncommon for adolescents to wonder, if only in passing, about their heterosexuality.

homosexuality: Preference for sex partners of the same sex.

heterosexuality: Preference for sex partners of the opposite sex.

The concern about homosexuality greatly overshadows the amount of homosexual activity that occurs among adolescents. Homosexual contact tends to occur in a relatively small number of adolescents. Also, it tends to occur rather early in adolescence, usually before 15 years of age, and is more common among boys than girls (Dreyer, 1982). In one sample of 600 adolescents 15 to 18 years of age, only 14 percent of the boys and 11 percent of the girls reported having had at least one instance of homosexual activity. When asked about their attitude toward homosexual activity among other adolescents, a majority of both sexes approved. Yet, boys were more accepting of homosexuality among girls than among boys, while girls were equally accepting of homosexual activity among both sexes. At the same time, in response to open-ended questions about homosexuality, these same adolescents expressed more negative, judgmental attitudes toward homosexual activity, reflecting some of their underlying anxiety and insecurity in this area (Haas, 1979).

In most instances, homosexual experimentation in adolescence is a passing phase of development and does not lead to a homosexual identity. Masters, Johnson, & Kolodny (1985) report that while about 2 in 10 men and 1 in 10 women experience homosexual activity in some phase of their lives, relatively few individuals become exclusively homosexual. Only about 4 percent of the men and 2 to 3 percent of the women ackowledge a lifelong preference for homosexuality. An extensive study of sexual preferences among heterosexuals and homosexuals by Bell, Weinberg, and Hammersmith (1981) has cast doubt on the familiar theories regarding the cause of homosexuality. These researchers found little evidence that male homosexuality is caused by having a dominant mother and a weak father. Nor is female homosexuality caused by girls choosing their fathers as role models. Also, the stereotype that homosexuality is often caused by being seduced by an older person of the same sex is untrue. Instead, it appears that sexual preference is firmly established by late adolescence. As children and adolescents, homosexuals have as many heterosexual experiences as their heterosexual counterparts, but tend to find these experiences less satisfying or frustrating. Gender noncomformity, such as boys avoiding sports like football while playing house or hopscotch, is a significant but not absolute predictor of a homosexual identity. All things considered, the lack of solid support for any of the usual theories of homosexuality raises the probability of a biological basis for homosexuality.

PROBLEMS IN SEXUAL BEHAVIOR

The fact that teenagers are becoming sexually active at an earlier age than in the past has intensified some of the perennial problems of adolescent sexuality. One of these is the failure to use contraceptives, a matter of increasing concern to parents and sex educators alike. A related problem is the incidence of premarital pregnancies, with American teenagers having one of the highest rates of teenage pregnancies in the world. Still another problem is the dramatic rise in sexually transmitted diseases in young people, which has become even more serious in recent years.

Use of Contraceptives

A study of sexually active couples, with young women ranging in age from 15 to 18, showed that the majority of them had discussed birth control on at least one occasion. But such discussions usually occurred *after* their first intercourse. In only one-fourth of the couples did partners agree that they had discussed birth control prior to their initial act of intercourse. Among couples who disagreed, girls more commonly than boys denied prior conversation regarding birth control. Couples who agreed they had good communication were more likely to practice effective contraception. Yet, one-fourth of all the respondents felt that the use of contraceptives had not been adequately discussed. And these individuals were found to be most at risk for an unintended pregnancy (Polit-O'Hara & Kahn, 1985).

Fewer than half the teenagers use a contraceptive during their first sexual intercourse, with a slightly higher proportion of women (48.9 percent) than men (44.1 percent) doing so. It's also possible that some of the men, especially the younger males, may not be aware what method their partners are using, so that contraceptive use may be greater than reported by men. The older the teenager, the more likely he or she is to use a contraceptive. Overall, whites are somewhat more likely than blacks to use contraceptives. But black women are more likely than white women to use a medically prescribed method. (See Table 10–3.) Young women rely mostly on the pill and to a lesser extent on the IUD and diaphragm. Young men tend to use the condom and to a lesser extent the withdrawal method. Men who plan to have intercourse are especially likely to rely on the condom. Overall, 5 out of 10 young women who use a prescription method obtain that method from a private physician. Another 4 out of 10 young women get prescriptions from a clinic, including hospitals as well as freestanding clinics. And 1 out of 10 women obtain contraceptives from some other source. Whites are much more likely than blacks to go to a private physician rather than use a clinic, 71 percent versus 29 percent (Zelnik & Shah, 1983).

A regulation proposed by the U.S. Department of Health and Human Services would have required that all family planning clinics supported by

Table 10–3 PERCENTAGE OF WOMEN AND MEN WHO USED A CONTRACEPTIVE DURING THEIR FIRST INTERCOURSE, BY AGE AT FIRST INTERCOURSE, ACCORDING TO RACE

Age at first intercourse	Percent who used any method			Percent who used a prescription method		
	Total	White	Black	Total	White	Black
Women	**48.9**	**51.2**	**40.8**	**19.9**	**15.3**	**40.7**
<15	31.0	33.2	26.9	10.2	4.7	22.9
15–17	52.1	52.9	48.8	18.8	12.1	48.0
≥18	62.3	63.7	47.7	29.4	28.9	*
Men	**44.1**	**46.0**	**34.0**	**22.0**	**21.6**	**25.1**
<15	34.0	36.4	27.8	16.8	15.6	20.8
15–17	48.5	48.9	44.7	20.5	19.2	34.1
≥18	59.1	59.4	*	38.3	40.0	*

*Fewer than 20 cases; in all others, N≥30.

Source: Melvin Zelnik and Farida K. Shah, "First Intercourse among Young Americans" *Family Planning Perspectives* (March/April 1983): 67.

Title X funds notify parents when teenagers receive prescription contraceptives from the clinic—popularly dubbed the "squeal rule." Opponents claimed that instead of encouraging family communication and contraceptive use, this rule would force teenagers to turn away from the clinics and to use less reliable methods, or none at all. Eventually, the squeal rule was dropped in the face of several court decisions that prohibited its use. At the same time, clinics have the responsibility to encourage young patients to involve their parents in decisions about the use of contraceptives. Surprisingly, there appears to be little relationship between family communication and contraceptive use. Frank Furstenburg and his colleagues (1984) found that teenage girls who communicated little with their mothers are as likely to use effective birth control as those who communicate well. Apparently, adolescents' intense desire for privacy in sexual matters and the embarrassment on the part of teenagers and parents alike in discussing birth control suggest that teenagers may benefit from professional guidance about contraceptives on a confidential basis.

Young people who plan to have sex are more apt to use a contraceptive than those who engage in sex spontaneously. And this holds true for males and females, blacks and whites. At the same time, a sizable group of those who plan to have sex go unprotected, including almost half the men and one-fourth of the women. And a large fraction of those who have not planned to have sex, about 4 in 10, use contraceptives (Zelnik & Shah, 1983). Generally, the use of contraceptives rises for subsequent acts of intercourse. Also, as we might expect, there is a strong positive association between contraceptive use and the couple's relationship. Couples who are going steady are the most likely to use contraceptives, with dating couples somewhat less inclined to do so. Individuals who have just met or have a "weak" relationship are the least likely to use contraceptives. Furthermore, the partners' views about whose responsibility it is to use birth control have a strong impact on contraceptive use. The great majority of both sexes, expecially among older teenagers, believe that both partners are equally responsible for practicing birth control. Not surprisingly, these teenagers are somewhat more likely to use a contraceptive than those who say one partner or the other should bear the entire responsibility. Just 1 percent of sexually experienced adolescents believe neither partner is responsible, and half of these teenagers say they used no contraceptive at their last intercourse (Zabin et al., 1984).

Why do young people fail to use contraceptives? The most frequently given reason is ignorance of which device to use or where to obtain them. Another common reason is the feeling that use of a contraceptive may spoil the spontaneity and pleasure of sex (Zelnik & Shah, 1983). A variety of other reasons may be seen in the study of 1200 young women 12 to 19 years of age, which inquired into why they had delayed coming to a family planning clinic. (See Table 10–4.) Many of the reasons reflect anxieties and fears as well as widespread ignorance about birth control. The sizable number who

Table 10—4 REASONS GIVEN FOR DELAYING VISIT TO FAMILY PLANNING CLINIC FOR BIRTH CONTROL HELP

REASON	PERCENT
Just didn't get around to it	38.1
Afraid my family would find out if I came	31.0
Waiting for a closer relationship with partner	27.6
Thought birth control dangerous	26.5
Afraid to be examined	24.8
Thought it would cost too much	18.5
Didn't think having sex often enough to get pregnant	16.5
Never thought of it	16.4
Didn't know where to get birth control help	15.3
Thought I had to be older to get birth control	13.1
Didn't expect to have sex	12.8
Thought I was too young to get pregnant	11.5
Thought birth control was wrong	9.2
Partner opposed	8.4
Thought I wanted pregnancy	8.4
Thought birth control I was using was good enough	7.8
Forced to have sex	1.4
Sex with relative	0.7
Other	9.7

Source: Adapted from L. S. Zabin and S. D. Clark, Jr., "Why They Delay: A Study of Teenage Family Planning Clinic Patients, "Family Planning Perspectives" (September/October 1981): 214.

responded that they "just didn't get around to it" also reflects immaturity and lack of responsibility in regard to sex. Consequently, there is increasing concern that sex education must include help in decision making as well as information on the availability of contraceptives.

Premarital Pregnancy

The failure to use contraceptives correctly or at all greatly increases the risk of premarital pregnancy. Over half of the sexually active young women who never use contraceptives have at least one unintended pregnancy. In one study, about two-thirds of those who became pregnant without trying to reported that they believed, for some reason, that they wouldn't get pregnant (Zelnik & Shah, 1983).

Some background statistics on premarital pregnancy may help to put this problem in perspective.

- 4 out of 10 girls under 15 years of age will become pregnant in their teens.

- 6 out of 10 teenage girls, including 1 out of 3 of those who are sexually experienced, will have at least one premarital pregnancy.

- 6 out of 10 teenage girls who have a child before 17 years of age will become pregnant again before the age of 19.

- 1 in 20 American teenage girls has a baby each year, one of the highest rates in the world. Although this is partly due to the exceptionally high (but declining) fertility rates among black teens, fertility rates among white teens are consistently higher than those of most European countries, Canada, and Australia (Zelnick, Kantner, & Ford, 1981; Westoff, Calot, & Foster, 1983).

miscarriage: The spontaneous premature termination of pregnancy; spontaneous abortion.

Although these statistics show how widespread unintended premarital pregnancy is, they don't fully convey the devastating consequences of the problem. To begin with, teenage mothers, especially those under 16, face increased health risks. They are apt to have more medically complicated pregnancies, including miscarriages and toxemia, as well as a higher risk of maternal death than mothers in their twenties. Also, babies of teenage mothers are more likely to be underweight, with twice as many of them dying in infancy as those born to women in their twenties (McCormick, Shapiro, & Starfield, 1981).

toxemia: A dangerous condition during pregnancy in which high blood pressure occurs.

Even more disturbing are the psychological and social consequences of unintended teenage pregnancies. Although it is illegal to expel pregnant teenagers from schools, most teenage mothers who keep their babies drop out of school and don't return. As a result, a large proportion of them never gain regular employment and are more likely to become dependent on government services and support (McGee, 1982). Males and females most at risk for becoming involved in teenage pregnancies tend to come from homes in the lower socioeconomic groups and have difficulty in school. Girls at greater risk usually come from large, father-absent families under a lot of stress, with black girls being especially at risk (Robbins, Kaplan, & Martin, 1985).

abortion: The spontaneous or medically induced removal of the contents of the uterus during pregnancy.

Unmarried teenage girls who find themselves pregnant face a series of difficult choices. They must decide how to deal with their pregnancy, often with little or no support—emotionally or financially—from the child's father. Some young women may decide to end the pregnancy by getting an abortion, though this often carries the risk of guilt and anguish. Most abortions are obtained by young unmarried women in the first 6 weeks of their pregnancy, with white women more likely to seek an abortion than black teenagers. A larger proportion of black teenagers choose to keep their child (Henshaw et al., 1985). Yet, as we discussed in an earlier chapter, the high rate of out-of-wedlock births combined with the greater proportion of black women who never marry contributes to the prevalence of single-parent families among blacks. About half of all black children under 18 years of age live in single-parent homes, with most of them living on incomes below the poverty level (Staples, 1985). Overall, less that 5 percent of young women put their babies up for adoption, with the proportion of unwed mothers giving up their babies for adoption being much higher among whites than blacks (McGee, 1982).

adoption: The voluntary and legal act of taking a child of other parents as a member of one's own family.

In about half the cases, teenage girls with a premarital pregnancy choose to marry the father of their child. Yet, as we've seen earlier, marriage reduces the chances that the girl will continue with school after the birth of the child. Then too, many of these marriages are entered into under pressure, with uncertainties on the part of both partners. Not surprisingly, couples who marry in their teens are more likely to have marital problems and to separate and divorce in later years than those who delay marriage until their twenties. Those who delay marriage until after the birth of the child are at even greater risk of marital failure (McLaughlin, 1986). At the same time, there are exceptions to the rule. Frank Furstenberg (1976) found that young women who made the best adjustment were those who married the father of the child, returned to school, and delayed having additional children. Within 5 years, these womens' lives resembled those of their counterparts who had delayed marriage and pregnancy until their twenties.

Sexually Transmitted Diseases

Another consequence of increased sexual activity among youth is the rise in sexually transmitted disease (STD), a broader and less value-laden phrase than the older term "venereal disease." The increase in sexually transmitted diseases probably reflects the increased sexual activity among young people as well as the tendency to have multiple sex partners. It is also believed that the widespread use of birth control pills reduces the use of two other contraceptive methods, vaginal spermicides and the condom, which offer some protection against these diseases. In any event, the incidence of sexually transmitted diseases continues to be highest among the 20 to 24 year olds, followed by the 15 to 19 years olds, and the 25 to 29 year olds (NIAID, 1981). Although young people frequently exhibit a casual attitude toward these diseases—that is, they're willing to seek treatment if necessary but aren't especially concerned about prevention—the dramatic rise of herpes and AIDS in recent years has generated a new caution among many youths, and rightly so (Leo, 1984).

gonorrhea: An infectious bacterial disease usually transmitted through sexual intercourse.

Gonorrhea continues to be one of the most common sexually transmitted diseases. Although about 1 million new cases of gonorrhea are reported each year, the true incidence is estimated to be closer to 2 million (Amstey, 1983). Gonorrhea may be transmitted by any form of sexual contact, from kissing to sexual intercourse. A woman who has intercourse once with an infected man has a 50 percent chance of catching gonorrhea, while a man who has intercourse with an infected woman has a lower risk, around a 25 percent chance of catching the disease. The old excuse "I caught it from a toilet seat" has been shown to be theoretically possible, though rare (Masters, Johnson, & Kolodny, 1985). Despite the fact that gonorrhea is readily treatable by antibiotic drugs, many young people do not seek treatment because they have so few symptoms that they fail to realize they are infected. This is especially true among women, with as many as two out of three women infected with gonorrhea not being aware of their condition. Yet, women with untreated gonorrhea may suffer from inflammation of the fallopian tubes, infertility, birth malformations, of menstrual disorders. Although a much smaller proportion of men remain relatively free of symptoms, untreated gonorrhea may be equally devastating to males. Gonorrhea is the single most important cause of sterility among males (NIAID, 1981).

syphilis: An infectious disease transmitted by sexual intercourse, which, if left untreated, may lead to the degeneration of bones, heart, and nerve tissue.

Syphilis, though much less common than gonorrhea, is a more serious disease. Men are twice as likely as women to have syphilis, with half of the men infected with syphilis being homosexual or bisexual (Masters, Johnson, & Kolodny, 1985). Although syphilis is generally transmitted by sexual contact, it can also be acquired from a blood transfusion or transmitted from a pregnant mother to the fetus. Fortunately, syphilis is readily treatable, usually with injections of penicillin. But if left untreated, the disease continues into an advanced stage in which it can cause brain damage, heart failure, blindness, or paralysis.

genital herpes: A sexually transmitted disease that, in addition to the discomfort of the symptoms, may lead to serious medical consequences.

Genital herpes, one of several herpes viral infections, has increased dramatically in recent years. Over a half million new cases of genital herpes are reported each year, with an estimated 20 million Americans now suffering from this disease. Furthermore, because of the frequent recurrence rate and lack of effective treatment, the number of cases continues to increase (Holmes, 1982). Although this disease is more common among young adults, an estimated 1 in 35 adolescents has herpes (Oppenheimer, 1982). Genital herpes is generally transmitted through direct sexual contact, such as sexual intercourse or oral-genital contact. At the same time, oral herpes can be transmitted through kissing, sharing towels, or drinking out of the same cup. And, contrary to earlier notions, if someone with cold sores or fever blisters in the mouth performs oral sex on another person, the latter person may develop *genital* herpes (Peter, Bryson, & Lovett, 1982). Although there is no known cure for genital herpes so far, there are drugs that may lessen the severity of the symptoms and shorten the time for healing (Corey et al., 1983). In addition to the intermittent discomfort of herpes sores, genital herpes can lead to serious complications. Newborn infants exposed to herpes sores in the birth canal may be infected with herpes and can suffer physical damage and even death. Furthermore, women infected with genital herpes are eight times more likely to have cervical cancer than women not infected (Harvard Medical School Health Letter, 1981).

AIDS (Acquired Immune Deficiency Syndrome): A sexually transmitted disease that is communicated through blood products and is ultimately fatal.

AIDS (Acquired Immune Deficiency Syndrome) is one of the newest and most frightening sexually transmitted diseases to come to public awareness. AIDS is caused by a virus that destroys the body's immune system. Thus, many germs which would ordinarily be harmless to someone with a normal immune system can produce devastating results, and ultimately death. AIDS is transmitted primarily through blood or blood products containing the virus—notably semen and blood. Although homosexual men make up the largest fraction of AIDS patients in the United States, the virus can also be transmitted through vaginal intercourse, and by intravenous drug users who share needles. The risk of getting AIDS through blood tranfusions has diminished considerably because of more careful screening of donors. But the risk of contracting AIDS through sexual intercourse between heterosexuals is expected to rise dramatically. One of the most troubling aspects of AIDS is the relatively long incubation period from the time of exposure to the appearance of symptoms: The incubation period is generally 12 to 18 months, but may last up to several years. During this time a person is probably contagious to others. It is estimated that as many as 1 million Americans may be infected with the AIDS virus, though more than 90 percent of them do not know it (Smilgis, 1987). Although present tests can identify people who have been infected with the AIDS virus, they cannot distinguish between those who carry the virus but are not affected by it and the much smaller number who will develop the full-blown AIDS syndrome itself. Since there is presently no cure for the AIDS syndrome, the disease is usually fatal. Marked public awareness of AIDS in recent years may well alter the attitudes toward casual sex among heterosexuals as well as homosexuals, with "safe sex" becoming the watchwords of the '80s. (Harvard Medical School Health Letter, 1985).

SEX EDUCATION

Many of the problems in adolescent sexuality can be traced to the lack of an adequate sex education in our society. Despite the openness and honesty in sexual matters and relative sophistication of today's teenagers in regard to sex, sex education continues to be mostly a hit-or-miss affair, varying widely in content and quality in different families and schools.

Sources of Sex Information

Teenagers continue to get most of their information about sex from the most unreliable source—their peers. In one study, Hershel Thornburg (1981) asked all the students in a midwestern high school where they had initially learned about a variety of sexual topics. As expected, peers were the most commonly cited source of sex information, followed in order by literature (pamphlets and books), mothers, and the school. Learning about sex from personal experience, fathers, physicians, and the clergy constituted a relatively small part of sex education. At the same time, initial sources of sex information varied somewhat depending on the particular topic. For instance, peers were the most apt to supply information about sexual intercourse, contraceptives, and homosexuality—topics not usually discussed elsewhere. Mothers were more likely to discuss menstruation and conception, usually with their daughters, while schools were usually the major source of information about sexually transmitted diseases. (See Table 10–5.)

Table 10–5 SOURCES OF FIRST SEX INFORMATION

	Abor-tion	Con-ception	Contra-ception	Ejacu-lation	Homo-sexuality	Inter-course	Mastur-bation	Menstru-ation	Pet-ting	Prosti-tution	Semi-nal Emis-sions	Vene-real Dis-ease	Totals
Peers	20.0	27.4	42.8	38.9	50.6	39.7	36.3	21.5	59.7	49.7	35.2	28.2	37.1
Literature	32.0	3.2	23.8	22.1	19.4	15.2	25.0	11.2	10.0	26.8	37.4	21.2	21.9
Mother	21.5	49.4	13.1	8.9	7.5	23.8	11.1	41.5	4.5	7.5	4.2	9.4	17.4
Schools	23.7	16.4	16.7	20.7	16.1	7.6	17.5	15.7	9.0	11.7	21.1	36.8	15.2
Experience	.5	.8	1.0	5.2	2.1	7.5	8.0	7.6	14.0	2.0	.7	1.1	5.4
Father	1.0	1.2	2.4	2.6	4.3	3.9	1.3	1.1	2.2	1.0	1.4	2.1	2.2
Minister	1.0	.9	.0	.7	.0	1.0	.0	.7	.2	1.0	.0	.0	.5
Physician	.3	.7	.2	.9	.0	1.3	.8	.7	.4	.3	.0	1.2	.3
Totals	100.0	100.0	100.0	100.0	100.0	100.0	100.0	100.0	100.0	100.0	100.0	100.0	100.0

N=1152

"Don't know" responses were eliminated from this table. Percentages are on all terms listed as known.

Source: Hershel D. Thornburg, "The Amount of Sex Information Learning Obtained During Early Adolescence," *Journal of Early Adolescence,* 2 (1981): 174.

Sex Education in the Home

sex education: Teaching the attitudes, information, and skills needed to become sexually competent, including formal programs of sex education in the schools.

It is generally agreed that informal sex education begins in the home. Children learn about sex the same way they learn about most other things—by observing their parents and talking with them. The manner in which parents hug, kiss and touch each other, or fail to do so, shows the degree to which they have accepted themselves and each other as sexual beings. Then too, the way parents handle their child's or their adolescent's questions about sex helps to shape the latter's attitudes toward sex. Parents who are embarrassed to talk about sex may unwittingly encourage an attitude of reticence or shame toward the teenager's own sexuality. When teenagers in one study (Haas, 1979) were asked if they had tried to discuss sex openly with their parents, about half of them said they had tried to do so at one time or another. Asked how their parents had responded to their questions, only a minority gave a positive response. More often than not, teenagers felt that their parents had responded negatively, either by avoidance, denial, teasing, or disapproval. As a result, about half of the teenagers either shared nothing about sex with their parents or only what they thought their parents would approve of. Only about 1 in 10 teenagers (9 percent of the boys; 11 percent of the girls) reported that they shared everything about their sex lives with their parents.

Many parents are well aware that they are not doing an adequate job with sex education in the home and favor some type of sex education in the schools. When asked if they think sex education will encourage sexual activity among their adolescents, most adults express the view that young people are likely to become involved anyway (Alan Guttmacher Institute, 1981). However, not all parents feel this way. A vocal minority of parents feel that exposing children and adolescents to information about sex will stimulate their curiosity and prematurely draw them into sexual activity. Some parents feel strongly that sex instruction is so closely bound up with moral and religious values that it should be done at home or in a religious setting. Still other parents hold the naive view that today's teenagers are so well-informed about sex that they have little if anything to learn about sex. But the facts show otherwise. While teenagers tend to be more sophisticated about sex in ways that their parents were not at the same age, much of their information is sketchy or just plain wrong. Yet, adolescents frequently do not appear to realize that their information is incomplete or inaccurate. For instance, it has been shown again and again that the majority of young high school students do not know when the most fertile part of the woman's menstrual cycle is likely to occur. Some believe you cannot get pregnant the first time you have sex or if you have sex infrequently—all of which helps to account for the failure to use contraceptives (Kisker, 1985).

Sex Education in the School

Parents and adolescents alike tend to favor sex education in the schools. Studies have shown that more than three-fourths of American adults believe sex education should be taught in school. And when such courses are taught, fewer than 5 percent of parents ban their children from attending (Kirby, Alter, & Scales, 1979; Alan Guttmacher Institute, 1981). Adolescents are even more supportive of sex education in the schools, with more than 8 out of 10 teenagers desiring such courses (Gallup, 1980/1985).

By contrast, barely one-third of American junior and senior high schools offer sex education, with only a handful of states requiring it, including New Jersey, Maryland, Kentucky, and the District of Columbia. Fortunately, an increasing number of schools are offering some type of sex education—ofen called "family life education." But the content and quality of sex education programs vary widely from one school to another, with many programs being remarkably incomplete (Orr, 1982). A survey of secondary schools has shown that sex education generally begins in high school, in the tenth grade on the average, with courses much more likely to be offered in grades 9 through 12 than in grades 7 and 8. The most popular topics are anatomy and physiology, reproduction, pregnancy, childbirth, birth control, and sexually transmitted diseases. Less likely to be included are topics such as homosexuality and prostitution. An encouraging trend is the number of schools that offer some instruction on sex roles, love, and marriage (Newton, 1982).

Educators and parents alike have a number of concerns about sex education. First, many educators feel that waiting until high school for sex education may be waiting too long. Thornburg (1981) found that over 80 percent of sex information had been acquired by the time teenagers began high school, a time when many teenagers have already begun sexual experimentation. As a result, many educators believe that sex education should begin in elementary school and be taught in an age-appropriate fashion throughout a child's formal education (Masters, Johnson, & Kolodny, 1985). Yet, only slightly over half (54 percent) of public school parents favor starting sex education in elementary school (Gallup, 1985). A second concern is what matters should be dealt with in sex education classes. Sex education should address more than sound factual knowledge, as important as this is. It should also deal with students' attitudes and help them to develop a sense of responsibility about sex, especially in regard to the use of contraceptives. Third, there is the issue of who should teach sex education classes. Ideally, sex education teachers should be highly qualified and should have a healthy attitude toward sex. At the least, teachers should have some special training in the field. But teachers' attitudes toward sex may be just as important as their knowledge of sex. Teachers who can answer students' questions and deal with controversial issues in a frank, fair manner encourage the development of similar attitudes among students. Fourth is the issue of teens' sexual knowledge in relation to their peers. Since teenagers tend to learn about sex from their peers, some programs have used teenage discussion groups led by skilled moderators with positive results (Kisker, 1985). Finally, there is the issue of parental involvement. Some evidence shows that sex education programs that have been the most successful have involved parents in one way or another (Scales & Everly, 1977). In one

**Table 10–6   PERCENTAGE OF UNMARRIED, SEXUALLY ACTIVE
TEENAGE WOMEN WHO HAVE BEEN PREGNANT, BY
WHETHER THEY HAD PREVIOUSLY TAKEN A SEX
EDUCATION COURSE IN SCHOOL**

	WHITE		BLACK	
	Age 15–17	Age 18–19	Age 15–17	Age 18–19
Had Sex Education	15.4	25.0	28.7	48.2
No Sex Education	25.0	31.9	49.2	54.0

Source: Adapted from survey data of Melvin Zelnik and Y. J. Kim, "Sex Education and Its Association with Teenage Sexual Activity, Pregnancy, and Contraceptive Use," *Family Planning Perspectives,* 14 (1982): 124.

approach, a broad coalition of community-oriented programs, such as the Girls Clubs of America, 4-H, Campfire, Inc., and the Salvation Army, have begun to implement sex education programs that involve both children and their parents (Gregg, 1982). At the same time, there is some evidence that parental communication about contraceptives has little effect on teenagers' use of contraceptives, suggesting that programs aimed at encouraging parents to communicate with their children about contraceptives have only limited potential for improving the contraceptive practices of teenagers (Kisker, 1985).

A final matter pertains to the impact of sex education courses. Contrary to the fears of some adults, there is ample evidence that formal sex education does not usually lead to premature or greater sexual activity among adolescents. Instead, there are generally fewer premarital pregnancies among those who have received formal sex instruction. (See Table 10–6.) Other positive gains associated with sex education programs include greater factual knowledge and accuracy of information about sexual matters, decline of fears and doubts about sex, greater awareness and use of contraceptives, better communication about sex, and, where specifically focused on, greater skill in decision making about sexual matters such as the use of contraceptives (Zelnik & Kim, 1982).

SUMMARY

Sexual Attitudes

1. We began the chapter by noting that attitudes toward sex generally became more liberal during the late 1960s and 1970s, leading to greater openness and honesty and personal choice in sexual matters.

2. Although today's teenagers are somewhat more conservative toward sex than teenagers in the 1970s, they have not rejected the changes of the recent past as much as they have put them in perspective, and they tend to be more sophisticated but wary about sexual relationships.

3. The majority of youth hold sexual attitudes and values that are consistent with responsible conduct, but do not always apply these attitudes to their personal behavior, with more than three-fourths of teenagers reporting that they had their first intercourse at a younger age than desirable.

4. Although the relationship between love and sex depends largely on the individual's age, sex, and sexual values, most youths favor sex within affectionate relationships.

Sexual Behavior

5. Masturbation is commonly practiced at adolescence, especially among boys, with the greatest change in masturbation practice being the reduced anxiety and guilt associated with it.

6. Adolescents engage in more intimate forms of erotic contact and at earlier ages than in the past.

7. Women generally have their first sexual intercourse at age 16 with a partner several years older than themselves; men tend to have their first intercourse at age 15 with a partner less than a year older than themselves.

8 In most instances, sex is a spur-of-the-moment decision, with fewer than one out of four teenagers having planned it.

9. Homosexual activity tends to occur relatively early in adolescence and involves more boys than girls.

10. Homosexual experimentation in adolescence is generally a passing phase of development that does not lead to a lifelong preference for homosexuality.

Problems in Sexual Behavior

11. A major problem in sex has to do with contraceptive use, with only half the teenagers using a contraceptive method during their first intercourse.

12. Contraceptive use generally rises in subsequent acts of sexual intercourse and is higher among couples going steady.

13. Over half the sexually active young women who never use contraceptives have at least one unintended pregnancy.

14. Although young women have several options for dealing with a premarital pregnancy, about half marry the father of the child, with such marriages being even more likely to end in divorce than other marriages.

15. Another possible consequence of sex is a sexually transmitted disease, such as gonorrhea, syphilis, herpes, and AIDS, with the highest incidence of these diseases occurring among 20 to 24 year olds, followed by the 15 to 19 year olds.

16. The dramatic rise in genital herpes and AIDS has already begun to alter young people's attitudes toward casual sex.

Sex Education

17. Teenagers continue to learn much of what they know about sex from the most unreliable source, their peers.

18. Sex education in the home leaves much to be desired, with most parents and teenagers sharing only limited information.

19. Although a majority of adults and adolescents alike favor sex education in the school, only about one-third of junior and senior high schools offer programs in sex education.

20. A major drawback to existing sex education programs is that most courses begin in high school, after many teens have already begun learning and experimenting with sex.

21. Contrary to adult fears, teenagers who receive formal instruction in sex do not engage in more premature sexual activity than teens who receive no instruction; in addition, teens who have been taught sex education have a lower rate of premarital pregnancy.

REVIEW QUESTIONS

1. How would you characterize the attitudes of today's adolescents toward sex?

2. What are some of the glaring inconsistencies between adolescents' sexual attitudes and their sexual behavior?

3. To what extent does the relationship between love and sex vary by the individual's age, sex, and sexual values?

4. In what sense do adolescents engage in more intimate physical contact at earlier ages than in the past?

5. How much basis is there to the criticism that teenagers begin having sex too early?

6. Is it true that homosexual play among adolescents generally represents a passing phase of sexual development?

7. What are some of the reasons teenagers fail to use contraceptives on their first intercourse?

8. How much of a problem is premarital pregnancy?

9. Select one sexually transmitted disease and describe the risk it represents to teenagers and what they can do to avoid it.

10. Describe the type of sex education program you would like to see, including school courses and parent involvement.

11 Work and Career Choice

- **ADOLESCENTS IN THE WORKPLACE**
  Teenage Employment
  Combining School and Work
  The Value of Work Experience
- **THE PROCESS OF CAREER CHOICE**
  Stages of Career Choice
  Career Maturity
  Identifying a Compatible Career
- **INFLUENCES ON CAREER CHOICE**
  The Family
  Socioeconomic Influences
  Sex Differences
- **PREPARING FOR A CAREER**
  Career Education
  Going to College
  Career Outlook
- **SUMMARY**
- **REVIEW QUESTIONS**

Looking around at parents and teachers already settled in their careers, young people may sometimes feel adults don't understand the perplexity and frustration one experiences in choosing a career. But such is usually not the case. When asked, many adults will frankly admit that settling on a career isn't as easy as it appears. Even people who achieve fame in their fields sometimes must struggle before finding their life work. Some people, like Charles Darwin, flounder throughout adolescence and give their parents a lot of worry before arriving at a firm career goal. Others, like Margaret Mead, entertain a wide variety of vocational aspirations before choosing their career. As you may recall from an earlier chapter, at one time or another, Margaret Mead wanted to be a lawyer, a nun, a writer, and a minister's wife with six children. As it turned out, she went on to get a master's degree in psychology, and later became a famous anthropologist. Still other people like Erik Erikson settle on their lifework later in life. After spending much of his youth as a traveling Bohemian or "hippie," Erikson started out as an art teacher and only later in his career became a psychoanalyst.

ADOLESCENTS IN THE WORKPLACE

career: Sequence of jobs or occupations; a purposeful life pattern of work.

Such experiences remind us that the choice of a career is an extended developmental process that involves a great deal of exploration and change before individuals settle on a firm career goal. Compared to youth in the past, young people today enjoy many advantages in selecting what they want to do in the workplace, such as greater educational opportunities, a wider variety of careers, and greater freedom of choice in career selection. At the same time, many careers in our technology-oriented society require a high degree of specialization and many years of expensive education, such that young people often feel under a lot of pressure to choose wisely. Yet, they spend so many years in school isolated from career-related work experience that they usually have little basis for knowing what type of work they want to do. Furthermore, rapid shifts in the need for particular specialized skills as well as the economy generally may make one's hard-won skills obsolete rather quickly, forcing people to change careers more often. All of this makes the choice of a career a major challenge for young people.

We'll begin by taking a look at the changing patterns of teenage employment and the pros and cons of their work experience. Then we'll examine the developmental process of career choice along with the various influences on young people's decisions, such as the family, socioeconomic background, and sex differences. Finally, we'll discuss the matter of preparing for a career, including the value of a college education.

Teenage Employment

workplace: Employment on a paid basis outside the home.

job: Position of employment.

The individual's transition to adult work roles in the United States generally occurs throughout a sequential process that begins with household chores in childhood and culminates in a full-time job in the formal workplace by late adolescence or the early twenties. During childhood, most boys and girls are assigned jobs around the house. Although younger children are equally likely to do the same type of chores around the house, such as clearing the table after a meal, household chores become more sex-typed as children grow older. However, as we mentioned in an earlier chapter, children from dual-career families tend to be more flexible about the particular tasks they perform at home than youngsters from the traditional, one-wage-earner homes. By early adolescence, teenagers take on informal jobs outside the home, frequently for neighbors. The most common jobs at this age are babysitting, newspaper delivery, house cleaning, and lawn and garden work. As adolescents reach the age required for employment in the formal work force—anywhere from 14 to 16 years of age depending on the state—they begin taking part-time jobs in the formal workplace. Teenagers may work after school, in the evenings, on the weekends, and during the summer. By their senior year in high school, more than three-fourths of the teenagers in school are working more than 15 hours a week on an outside job (Grant & Snyder, 1984). Full-time employment usually begins after graduation from high school or college.

The range of jobs available to adolescents in the formal workplace is rather limited. One-half of teenagers who have jobs either work in restaurants or in the retail trades. About one in five working teenagers is employed in restaurants, either preparing food, waiting on customers, clearing tables, or cleaning dishes. Many of them work in the fast-food restaurants. Another one in five teenagers works in a retail business, often as a cashier or a sales clerk. The remainder of adolescents work in a variety of jobs, such as clerical work, service station attendants, maids, and other types of service jobs or manual labor. Relatively few teenagers now work on farms or in factories. And only a small proportion of teenagers have jobs that are interesting and challenging (Lewin-Epstein, 1981).

Occasionally, one sees a girl delivering newspapers and a boy babysitting. But these are still exceptions to the rule. In one study of several thousand teenagers, only 9 percent of the newspaper carriers were girls and only 5 percent of the babysitters were boys. It appears that part-time jobs for teenagers are no less sex-typed than those for adults. Boys generally work as manual laborers, gardeners, newspaper carriers, service station attendants, and busboys. Girls are more likely to work as maids, babysitters, food counter workers, and waitresses. Furthermore, boys work longer hours than girls and boys' jobs generally pay better than girls' jobs (Greenberger & Steinberg, 1983).

Combining School and Work

work: Generally taken to mean economic employment.

One of the most remarkable changes in the past couple of decades is the increased proportion of teenagers who work during the school year. Prior to the 1950s, it was unusual for students to combine school and work. Today, it's unusual for students *not* to work after school. About one in three public school students in the ninth grade hold part-time jobs after school. By the senior year, more than 9 out of 10 students work after school (Grant & Snyder, 1984). A major reason for this change is the rapid expansion of those aspects of the American economy that need a large number of part-time workers, namely the retail trades and service sectors. Employers, such as the fast-food chains, need workers who are willing to work for relatively low wages and short shifts, requirements readily filled by teenagers. Another factor has been the dramatic increase in inflation since the 1970s—it cost more to be an American teenager than it did in the past. Practically everything teenagers buy costs more, including candy, gum, toys, records, fast food, tickets to games and movies, cars, and gas, not to mention alcohol and drugs.

Surprisingly, students from middle-class families are even more likely to work after school than those from less privileged homes, which is just the opposite from the practice in earlier eras. The most likely explanation is that

Students who work 20 hours or more each week at an outside job are often too tired to study. (National Archives)

middle-class teenagers have an easier time finding jobs because they live in the suburbs where many of the newer jobs are located. One study of 4587 high school students found that employed students were more likely to come from families in the higher socioeconomic groups with a mother or father in the higher-status careers. Unemployed students were more apt to come from families in the lower socioeconomic groups and to have parents in the lower-status, low-paying careers (Schill, McCartin, & Meyer, 1985).

Students are also working more hours than ever before. The average high school sophomore works about 15 hours a week. Seniors average close to 20 hours a week. Almost half the seniors in public high schools work more than 20 hours a week (Grant & Snyder, 1984). (See Table 11–1.) The sum of 30 hours per week in class (at 6 hours per day) plus 6.5 hours of homework a week on the average, combined with 20 hours a week at a part-time job makes the student's total work load over 50 hours a week. And many students work even more. As a result, students are often too tired to study, with school grades more likely to suffer after about 15 to 20 hours a week. Mar-

**Table 11—1   PART-TIME EMPLOYMENT BY
HIGH SCHOOL SENIORS (in percents)**

Hours worked per week	Boys	Girls	Total
None	2.9	6.1	4.5
1 to 4 hours	7.5	11.9	9.8
5 to 14 hours	17.3	20.4	18.9
15 to 21 hours	24.6	29.2	27.0
22 to 29 hours	18.1	15.9	17.0
30 to 34 hours	9.7	6.8	8.2
35 hours or more	19.1	9.0	13.8
Not reported	.8	.8	.8

Source: W. Vance Grant and Thomas D. Snyder, *Digest of Educational Statistics, 1983–1984* (Washington, DC: National Center for Education Statistics, Government Printing Office, 1984), p. 74.

ginal students' grades are especially likely to suffer from part-time jobs. The Schill, McCartin, & Meyer study showed a curvilinear relationship between hours on the job and the students' grade-point averages (GPAs). Grades tended to increase somewhat for students with part-time jobs up to about 20 hours a week. After that, grades steadily decreased. Furthermore, students who work long hours are less likely to participate in extracurricular activities and tend to become more distant from their families and friends.

The Value of Work Experience

Such findings raise questions about the impact of part-time jobs on adolescent development. All too often, adults have assumed that work is "good" for teenagers. It teaches them about the value of money and the real world. Work experience also helps prepare them for adulthood. Yet, it is also apparent that teenage work often exacts a price from teens' social development. More realistically, it's a matter of balancing the advantages and disadvantages of work experience.

The biggest gain from part-time work is the practical knowledge acquired. Young people learn how to find and hold a job, a valuable lifetime skill. They also learn how to budget their time. Work also develops the individual's sense of responsibility and work maturity, that is, the ability to complete a task and feel pride in a job well done. Although most adolescents work primarily for money, getting paid for what they do also brings self-reliance. But one of the most valuable lessons of all is learning how to get along with other people. One girl said, "My job as a cashier in a supermarket has helped me to be less shy and to talk to people" (Cole, 1981).

On the minus side, most adolescent jobs are dull, monotonous, and stressful. In contrast to adolescents in more traditional societies, who come into contact with adults who prepare them for their career goals—as in an apprenticeship program—today's teenagers spend most of their time with other adolescents in jobs that are unrelated to their career goals. Adolescents have few opportunities to make decisions on their own. They generally get little instruction from their supervisors, who are often only a few years older. Furthermore, adolescents in their work seldom make use of the skills they are learning in school, such as reading and writing. Also, most part-time jobs do not develop teenagers' sense of social responsibility or their concern for others. Instead, students often acquire cynical views about the business world, such as the view that there is something wrong with people who work harder then they have to. Teens are also more likely to go along with unethical business practices, such as supporting the idea that poorly paid workers are entitled to take little things from their jobs as a way of making up for their unfair treatment (Steinberg et al., 1982).

Such findings raise questions about the value of adolescents' work experience during school. On the one hand, the advantages of having a part-time job in high school appear to be short-lived. That is, students who have worked during high school tend to have an easier time finding full-time jobs with good pay than those who have not worked after school. But within a few years, there is little or no difference between the two groups of students. Those who didn't work during high school are just as likely to be employed and earn just as much money as those who worked during high school (Freeman & Wise,

Most adolescents work in jobs that are unrelated to their career goals. (Courtesy NCR)

1982). On the other hand, students may achieve more lasting benefits from supervised work-study programs that integrate adolescents' learning experiences and their career goals. In such programs, special effort is made to match students' jobs with their interests and to provide appropriate instruction and supervision, thereby maximizing the learning experience of adolescent work. Such programs may also help adolescents to view work in a more meaningful way, increasing their sense of social responsibility and the value of feedback regarding their work performance.

THE PROCESS OF CAREER CHOICE

Because our society places such a high value on work, there is considerable pressure on teenagers to choose a career. At first casually and then more anxiously, parents begin asking their children, "What do you want to do when you grow up?" Schools, too, expect adolescents to make important decisions bearing on their choice of a career. Which courses do you want to take? Which activities are you most interested in? Students' choices and how well they do in their courses and extracurricular activities, in turn, help to shape their eventual choice of a career. Additional pressure to choose a career comes from within. Serious concerns about a person's life work appear by the age of 12 or so and generally increase with intensity into adulthood. Anxiety builds during the last year or so of high school, when students usually ponder such questions as: "Do I want to go to college?" "If so, which college?" "Or, if not, what am I going to do after graduation?"

Much of the anxiety over career choice might be alleviated among adolescents and parents alike if they were more aware that career choice is a process that occurs over a period of many years and involves a great deal of exploration and change. Here in this section, we'll take a look at the developmental process of career choice, the matter of career maturity, and the task of identifying a compatible career.

Stages of Career Choice

Eli Ginzberg (1972/1977) regards career choice as a developmental process that begins in childhood and culminates in late adolescence or youth; it includes three stages: the fantasy, tentative, and realistic stages, respectively.

The *fantasy stage of career choice* lasts through childhood up to the age of 11. During this stage children base their career interests mostly on imagination, without regard for ability, opportunity, or training. Children may imagine themselves as astronauts one week and doctors the next. Parents generally don't worry too much about their childrens' notions of careers at this age, intuitively realizing that such childhood fantasies generally have little relationship to adult choice.

The *tentative stage* begins at puberty and lasts through high school, from about 11 to 18 years of age. It includes several substages, during which adolescents are apt to make career choices successively on the basis of their interests, their abilities, and their values. Then in a transitional substage at about 17 or 18 years of age, adolescents move toward a more realistic choice in response to pressures from the school, parents, peers, colleges, and graduation.

The *realistic stage,* from 18 years of age on, is a time for resolving the choice of a career more realistically on the basis of a candid self-appraisal and viable career options. It too is divided into substages. During the exploration period individuals gain greater knowledge of their strengths and weaknesses, as well as the appropriate career options. Then, in the crystallization period, they make a commitment to a particular field or career. During the specification period they may further refine their choices to an area of specialization. For example, a nurse may decide to work in intensive care units and seek the appropriate training for it.

Don Super (1980) has also proposed a developmental theory of career choice, but emphasizes the relationship between the individual's self-concept and his or her choice of a career. Thus, career choice involves exploring and affirming one's overall identity in the world of work. The basic idea is that individuals tend to choose those careers that allow them to affirm their personal identities, including their needs, interests, and abilities.

vocational identity: A way of thinking about one's self in relation to the kind of life work desired.

Although adolescence is a critical time in the choice of a career, the process of vocational identity lasts throughout the lifespan and includes five broad stages. The stages, together with the approximate ages, are as follows:

Growth (up to 14 years)
Children and early adolescents get ideas about the nature and meaning of work through their play, interests, and activities.

Exploration (15 to 25 years)
Adolescents and youth become more aware of the need to make career-related decisions, including the exploration of interests, information, and educational programs, and eventually the choice of a career.

Establishment (26 to 45 years)
Adults generally settle down into an appropriate career pattern, and acquire experience and further specialization in their chosen fields.

Maintenance (46 to 65 years)
Individuals acquire further expertise, seniority, and status in their chosen careers.

Decline (66 years and over)
People normally retire and explore new roles and new ways of understanding themselves in a way that emphasizes satisfaction outside their careers.

According to Super, problems in career choice and adjustment generally reflect the difficulties a person has in achieving a clear self-concept and positive identity.

Career Maturity

career maturity: How well individuals are progressing toward a career or vocational identity.

Don Super has proposed the concept of career maturity as a way of determining how well individuals are progressing toward their vocational identities. Career maturity varies somewhat from one stage to the next but can be measured with instruments such as the Career Maturity Inventory. In one study, the stage-related career maturity was measured in over 600 salespeople from 21 to 60 years of age. Individuals aged 21 to 30, who were in the exploration and early part of the extablishment stages, were characteristically exploring their career identities and shifted jobs more frequently than workers at older ages. Individuals aged 31 to 44, who were well into the establishment stage, had typically stopped exploring career choices, but were moving between jobs and companies as a way of advancing their present careers. Workers in the 45 to 60 age bracket, who were in the maintenance stage, were less willing to relocate and had little desire to be promoted. People this age were generally more concerned with making the transition to retirement (Slocum & Cron, 1985).

According to Super, the individual's self-concept is a major factor in achieving maturity. Thus, high school and college students who have a firm personal identity and high esteem will readily explore career options and choose challenging careers that demand hard work, risk, and growth. But students who are plagued by self-doubt, feelings of unworthiness, and a fear of failure tend to choose less challenging careers that are beneath their abilities or may have a lot of difficulty making a choice. As you might expect, there is some support for this view. For instance, in one study college students were administered instruments measuring their self-concepts, career maturity, and work achievement. The results showed that individuals with high self-esteem generally had more mature career attitudes and higher work achievement scores (Crook, Healy, & O'Shea, 1984). At the same time, most experts believe that career maturity depends on many other factors as well, such as intelligence, socioeconomic differences, and education, including career education.

Identifying a Compatible Career

Students generally need assistance in identifying a compatible career. In the first place, youths spend long years in school, with their part-time and summer employment generally unrelated to their career goals. Then too, with more than 20,000 careers to choose from, it's hard to know where to begin. As a result, many students do not know what information to seek about careers, how to go about it, or where to look. Most students would benefit from talking over their educational plans and career goals with an interested teacher, school or career counselor, or someone in their fields of interest. Yet, the average high school student spends less than 3 hours a year with a counselor at school (Super & Hall, 1978). Counselors at school and career guidance centers can refer students to the information they need in order to explore the various career options and the appropriate educational requirements. Counselors also have access to a wide assortment of career inventories that may help students identify the most compatible careers for them. Two familiar inventories are Holland's Self-directed Search (SDS) and the Strong-Campbell Interest Inventory (SCII). Both of these instruments make use of Holland's personality-occupational types, which match different personality characteristics with the requirements of various careers. Although no one's

---

Holland's Six Personality-Occupational Types

The following are descriptions of Holland's six personality-occupational types. These descriptions are, most emphatically, only generalizations. None will fit any one person exactly. In fact, most people's interests combine all six themes or types to some degree. Even if you rate high on a given theme, you will find that some of the statements used to characterize this theme do not apply to you.

The archetypal models of Holland's six types can be described as follows:

Realistic: Persons of this type are robust, rugged, practical, physically strong, and often athletic; have good motor coordination and skills but lack verbal and interpersonal skills, and are therefore somewhat uncomfortable in social settings; usually perceive themselves as mechanically inclined; are direct, stable, natural, and persistent; prefer concrete to abstract problems; see themselves as aggressive; have conventional political and economic goals; and rarely perform in the arts or sciences, but do like to build things with tools. Realistic types prefer such occupations as mechanic, engineer, electrician, fish and wildlife specialist, crane operator, and tool designer.

Investigative: This category includes those with a strong scientific orientation; they are usually task-oriented, introspective, and asocial; prefer to think through rather than act out problems; have a great need to understand the physical world; enjoy ambiguous tasks; prefer to work independently; have unconventional values and attitudes; usually perceive themselves as lacking in leadership or persuasive abilities, but are confident of their scholarly and intellectual abilities; describe themselves as analytical, curious, independent, and reserved; and especially dislike repetitive activities. Vocational preferences include astronomer, biologist, chemist, technical writer, zoologist, and psychologist.

Artistic: Persons of the artistic type prefer free unstructured situations with maximum opportunity for self-expression; resemble investigative types in being introspective and asocial but differ in having less ego strength, greater need for individual expression, and greater tendency to impulsive behavior; they are creative, especially in artistic and musical media; avoid problems that are highly structured or require gross physical skills; prefer dealing with problems through self-expression in artistic media; perform well on standard measures of creativity, and value aesthetic qualities; see themselves as expressive, original,

personality exactly fits these types, finding the combination of those types that most resembles one's personality can suggest representative compatible careers. (See the boxed item on Holland's personality-occupational types.)

Students often wonder whether such inventories can really help. A lot depends on how these inventories are used. If a person relies on them as a substitute for making a career choice, the answer is no. But if the results are used as a guide in reaching one's own decision, then the answer is decidedly more positive. Talking over the results of such instruments may furnish valuable leads for the most compatible careers for each person. These inventories will not tell people how happy or successful they will be. Such outcomes depend more on people's abilities, motivation, and available opportunities. But these inventories may provide a valuable aid in determining whether people will persist or drop out of a given career. Individuals who enter a career

---

intuitive, creative, nonconforming, introspective, and independent. Vocational preferences include artist, author, composer, writer, musician, stage director, and symphony conductor.

Social: Persons of this type are sociable, responsible, humanistic, and often religious; like to work in groups, and enjoy being central in the group; have good verbal and interpersonal skills; avoid intellectual problem-solving, physical exertion, and highly ordered activities; prefer to solve problems through feelings and interpersonal manipulation of others; enjoy activities that involve informing, training, developing, curing, or enlightening others; perceive themselves as understanding, responsible, idealistic, and helpful. Vocational preferences include social worker, missionary, high school teacher, marriage counselor, and speech therapist.

Enterprising: Persons of this type have verbal skills suited to selling, dominating, and leading; are strong leaders; have a strong drive to attain organizational goals or economic aims; tend to avoid work situations requiring long periods of intellectual effort; differ from conventional types in having a greater preference for ambiguous social tasks and an even greater concern for power,

status, and leadership; see themselves as aggressive, popular, self-confident, cheerful, and sociable; generally have a high energy level; and show an aversion to scientific activities. Vocational preferences include business executive, political campaign manager, real estate sales, stock and bond sales, television producer, and retail merchandising.

Conventional: Conventional people prefer well-ordered environments and like systematic verbal and numerical activities; are usually conforming and prefer subordinate roles; are effective at well-structured tasks, but avoid ambiguous situations and problems involving interpersonal relationships or physical skills; describe themselves as conscientious, efficient, obedient, calm, orderly, and practical; identify with power; and value material possessions and status. Vocational preferences include bank examiner, bookkeeper, clerical worker, financial analyst, quality control expert, statistician, and traffic manager.

Reprinted from David P. Campbell and Jo-Ida C. Hansen, *Manual for the Strong-Campbell Interest Inventory*. Form T325 of the *Strong Vocational Interest Blank*, 3rd ed., with the permission of the distributors, Consulting Psychologists Press Inc, for the publisher, Stanford University Press. c 1974, 1977, 1981, by the Board of Trustees of the Leland Stanford Junior University.

that is very similar to their career profile tend to remain in that career. By contrast, individuals who enter a career that is highly dissimilar to their career profile are more apt to drop out eventually. Considering all the time and money people invest in preparing for their careers, such information can be extremely helpful.

INFLUENCES ON CAREER CHOICE

Although young people are generally encouraged to choose their own career, they rarely make such an important decision entirely on their own. In discussing their plans with a school counselor, they may be encouraged or discouraged by what the counselor says. Young people are even more likely to share their educational and career plans with their parents, and parental support—or the lack of it—becomes an important factor in whether they pursue their career goals or change them (Sebald, 1986). Then too, young people tend to choose careers that are familiar to them, usually on the basis of the particular people they've known in those careers, rather than out of an objective consideration of unfamiliar careers. Sometimes they are well aware that their career preferences have been influenced by others, such as the girl who goes ahead with her plans of applying to medical school largely because her uncle, a physician, has promised to lend her part of the necessary money. But more often than not, young peoples' choices of a career have been influenced by others in more indirect, subtle ways. The degree to which young people's career choices have been influenced by their families, socioeconomic backgrounds, and sex roles has been the subject of extensive studies.

The Family

The family's influence on young people's career choices has been attributed to such factors as the parents as role models, that is, the parents' education, career, income, and degree of success. But numerous studies suggest that the impact of parental models is mediated through a variety of processes and relationships within the family, such as how satisfied parents are with their careers, how close young people are with a given parent, childrearing practices, and values in the home. As a result, the influence of the family on career choice is more complex than suggested by the findings on any one study, which is well to keep in mind throughout this section.

The father's career usually has an important bearing on the career choices of his sons, though to a lesser extent for his daughters. Although this pattern holds true across many fields, it is especially marked for sons of professionals, technical workers, and those who enter public service. Even when sons do not choose their father's career, they often choose fields with similar career characteristics, such as the extent of work autonomy involved, the complexity of

intrinsic career values: Valuing a career mostly for the work activity itself.

extrinsic career values:
Valuing a career more for external rewards, such as money or status, than for the work activity itself.

work activities, and whether the rewards are intrinsic or extrinsic. At the same time, the father's influence tends to be greater when he has a high-status career and a close relationship with his son. Furthermore, sons of fathers in the professions, such as law or medicine, are more likely to hold intrinsic career values, while sons of fathers in the business world are more apt to hold extrinsic values. Sons who have a close, positive relationship with their mothers are also more likely to choose a people-oriented career than sons who remain more distant from their mothers (Mortimer & Kumka, 1982).

Although girls generally identify more strongly with their mothers, their choice of a career is also influenced by their father, especially the father's attitude toward women's roles. Young women whose fathers have a negative attitude toward the homemaker role are more likely to plan on a college education and to work outside the home. Furthermore, the type of career they choose is influenced by their father's values. For instance, Tenzer (1977) studied women in male-dominated careers (for example, lawyers, doctors, and managers) and female-dominated careers (such as social workers and nurses). She found that, while both groups of women identified more strongly with their mothers than their fathers, women in the male-dominated careers perceived their fathers as stressing extrinsic values and behaviors more than expressive values. And the opposite was true for women in the female-dominated careers. As you might imagine, young women's career choices are also greatly influenced by their mother's example. Young women from families in which the mother works outside the home generally have a high regard for women as competent people in the workplace, believe in more flexible sex roles, are more likely to plan on working outside the home, and aspire to higher-status careers than women whose mothers remain in the home (Fox & Hesse-Biber, 1984). At the same time, the impact of the mother as a role model is mediated by the same type of influences discussed in relation to the father, such as the status of the mother's career, whether the mother works full- or part-time, how successful she is, and how satisfied she is in her role. When the mother is satisfied with her role, whether she is employed outside the home or not, there is a greater chance the daughter will follow the mother's role. But the opposite is true if the mother is dissatisfied with her role (Altman & Grossman, 1977).

Young people are especially influenced by the way they are treated in the home, by the childrearing practices and values stressed in the home. Parents with higher education levels, higher-status careers, and higher incomes tend to treat their adolescents more as equals, to explain parental rules, and to be concerned about why young people behave the way they do. And the parents generally expect their young people to be concerned about the same matters. By contrast, parents with lower levels of education, lower-status careers, and lower incomes tend to do the opposite. That is, they are inclined to treat their adolescents in a more authoritarian way, do less explaining of parental rules, and stress conformity of external behavior. Also, parents from lower socioeconomic groups often work as part of a group or team, and are more likely to stress being able to work as part of a crew, while parents from higher socioeconomic groups are more likely to work in one of the professions and stress the value of individual initiative (Schulenberg, Vondracek, & Crouter, 1984).

Socioeconomic Influences

The socioeconomic status of the adolescent's family continues to be a powerful predictor of career choice, primarily because it includes so many other important factors such as parental education, income, career status, childrearing practices, and values. As a result, a vicious cycle often occurs, with youths tending to remain in the same broad socioeconomic group in which they are reared. In the first place, socioeconomic status limits the opportunities available to those with different socioeconomic backgrounds. Thus, youths from the lower socioeconomic groups tend to have fewer opportunities for higher-status careers because of a combination of negative influences in their homes, neighborhoods, and schools, and the limited job market, with the reverse being true for youth from the middle and higher socioeconomic groups. Even when they are exposed to opportunities for higher-status careers, youths from the lower socioeconomic groups are not as likely to take advantage of them because of the way they have been socialized. That is, having been given less encouragement for educational or social mobility, they either don't aspire to higher-status careers or think such careers are not possible for them. Consequently, there is a tendency for socioeconomic status to perpetuate itself in the next generation, with career mobility occurring more frequently among those in the middle ranges than among those in the extremely high or low socioeconomic statuses (Schulenberg, Voncracek, & Crouter, 1984).

There is a strong association between young people's career aspirations and achievements and their socioeconomic background. In one study, Bachman, O'Malley, and Johnston (1980) have been following a national sample of more than 2000 males since they were tenth graders in the mid-1960s. They have found that career aspiration and career achievement generally reflect the socioeconomic status of the individual's family and remain fairly stable across time. As a result, students from the middle and upper socioeconomic brackets aspire to higher-status careers more often than would be expected by chance, whereas students from the lower socioeconomic brackets aspire to such careers less often than expected by chance. Although some working-class and middle-class youth aspire to similar higher-status careers, middle-class youth are more likely to feel such goals are attainable and to choose them. As students progress through high school, they generally become more realistic about their career aspirations and modify them according to their educational attainments. Thus, students who achieve their educational aspirations tend to have their career aspirations reinforced, while those who fail to attain their educational goals generally revise their career expectations downward. In many, though not all, instances, youths feel most comfortable about aspiring to and achieving levels of career status that are roughly comparable to those of their parents (Kohn & Schooler, 1983).

What about the traditional notion that young people expect to do better than their parents? That is, given greater advantages than their parents had in the home, school, and workplace, youths naturally expect to get a better education and choose a higher status career than their parents. This is part of the American dream: Through education and hard work, each generation can better itself. There is still a great deal of truth in this view. But there are also other forces at work in society and the workplace that are modifying career mobility. In the first place, it is especially difficult for youth from affluent homes to surpass their parents. Actually, faced with the highly competitive job market of the past few years, many youths will have difficulty matching the lifestyle of their parents, much less improving upon it. Then too, the increased importance of self-fulfillment values in the past decade or so has brought about changes in career mobility. Thus, as what is expected from a job shifts from money and status to work that is intrinsically rewarding, there may be less emphasis on upward mobility. That is, many young people may feel successful when they find a career that is interesting and satisfying, even though it may be at the same social and economic level of their parents, or in some cases, lower (Yankelovich, 1981).

Sex Differences

Traditionally, the career choices of males and females have been highly sex-typed, with sex differences often outweighing educational, cultural, and racial differences in career aspirations and achievements. Boys have chosen from a much wider range of careers than girls. Furthermore, boys have been more likely to choose challenging, high-status careers that offer opportunities for independence and leadership as well as high pay. Girls have been more inclined to shun competitive, high-status careers in favor of lower-status, service-oriented careers with low pay (Grotevant & Thorbecke, 1982).

The traditional pattern of sexual inequality in career choice stems from a variety of factors in American society, especially the stereotyped sex roles and restricted opportunities for women in the workplace. From infancy, boys and girls are treated differently by significant others—parents, teachers, friends, and supervisors—so that young men and women come to think of themselves and their aspirations differently. A man is regarded as the primary breadwinner; a woman is viewed as the homemaker. Consequently, girls who aspire to a career generally experience a conflict between career and family roles, whereas men do not. Even when women marry and work outside the home, as an increasing number of young women are doing, they continue to assume the major responsibilities for nurturing the young and for household chores, as we discussed in an earlier chapter. Such role expectations on the part of men and women alike

Young women in high status careers usually have received a lot of support in their homes and schools. (Laima Druskis)

make it difficult to combine career and family roles, so that while many women work outside the home, their primary career is that of homemaker.

Then too, career-oriented women have had fewer opportunities in the workplace. Even with the extensive changes in the past decade or so, the majority of women are crowded into just 20 of the Labor Department's 427 job categories. Women are overrepresented in such jobs as secretary, waitress, house servant, nurse, and elementary school teacher. But they are underrepresented in such jobs as auto mechanic, police officer, miner, and construction worker. Only a small proportion of women belong to unions or apprenticeship programs in the trades. Furthermore, the traditionally feminine careers are unevenly distributed throughout the socioeconomic hierarchy of careers, with most of them in the low-paying service sector. Consequently, women who choose traditionally feminine careers not only limit their career mobility but only earn about two-thirds as much as men with the same education and job (Lewin, 1984).

Fortunately, the outlook for women is improving because of more flexible sex roles, social practices, and greater sexual equality in the workplace. As a result, surveys of high school students show that girls generally have higher educational and career aspirations than they did a decade ago (Dunne, Elliot, & Carlsen, 1981). Also, college enrollment has increased faster for women than for men in the past decade, with the number of women attending graduate and professional school increasing by 75 percent (Dearman & Plisko, 1981). More women are also entering the workplace than ever before. More than 9 out of 10 women work outside the home sometime during their lives, with about half of them employed in the workplace at any one time. More young women are working before marriage, delaying marriage, combining work and marriage, having fewer children, and returning to work sooner after marriage. Women are also choosing from a broader range of careers, including many nontraditional careers for women. Although the *proportion* of women in the nontraditional careers remains small, the *rate* of increase is dramatic. For instance, since 1960, the percentage of college professors who are women has increased by two-thirds, the percentage of physicians who are women has more than doubled, the percentage of women workers in the life and physical sciences has increased two and one-half times, women engineers have increased four times, and the percentage of lawyers and judges who are women has risen nearly five times (U.S. Department of Labor, 1982).

Table 11–2   DISTRIBUTION OF THE SEXES IN SELECTED CAREERS

MALE-DOMINATED CAREERS	Percent men	
Auto mechanics	99.3	
Carpenters	98.8	
Truck drivers	97.9	
Dentists	93.5	
Police	89.9	
FEMALE-DOMINATED CAREERS	Percent women	
Secretaries	98.4	
House servants	96.2	
Registered nurses	95.1	
Bookkeepers	91.5	
Telephone operators	88.8	
EVENLY DISTRIBUTED CAREERS	Percent men	Percent women
Secondary school teachers	46.0	54.0
Real estate sales	48.4	51.6
Psychologists	49.6	50.4
Personnel managers	55.5	44.5
Accountants	55.9	44.1

Source: U.S. Bureau of the Census, *Statistical Abstract of the United States, 1987,* 107th ed. (Washington, D.C.: U.S. Government Printing Office, 1986), pp. 385-386

Young women who choose high-status careers generally have high self-esteem and have received a lot of encouragement and support in their homes, schools, and communities. Girls whose mothers are college educated and work outside the home are especially apt to aspire to a career (Fox & Hesse-Biber, 1984). The father's encouragement of his daughter's educational and career aspirations is also an important factor, as we mentioned earlier, and is especially crucial when the mother works in a traditionally feminine career (Weishaar, Green & Craighead, 1981). On the other hand, girls whose fathers are in low-status careers and whose mothers don't work outside the home are much more likely to plan on being a full-time homemaker (Falkowski & Falk, 1983). Women who choose nontraditional careers generally come from the higher socioeconomic groups, partly because of the combination of favorable influences mentioned above. But socioeconomic factors appear to have less impact on women than on men, partly because of the extensive changes in sex roles and social practices in recent years (Schulenberg, Vondracek, & Crouter, 1984). The increasing proportion of women attending college is also a factor, with a college education generally increasing career aspirations among women. (See Table 11–2.)

PREPARING FOR A CAREER

Occupational Outlook Handbook: A resource for career choice, revised and published every two years by the Department of Labor Statistics.

The mention of college brings us to the important matter of preparing for a career. Actually, there are a variety of ways young people may prepare for a career. Some careers are normally entered through an apprenticeship program, vocational school, or an on-the-job training program. Other careers require a 2-year or a 4-year college degree. Most professions like medicine and law also require an advanced degree and professional training. A very important part of the decision-making process is learning what type and how much education is needed, the entrance requirements, the expected expenses, and the job prospects. Much of this information may be acquired during talks with the school counselor or from resources such as the *Occupational Outlook Handbook*. But in many instances, students remain ill-informed and have educational plans that are not consistent with their career goals. In one study, a comparison of educational plans and career goals among high school seniors found that one-fifth of them were planning on too little education and one-third were planning on too much (Grotevant & Durrett, 1980).

Career Education

Many people feel that such misunderstanding could be avoided or minimized if students had more exposure to careers earlier in their education, as in career education. The general purpose of career education is to bridge the gap

Vocational high school students examining an engine. (Cities Service Company by Nelson Morris)

career education: Helping people to develop a sense of direction and make responsible career choices, including academic and supervised work-study programs.

between the individual's learning experience and the workplace. Career education is especially needed among today's students because of the long years spent in school isolated from the workplace, except for the part-time jobs unrelated to career goals so characteristic of teenage employment. A Gallup poll (1985) showed that over half (57 percent) of public school parents believe some type of career education should be required for all high school students, college-bound youth as well as those not planning to attend college. Some of the components of a comprehensive career education include providing information on: the career implications of every learning experience, attitudes and values of a work-oriented society, career awareness, career exploration, decision-making skills, career choice, entry-level work skills, specific vocational training, and supervised work experience.

A master plan for school-based career education would involve students at all age levels. Throughout elementary school, the emphasis would be on career awareness. Children would be exposed to the different kinds of work done by adults in the community. Teachers would point out the career implications of various learning experiences, including practical problems in arithmetic, and would simulate appropriate work experience in the classroom. Beginning in middle school or junior high school and continuing into high school, the emphasis would shift to career exploration. Students would explore career clusters, or groups of related careers. After becoming familiar with all of the clusters, they would explore in more depth one or two clusters of the greatest interest to them, such as health careers. Students would then become aware of the many different jobs in these clusters and how each job relates to the larger society. As they move into high school, students would be assisted in formulating tentative career goals and become aware of the necessary preparation to achieve them. Students who are not planning to go to college would also be encouraged to develop entry-level skills useful in a variety of fields, as well as basic vocational training in a field of their choice.

A valuable part of career education is providing appropriate experience in the workplace. In contrast to the usual teenage employment, such experience should include appropriate instruction and supervision and should help integrate what is learned in school and on the job. One approach is the familiar work-study programs, in which students spend half the day in class and the other half at work on the job. In addition, some private schools have initiated internships which provide students firsthand experience in a field related to their career goals. Examples would be helping in a lawyer's office, a television studio, a dentist's office, or at a newspaper. One school requires all juniors and seniors to spend their Fridays working on a semester-long internship instead of going to class. Students are also encouraged to start their own small businesses, such as a painting company (Woodall, 1984). Another option is providing students with human-services education, with supervised work experience for academic credit but without pay. This may be especially valuable for people considering professional fields such as teaching, nursing, and medicine as well as social service agencies in which entry-level, people-oriented jobs are not available on a paid basis. Still another option is volunteer work, including social service agencies, hospitals, community centers, tutoring in schools, and libraries. One way for organizations to make better use of volunteers is to arrange voluntary work around the free time in students' schedules, such as near lunch or at the end of classes. In one school, students organized a catalogue of service opportunities and made it available in a volunteer bureau as a kind of clearing house for student volunteers.

Despite the commendable goals of career education advanced since the 1970s, few comprehensive career education programs have been implemented. And now that the emphasis in educational priorities has swung toward improving basic skills and academic excellence, the funding for career education is likely to become even more competitive. Furthermore, the results of career education vary widely with the particular programs and students, and generally have been more successful in the area of increasing awareness of careers and how to go about selecting a career than substantial changes in unemployment. One study focused on the relationship between career maturity and career education practices among 2280 students in the ninth and eleventh grades in 38 secondary schools in metropolitan areas. The results showed that students in schools with career education programs achieved higher gains in their career maturity during high school than those in schools with no career education programs. There was also a strong association between the type of career education program and support given, such that the greatest gains in career maturity occurred among students in schools offering innovative career education programs strongly supported by the school and community (Trebilco, 1984).

Going to College

A college education continues to be an important factor in the choice of a career. According to a Gallup poll (1985), 9 out of 10 adults (91 percent) feel that a college education is very important or fairly important. When parents of public school students were asked if they would like their oldest child to attend college, 88 percent said yes. Yet, when asked if they thought their child would go to college, only 66 percent said yes. The biggest single reason cited was the high cost of a college education, a realistic consideration in view of the increasing costs of college and reduced federal funds for student grants and loans.

The proportion of high school graduates going on to college has also increased in the past two decades. While only one-third of the high school graduates went on to college in 1960, about half of them do so today. When high school seniors throughout the country were asked to indicate their plans for the next year, about half of them planned to attend college—38 percent at a 4-year college and an additional 14 percent at a 2-year or community college. Another one-third of the students planned to work full-time, including apprenticeships and on-the-job training programs as well as serving in the military. The rest of the students had a variety of other plans such as working part-time, being a homemaker, and undecided (Grant & Snyder, 1984) Interestingly, more college-bound youth are delaying their entry into college, with about half of them waiting several years before enrolling in college. Students are also more likely to interrupt their college education than in the past. About half of the students in 4-year colleges and more than two-thirds of those in 2-year colleges drop out, mostly in the first year. However, about half of those who drop out eventually return and complete their degree (Dearman & Plisko, 1981). Students drop out of school for a variety of reasons, but they often do so because they feel a need to work to save money for college and/or to clarify their educational and career goals.

One of the most significant changes in recent years has been the marked increase of career-minded students. Whereas barely half of the college students were enrolled in a career-oriented program in 1960, more than two-thirds of them were taking such a program by 1980. Much of the reason has to do with the tight job market, inflation, and increased cost of a college education. As a result, more students have been taking degrees in job-related fields such as business and management, computer science, health

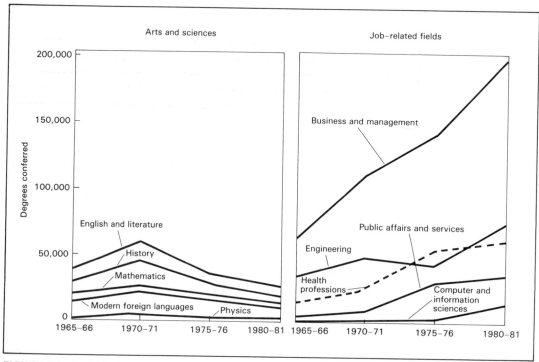

FIGURE 11–1 BACHELOR'S DEGREES CONFERRED IN SELECTED FIELDS IN THE UNITED STATES, 1965–1966 TO 1980–1981.

Source: W. Vance Grant and Thomas D. Snyder, *Digest of Education Statistics, 1983–84* (Washington, DC: National Center for Education Statistics, Government Printing Office, 1984), p. 119.

professions, and engineering, while fewer of them take degrees in the arts, humanities, and social sciences (Grant & Snyder, 1984). (See Figure 11–1.) Such changes also reflect the perceived value of a college education. When the respondents in the annual Gallup poll (1986) were asked about the advantages of a college education, 34 percent of them cited better job opportunities, 8 percent to get a better-paying job, and 9 percent to achieve financial security. Relatively few Americans mentioned preparation for life (23 percent), to acquire knowledge (10 percent), to think, learn, and understand (3 percent), to learn how to get along with others (4 percent), or to contribute to society (3 percent). Yet, in the long run, the latter advantages of a college education might be just as valuable, if not more so, than the more obviously career-related advantages of a college education. Given the rapid changes in the job market and the fact that roughly one in nine workers changes his or her career every year, one of the most valuable but overlooked rewards of a college education is the ability to continue learning on one's own and being able to adapt to continuing change (U.S. Department of Labor, 1986).

Career Outlook

career outlook: Projected trends in regard to careers and job openings in the workplace.

Individuals planning on a career should be aware of the rising educational level of the work force and the importance of a college education. Between 1970 and 1984, the proportion of the work force with some college increased from 26 percent to 41 percent. (See Figure 11–2.) The increasing level of education in the work force reflects both the retirement of older workers, many of whom had little formal education, and the entry into the workplace of young people who generally have a higher level of formal education. Among workers aged 25 to 34, nearly half have completed at least 1 year of college, and one-fourth have 4 or more years of college. The advantage of college-educated workers can be seen in their lower unemployment rates and higher lifetime earnings. In 1984, the unemployment rate among 20 to 24 year olds with 4 years of high school was 13 percent. For those with 1 to 3 years of college, the unemployment rate was 7.8 percent, but for college graduates the unemployment rate was only 4.9 percent (U.S. Department of Labor, 1986). Although earnings vary greatly between people in different careers as well as among those within the same career, college graduates generally enjoy higher earnings than those with less education. In the general population, the median income of college graduates is about 25 percent higher than that of high school graduates (U.S. Bureau of the Census, 1985).

About one out of every four jobs in the workplace requires a college degree. But the rate of increase in the projected job openings between now

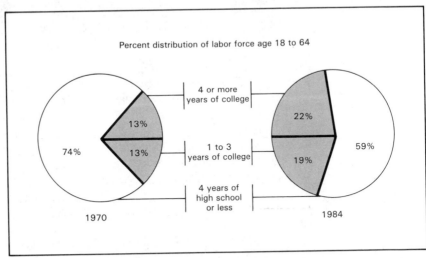

FIGURE 11–2 THE PROPORTION OF WORKERS WITH A COLLEGE BACKGROUND HAS INCREASED SUBSTANTIALLY SINCE THE 1970s.

Source: From *Occupational Outlook Handbook, 1986–87* (Washington, DC: U.S. Government Printing Office, 1986), p. 16.

Many of the fast growing careers are in the computer and health fields. (Courtesy IBM)

and 1995 varies somewhat from one field to another. For instance, fewer openings are expected for secondary school teachers and college professors, because of the decline in the youth population. On the other hand, an increase in job openings is expected for kindergarten and elementary school teachers, mostly because of the rise in the birth rate in recent years. Several of the major career groups having the largest proportion of college-educated workers are expected to increase faster than average. Executives, administrators, and managers are expected to increase because firms increasingly depend on trained management specialists. Also, many of the careers in the professional specialties are expected to increase, including computer-related careers, engineering, and health specialties, along with the related technicians and support workers. Almost half of the 20 fastest-growing careers are in the computer or health field, which will continue to be among those with the strongest future growth (Silvestri & Lukasiewicz, 1985).

Career prospects depend not only on the job openings but also on the number of people looking for these types of jobs. Generally, the labor force is expected to grow through the mid-1990s but at a slower rate than in the recent past. Much of the growth in the labor force will be the continued, though slower, rise in the number and proportion of women seeking jobs, with women accounting for nearly two-thirds of the labor force growth between now and 1995. While the number of young people in the 16 to 24 age bracket is expected to decline, the number of people in the 25 to 54 age group is expected to increase considerably because of the aging of the baby boomers. Both of these factors will keep the job market competitive into the 1990s, especially among those in the 25 to 54 age group. Furthermore, since the supply of college graduates has been increasing faster than the fields which employ college-educated workers, the job market is expected to remain competitive for college graduates. In recent years, one in five college graduates has had to take a job that does not require a college education. Yet, not all careers that require a college degree will be crowded. Good opportunities are expected in a number of fields, including computer-related careers, engineering, and the health-related careers. Despite the competitive job market for college graduates, a degree is still needed for most high-status, high-paying jobs. Individuals interested in these careers should not be discouraged from pursuing a career that they believe matches their interests and abilities. But they should be aware of the job market conditions (U.S. Department of Labor, 1986).

SUMMARY

Adolescents in the Workplace

1. The transition to adult work roles in the United States occurs gradually, with young people doing household chores and holding part-time jobs before assuming responsibility for a full-time job in the workplace.

2. One of the remarkable changes in the past decade or so has been the greater proportion of teenagers who work during the school year, with the average student-worker spending about 15 to 20 hours a week on the job.

3. Although students gain a lot of practical knowledge from their part-time jobs, their work tends to be dull and stressful and unrelated to their career goals, with school grades being adversely affected after 15 to 20 hours on the job.

The Process of Career Choice

4. The choice of a career is generally an extended developmental process that involves a great deal of exploration and tentative choices before settling on a firm career goal.

5. Individuals with high esteem who have appropriate opportunities for career exploration and preparation tend to move toward career maturity faster than those who are plagued by self-doubt and have little opportunity for career exploration.

6. Students generally need some type of assistance in identifying a compatible career, such as talking with a school counselor or taking a career inventory.

Influences on Career Choice

7. Young people's choice of a career is influenced by a variety of factors in their families, such as the parents' education, careers, childrearing practices, and values as well as the extent to which adolescents identify with their parents.

8. The socioeconomic status of the adolescent's family also continues to be a significant factor in career choice, with youths from affluent, educated families more likely to choose a high-status career than those from less affluent families.

9. The restricted career choices and opportunities that have limited women's role in the workplace in the past have resulted in the majority of women being crowded into just 20 of the Labor Department's 427 job categories.

10. At the same time, because of the extensive changes in sex roles and rules governing the workplace, young women now have higher educational and career aspirations and are choosing from a wider variety of careers than in the past.

Preparing for a Career

11. Students in schools with career education programs, especially the innovative and well-supported ones, tend to experience greater gains in career maturity in school than those who have little or no career education.

12. The proportion of high school graduates going on to college has increased in the past two decades, with about half of them enrolling in college, most in career-oriented degree programs.

13. Although the job market is expected to remain competitive for college graduates between now and 1995, a college degree is still needed for most high-status, high-paying careers.

REVIEW QUESTIONS

1. What are the most common part-time jobs for 16 and 17 year olds?

2. What are the pros and cons of spending 20 hours a week at a job while attending high school or college?

3. How valuable is the typical teenager's work experience?

4. Describe Ginzberg's three stages of career choice.

5. Describe Don Super's five stages of vocational identity.

6. How can a student improve his or her chances of choosing a compatible career?

7. What are some of the ways young people's career choices may be influenced by their families?

8. To what extent is the socioeconomic status of one's family a factor in career choice?

9. What are some of the ways sex differences influence career choice?

10. To what extent is a college education an advantage in the workplace?

12 Moral Development and Religion

- **THE DEVELOPMENT OF MORAL REASONING**
  Kohlberg's Stages of Moral Reasoning
  The Attainment of Moral Reasoning
  A Critique of Kohlberg's Theory

- **MORAL BEHAVIOR**
  Resistance to Wrongdoing
  Helping Behavior
  Personal and Moral Values

- **RELIGION**
  The Developmental Sequence
  Cults
  Religion and Moral Behavior

- **SUMMARY**

- **REVIEW QUESTIONS**

Lisa, a 16 year old who recently obtained her driver's license, has borrowed the family car for an errand at a local shopping mall. On the way over to the mall, she picks up her boyfriend, Brad. A short time later, while parking her car at the mall, Lisa accidentally scrapes the fender of a Mercedes sedan in the adjacent parking space. Inspecting the damage, Lisa discovers a small indentation in the fender of the other car along with a smear of blue paint from her parents' car. Lisa is very upset and says, "Maybe I should leave a note with my name and phone number on the windshield of the Mercedes." "I'd forget it," Brad chimes in. "The guy's insurance will cover it." Shaking her head, Lisa replies, "I'd feel better if I left a note." "Listen," says Brad, "anyone who drives an expensive car like that can afford to fix a small scratch in the fender. Come on, let's move the car to another parking place before someone finds out about this."

THE DEVELOPMENT OF MORAL REASONING

moral reasoning:
Thought or deliberation in regard to what is right and wrong.

moral development: The process by which people come to adopt the standards of right and wrong transmitted by their families and culture, including cognitive, emotional, and behavioral aspects of morality.

moral dilemmas:
Hypothetical situations in which one must choose between two equally unpleasant moral alternatives.

What do you think Lisa should do? Should she contact the driver of the other car? Or should she overlook the incident and move her car to another parking space? When individuals of various ages are faced with such moral dilemmas, their responses vary considerably, depending largely on their age and the maturity of their moral reasoning. Younger children tend to be concerned about the reactions of authorities, such as what would happen if Lisa were caught and punished. Older children and adolescents are capable of more complex moral reasoning and show more concern for the rules of society. Such differences reflect the characteristic types of reasoning associated with the various stages of moral development.

Lawrence Kohlberg's theory of moral development is one of the best-known approaches to how adolescents develop their sense of right and wrong. Building on Piaget's model of cognitive and moral development, Kohlberg (1984) has formulated a comprehensive, lifespan theory of moral reasoning. Essentially, Kohlberg holds that children and adolescents progress through a sequence of six distinct stages of moral reasoning that parallel cognitive development, with higher stages of moral reasoning presupposing the attainment of formal or abstract thinking. Each stage develops out of and includes the attainments of the previous stages, so that the individual's moral reasoning becomes increasingly complex with age and experience. Similar to Piaget, Kohlberg views the sequence of stages of moral reasoning as invariant, though the rate of development varies considerably among individuals, with those thwarted in their moral development remaining at a given stage of moral development, perhaps permanently. Kohlberg holds that the attainment of each successive stage of cognitive thought is a necessary but not a sufficient condition for reaching the corresponding level of moral reasoning. That is, many adolescents who have acquired the capacity for formal, abstract thought have not yet attained the corresponding level of moral reasoning. A primary reason is that the development of moral reasoning also depends on other factors, such as the individual's learning experiences and social interaction. As a result, moral reasoning tends to lag somewhat behind cognitive development.

Kohlberg's knowledge of moral development has been acquired through 20 years of extensive research using a method of personal interviews. Individuals of various ages are presented with various moral dilemmas involving moral concepts, values, and issues. Then the subjects are asked what the actor in the situation ought to do and—more important—the reasons or justifications for doing so. For example, in one such dilemma, a woman is near death from a very bad disease. There is one drug doctors think might save her. Yet, it is very expensive. After Heinz, the woman's husband, is unable to borrow the necessary money, he pleads with the druggist to let him pay for the drug in installments. But the druggist refuses. Now Heinz must decide whether to steal

Table 12–1 KOHLBERG'S LEVELS AND STAGES OF MORAL REASONING

Stage Descriptions	Examples of Moral Reasoning in Support of Heinz's Stealing	Examples of Moral Reasoning Against Heinz's Stealing
Preconventional Morality Morality is based on what one has to gain or lose personally		
1. Avoidance of punishment by authorities	"If you let your wife die, you'll get in trouble."	"You shouldn't steal the drug because you'll be sent to jail."
2. Acting in one's own self-interest	"If you get caught you could give the drug back and get a lighter sentence."	"You may not get much of a jail term, but your wife will probably die before you get out."
Conventional Morality Morality is based on upholding social conventions		
3. Gaining approval/avoiding disapproval	"Your family will consider you an inhuman husband if you don't."	"It isn't just the druggist but everyone else who will think you're a criminal."
4. Doing one's duty to maintain social and legal order	"If you have any honor, you won't let your wife die because you're afraid to act in a way that will save her."	"You'll always feel guilty for breaking the law."
Postconventional Morality Morality is based on ethical principles		
5. Affirming socially agreed upon rights	"If you let your wife die, it would be out of fear rather than thinking it through."	"You'd lose respect for yourself if you get carried away by your emotions and forget the long-range point of view."
6. Adhering to universal ethical principles	"If you don't steal the drug you have obeyed the external law, but you wouldn't have lived up to your own conscience."	"If you stole the drug, you would condemn yourself because you wouldn't have lived up to your own standards of honesty."

Adaptation of Table 1.6, "Motives for Engaging in Moral Action," pp. 52–53 from *The Psychology of Moral Development*, Volume II by Lawrence Kohlberg, Copyright © 1984 by Lawrence Kohlberg. Reprinted by permission of Harper & Row, Publishers, Inc.

the drug he cannot afford in order to save his dying wife. Subjects are asked what they would do in such a situation and to explain the reasons for their decisions. On the basis of the responses to such moral dilemmas, Kohlberg and his associates have concluded that the moral reasoning of children and adolescents tends to evolve through the same sequence of stages. (See Table 12–1.)

Kohlberg's Stages of Moral Reasoning

Kohlberg (1984) has identified six successive stages of moral reasoning, divided into three major levels: preconventional, conventional, and postconventional.

preconventional level: The lowest level of moral development, in which one's moral judgment is based on external authority, including rewards and punishment.

At the preconventional or premoral level, which parallels Piaget's pre-operational stage of thinking, the child remains highly egocentric and unable to adopt the views of others. The sense of right and wrong is understood in relation to external authority. In stage 1, designated variously as heterono-mous morality or the punishment and obedience orientation, the child obeys parental rules to avoid punishment. For example, Jeff avoids running into the street because he fears being punished by his parents. By stage 2, a type of instrumental hedonism comes into play, such that the rules are followed to meet the child's own interest and needs with the assumption that others will do the same. The child begins to take into consideration the views of others, but mostly in order to get what he or she wants. At this stage, Jeff tries not to run into the street mostly because of the reward promised by his mother.

conventional level: The second major level of moral development, in which one's moral judgment is based on internalized, socially acceptable moral standards.

At the conventional level, which roughly parallels Piaget's stage of concrete operational thinking and the early stages of formal thought, children and adolescents are better able to grasp the views of others, so that moral reasoning is based more on internalized, socially acceptable authority. At stage 3, characterized by mutual interpersonal expectations and relationships, children and adolescents regard right and wrong in relation to what others expect of them in terms of their roles—that is, as daughter or son, sister or brother, and friend. Motives and intentions also become more important at this stage. For example, Kathy doesn't tell her parents that her brother is smoking cigarettes because Kathy doesn't want to betray her brother's trust and friendship. Naturally, Kathy expects her brother would do the same. By stage 4, sometimes known as the social system or "law and order" stage, the individual's moral reasoning takes into account the larger social system such as the rules of the group or laws of society. At this stage, the sense of right and wrong is based on a more generalized approach than that assumed at the level of interpersonal relation-

postconventional level: The highest major level of moral development, in which one's moral judgment is based on universal, abstract moral principles.

ships and often involves consideration of "What if everyone does this?" An example would be Kathy's obeying the stop signs at street crossings when she drives because she feels this is necessary for everyone's safety.

The postconventional level of moral reasoning presupposes the attainment of formal operational thinking, such that moral judgment is based on abstract moral principles and a more generalized role taking. In stage 5, moral reasoning is based on the social contract or system of principles which takes into account the individual rights agreed upon by society as a whole, such as the Constitution of the United States. Thus, a person's conviction for stealing a car (violation of laws in stage 4) might be thrown out by a higher court because the defendant's rights had not been duly respected during the process of arrest and trial. By stage 6, the highest level of moral development, individuals follow self-chosen moral precepts based on universal ethical principles. At this level, particular laws or social agreements are held to be valid or not depending on whether they are based on certain universal principles, such as the equality of human rights and respect for the dignity of human beings as individual persons. For example, an individual might feel that people should not be denied their basic rights because of their skin color despite laws to the contrary in their particular society.

universal ethical princi-ples: The understanding of right and wrong based on general moral principles such as the respect for individual rights.

The Attainment of Moral Reasoning

Kohlberg (1984) and his colleagues have found a strong relationship between peoples' overall developmental stage, as reflected in their age, and their stage of moral reasoning. Most preschool-aged children and those in the early grades, who are in Piaget's stage of preoperational thinking and the early stages of concrete operational thought, characteristically exhibit preconventional moral reasoning. As older children and early adolescents reach the stage of concrete operational thinking and the beginning stages of formal thought, an increasing proportion of them can use conventional moral reasoning. Data from a 20-year longitudinal study of moral development by Kohlberg and his colleagues (Colby et al., 1980) chart the growth of moral reasoning with age, from late childhood on. During mid- to late adolescence there is a dramatic decline in the proportion of individuals using preconventional moral thought, that is, stages 1 and 2. At the same time, there is a marked increased in the proportion of adolescents in the early stages of conventional moral thinking, that is, stage 3 or the interpersonal orientation. Also, there is a slow but steady rise in stage 4 thinking, or the more advanced conventional moral reasoning reflecting the more generalized viewpoint and laws of society, so that by the time individuals have reached their thirties more than 6 out of 10 of them exhibit conventional moral understanding. Stage 5 thinking, or the early stages of postconventional moral thinking, which does not appear generally until after 20 years of age, grows much more slowly, and never characterizes more than 10 to 15 percent of the early adults. (See Figure 12–1.) Apparently, since formal thinking does not always emerge in adolescence, if at all, neither do the higher stages of moral reasoning. Individuals whose characteristic moral reasoning is at the postconventional level tend to be in the distinct minority.

The development of moral reasoning depends largely on the interaction

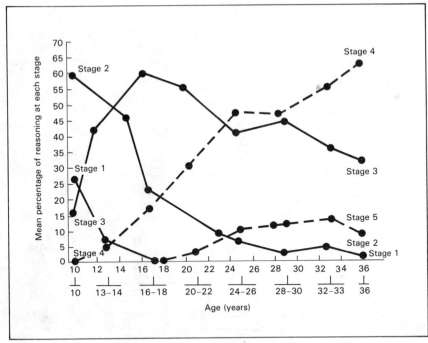

FIGURE 12–1 THE PERCENTAGE OF PEOPLE IN EACH AGE GROUP EXHIBITING THE VARIOUS STAGES OF MORAL REASONING.

Source: A. Colby, L. Kohlberg, J. Gibbs, and M. Lieberman, "A Longitudinal Study of Moral Judgment," *Monographs of Society for Research in Child Development.* © 1980 The Society for Research in Child Development, Inc.

cognitive-disequilibrium model: The concept that people progress in their moral understanding through confronting new moral issues, reexamining their former beliefs, and progressively reorganizing their moral reasoning.

between the individuals and their social environment, and involves both cognitive and social processes as well as maturation. According to Kohlberg's cognitive-disequilibrium model, individuals progress in their moral reasoning through the experience of cognitive disequilibrium. That is, when they confront new information or situations they do not fully understand, the resulting conflict or disequilibrium motivates them to reorganize their moral reasoning. Thus, individuals who continue to learn and grow in their moral development are progressively reorganizing their moral reasoning to deal with new problems and issues in their lives. Kohlberg estimates that at any time, about half of a person's moral thinking will be at a given stage, with the remainder reflecting the adjoining stages. Normally, individuals take the moral views of one stage above their own characteristic stage as their moral ideal. At the same time, there is increased awareness that the growth of moral reasoning involves more than cognitive development and is also dependent on the individual's social interactions, especially social role taking, that is, the ability to understand moral issues from another person's viewpoint. Consequently, efforts to facilitate the growth of moral reasoning must include more than the

Discussing moral issues and conflicts helps to deepen adolescents' moral understanding. (Ken Karp)

cognitive dimensions of development. For instance, in one experimental study, high school seniors 16 to 18 years of age enrolled in three psychology classes were randomly assigned either to one of two treatment groups or to a control group. Those in the treatment groups participated in a 3-week unit on character development that included classroom experiences involving the development of listening and communication skills, empathy, social role taking, and assertive behavioral skills as well as the discussion of moral dilemmas. The results showed that the students who were exposed to this enriched learning environment exhibited a significant improvement in their levels of moral reasoning compared to those in the control group (Kessler, Ibrahim, & Kahn, 1986).

A Critique of Kohlberg's Theory

Despite the helpfulness of Kohlberg's approach and the extensive research it has generated, his theory has been criticized on a number of points. A recurring criticism is that relatively few people ever attain the highest stages of moral thought even in American society. The fact that even fewer people in the less-developed societies, especially those not using a democratic form of government, reach the higher stages of moral thought has been taken to mean that Kohlberg's theory is ethnically and culturally biased on the one hand, and also that his stages of moral development are not universal and invariant. At the same time, the lack of empirical verification of Kohlberg's theory in other societies may have been due partly to earlier scoring procedures, and better support may be expected from research using Kohlberg's newer, improved methods (Hoffman, 1980).

Another criticism is that many of the moral dilemmas used in Kohlberg's studies fail to deal with the issues that are most important to adolescents. In one study, Yussen (1977) asked adolescents in the seventh, ninth, and twelfth grades to write a realistic moral dilemma of their own. When Yussen analzyed the adolescents' moral dilemma, he discovered that they were quite different from those used by Kohlberg. A central issue for all adolescents was interpersonal relationships, especially among friends. There were also age differences, with seventh graders more likely to be concerned about matters of physical safety, that is, harm from threats and physical violence, whereas ninth and twelfth graders were more concerned about sexual matters, and twelfth graders were beginning to struggle with job-related difficulties. Other issues that emerged in the adolescents' own moral dilemmas included the use of alcohol and drugs, civil rights, and stealing.

Still another criticism focuses on Kohlberg's emphasis on the abstract principles of moral reasoning. Carol Gilligan (1985) holds that Kohlberg's theory reflects a male bias, with the emphasis on the abstract, impersonal principles of right and wrong superseding the interpersonal aspects of morality which are often more developed among women. Thus, when boys score higher than girls on Kohlberg's measures of moral development, this may be more of an indication that the measures themselves are biased. A more comprehensive account of moral development should give greater consideration of right and wrong in relation to people's connectedness or relationships with others in a given situation.

Despite these criticism, Kohlberg's theory remains one of the most helpful explanations of moral development available, especially in an area that has lacked serious attention from the social sciences. Even though Kohlberg's stages of moral reasoning may not be universal or invariant, Kohlberg's theory is helpful in understanding the development of moral thought, especially in our society with its democratic form of government. But we should bear in mind that Kohlberg's theory is essential a *cognitive* view of moral reasoning, and that this must be seen in relation to other aspects of moral development as well.

MORAL BEHAVIOR

moral behavior: The conformity of behavior to commonly accepted standards of right and wrong.

The ability of youths to reason about right and wrong in a mature way in a given situation is one thing; whether they act, and how they act in the situation is something else. For instance, faced with the possibility of getting caught for classroom cheating, 15 year olds may act out of self-serving motives rather than in accordance with their more sophisticated moral reasoning. As a result, morality is considerably more complex than moral reasoning, as important as this is.

Here in this section, we'll examine some of the behavioral aspects of morality. We'll begin by discussing the resistance to wrongdoing, that is, the avoidance of such behaviors as cheating, lying, and stealing. Then we'll look at the development of empathy and prosocial behavior, such as coming to the aid of a stranded motorist. Finally, we'll consider the importance of personal and moral values, and the characteristic value-behavior conflicts that occur during adolescence. In each case, we'll note the personal and social factors which help to shape moral behavior, such as childrearing practices, sex roles, and situational influences.

Resistance to Wrongdoing

unethical: That which violates established norms of right and wrong.

Every day, adolescents and adults are faced with choices that involve doing the right thing versus taking the easy way out. That is, they may lie, cheat, steal, or do something to gain an immediate reward in an unethical or illegal way. The temptations are great, and, judging by newspaper headlines and surveys on the subject, so is the amount of unethical behavior. In one national survey, investigators (Hassett, 1981) polled over 24,000 people about a variety of moral issues. One-fourth of the subjects were between 13 and 25 years of age, another one-fourth were in their mid- to late twenties, and the rest were older. The results showed that during the year prior to the survey a majority of respondents had broken the law or engaged in some unethical behavior. (See Figure 12–2.) Furthermore, while 89 percent of them reported they feel it is unethical to drive away after scratching a car without telling the owner, 44 percent of them admitted they would probably do it. And if they were sure they wouldn't get caught, over half of them (52 percent) would be even more likely to do so. Similarly, while 85 percent of them feel it is unethical to keep $10 extra change at a local supermarket, one-fourth of them (26 percent) said they would probably do so. One-third of them (33 percent) indicated they would keep the change if they were sure they wouldn't get caught.

sociopathic: Lacking an adequate sense of right and wrong.

Some individuals are more likely than others to engage in wrongdoing. For instance, those who cheat on exams tend to be low in the ability to delay gratification, low in interpersonal trust, and low in self-esteem. Such people also tend to be high in sociopathic tendencies, high in the need for approval, and high in chronic self-destructive tendencies. Altogether, such individuals tend to be emotionally and morally immature and unable to sacrifice short-term gains in order to obtain future rewards (Baron & Byrne, 1984). At the same time, how individuals will behave at any given time also depends heavily on the particular circumstances. In nonviolent crimes, the situation seems to play a major role, with individuals deciding whether to act on the basis of

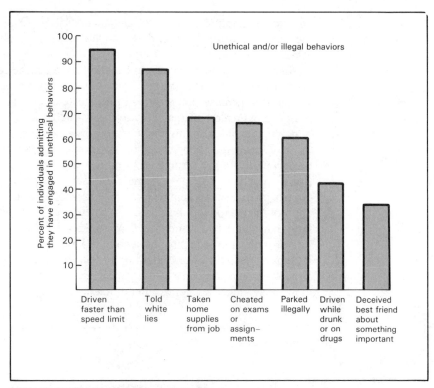

FIGURE 12–2 UNETHICAL BEHAVIOR.
Source: Adapted from data in the Psychology Today Survey Report on unethical behavior—James Hassett, "But that Would Be Wrong . . .," *Psychology Today* (November 1981): 41. Reprinted from *Psychology Today* Magazine, © 1981 The American Psychological Association.

the benefits to be gained, the probability of getting caught, and the costs involved if the misdeeds are discovered. For instance, with respect to stolen property, the possible benefits are often exceeded by the high risk of getting caught, convicted, and punished, such that the incidence of property crime is relatively high among young, risk-taking males. By contrast, with fraud there's often a high potential gain, with a low risk of indictment and a low penalty, so fraud is widely practiced by both men and women from age 20 to 60 (Lockhard, Kirkevold, & Kalk, 1980).

Have you ever copied another student's answers on an exam or passed off another person's paper as your own? If so, you have plenty of company. According to the survey cited earlier (Hassett, 1981), two-thirds of the respondents admit to having cheated on an exam or school assignment at one time or another. Cheating tends to be more common among males than among females, and more common among those with average grades or below than among those with higher than average grades (Baron & Byrne, 1984). When asked why they cheat, some teenagers say that since others are doing it they have to cheat to protect themselves (Gallup, 1979). Others frankly admit that they cheat because it's easier than studying and saves time (Norman & Harris, 1982).

Attempts to reduce classroom cheating have usually relied on one of two approaches or a combination of both. First, there are attempts to alter the situation to increase the perceived probability of getting caught, such as the presence of an alert examiner in the classroom. Unfortunately, the probability of getting caught is so low that cheaters are seldom deterred by this fear. Of those who say they have cheated, only one out of five admit they have ever been caught, about the same proportion as those who ever fear getting caught (Norman & Harris, 1982). There is also some evidence that when students are placed in a situation in which a high level of effort is rewarded fairly, they are more likely to work hard and less likely to cheat when performing similar tasks at a later date (Eisenberger & Masterson, 1983).

A second approach aims at altering the individual's moral awareness and feelings of guilt. Since anywhere up to half the students report some feelings of guilt after cheating on a test or paper, especially an important one, attempts to increase ethical awareness might be effective in reducing classroom cheat-

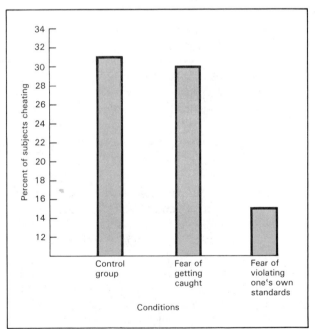

Figure 12–3 THE REDUCTION OF CHEATING BEHAVIOR: GUILT VERSUS FEAR OF GETTING CAUGHT.

Source: Based on data from R. A. Dienstbier, R. L. Kahle, K. A. Willis, and G. B. Tunnell, "The Impact of Moral Theories on Cheating: Studies of Emotion Attribution and Schema Activation," *Motivation and Emotion, 4* (1980):193–216. Used by permission of Plenum Publishing Corporation.

ing. In one experiment, investigators (Dienstbier et al., 1980) administered a vocabulary test to students followed by some material with moral theories. Students in the externally oriented condition read material that emphasized the threat of punishment and fear of getting caught. Those in the internally oriented condition read material that stressed the emotional tension that resulted from violating one's own moral standards. At this point, students had the opportunity to cheat. They were given the correct answers and warned not to change their answers. During the next couple of minutes the experimenter supposedly was busy with a telephone call, making it easy for students to cheat. But there was a sheet of pressure-sensitive paper in the test booklets so that changed answers could be identified easily. The results showed that less than one-sixth of the students in the internally oriented condition cheated, compared to almost one-third of the students in the externally oriented and control groups. (See Figure 12–3.) The investigators suggested that the internally oriented approach was more effective because most of the students already possessed the appropriate moral rule but needed a slight reminder to act on their moral views. At the same time, the effectiveness of this approach was probably enhanced because of other factors, such as the reliance on free choice versus external pressure and the appeal to a positive self-image in which students like to think of themselves as honest and ethical individuals— all of which reflect democratic socialization practices.

Helping Behavior

The adolescent's ability to resist wrongoing reflects socialization practices that stress the importance of subordinating the individual's impulses to the requirements of conventional authority or law. However, children and adolescents are also socialized in varying degrees in regard to helping or prosocial behaviors. Furthermore, increasing attention is now being given to the humanitarian aspects of moral behavior, especially to the development of empathy as a motive for helping others.

empathy: The ability to share another's emotions as one's own.

Martin Hoffman (1980) explains how a person's readiness to help people usually results from a combination of empathetic affect and the cognitive ability to take the view of others. Infants and young children may experience a kind of global empathetic distress response. Yet, lacking a clear distinction between themselves and others, they may be unclear as to who is feeling the distress. As children acquire a rudimentary sense of self and subjective perspective taking, they may experience a more specific empathetic distress response. But since they cannot yet distinguish accurately between their feelings and those of others, young childrens' efforts to help may consist in giving others what they themselves find most comforting, such as offering their teddy bear or security blanket to a troubled friend. Older children are capable of responding to signals of distress with more appropriate empathetic affect and objective perspective taking, such that they can recognize signs of happiness

helping behavior (prosocial behavior): Actions intended to benefit others that have no apparent selfish motivation.

and sadness in others in specific situations. Sometime during late childhood and early adolescence, largely because of the emergence of greater autonomy, self-identity, and social perspective taking, youths can not only empathize with people's feelings in transitory situations but can also imagine the other person's general condition. Thus, individuals can experience empathy together with a fairly realistic mental image of the general plight of others. One might draw a parallel here between the emergence of the capacity for mature empathetic arousal and Kohlberg's conventional level of moral reasoning (Muuss, 1986).

Hoffman points out that individuals may continue to respond to others' distress in a purely empathetic manner, that is, to feel uncomfortable and highly distressed themselves. Or they may also experience a sense of compassion and a desire to help the other person, which Hoffman designates as the response of "sympathetic distress." Ordinarily, it is only as individuals experience sympathetic distress that they actually engage in helping behaviors. Those who feel empathetic or sympathetic arousal and do nothing about it, and who then attribute criticism or blame to themselves for not helping, are subject to the sense of guilt. Guilt over inaction has much in common with sympathetic distress, with the primary difference being that in the case of guilt observers are aware of something they could have done but didn't.

Empathy is such an important component of helping behavior that considerable attention has been directed at how it develops or fails to develop. Since identical twins are more similar in empathetic concern than fraternal twins, the disposition to be empathetic may be partly hereditary. The child's early experiences also play an important role in the development of empathy. For instance, empathy tends to be very low in abused children as well as in those who are spanked a great deal by their parents. On the other hand, females who are high in empathy report receiving more affection from their parents and are more apt to discuss their feelings with their mothers than females low in empathy. Prosocial behavior is also influenced by the type of play children engage in, with those exhibiting high empathy more likely to have played cooperative games than competitive ones. Furthermore, the motive to help others seems to increase with age, so that late adolescents and adults value altruistic behavior more than children do. In part, this may be due to older individuals' feeling a greater sense of personal responsibility and knowing what to do in situations of human need (Baron & Byrne, 1984).

At the same time, the inclination to help people also depends on a variety of other factors. For instance, sex-role differences and the kind of behavior required to help often determine who will help. Among children in the elementary grades, girls are often perceived as more empathetic than boys, though girls are only slightly more altruistic in their actual behavior (Shigetomi, Hartmann, & Gelfand, 1981). Among older adolescents and adults, males are more likely to help in certain situations, such as offering to help pick up dropped coins and hitchhikers. With a stalled car on the highway, the vast

A VISTA volunteer helps an elderly woman with
home repairs. (ACTION)

bystander effect: The
finding that people are
much less likely to help
in an emergency when
they are with others than
when they are alone.

majority of people who stop are males, especailly when the stranded motorist
is a female. Consequently, helping behavior in different situations is affected
by sex roles, such as the nurturing role in females and the risk-taking, protective role in males. A number of other factors also modify the probability that
individuals will help. Individuals who have a strong fear of embarrassment or
social blunders are less likely to take the initiative to help. On the other hand,
those with a strong need for approval may be more inclined to help, especially
when there are witnesses to their good deeds. However, when the situation
poses the threat of danger, the opposite is often true. That is, people are less
likely to intervene in a dangerous situation, such as coming to the aid of
someone being mugged; this is labeled the bystander effect. A major explanation is that the presence of others leads to a diffusion of responsibility, so that
there is less likelihood of anyone intervening. Conversely, when a lone individual observes a mugging, this person is more apt to feel responsible and come
to the aid of the victim. At the same time, the willingness to help others even in
a dangerous situation is affected by many other factors. An important factor is
feeling competent and knowing what to do in a given situation. For instance, a
comparison of individuals who had intervened in actual instances of violent
crime and those who failed to intervene disclosed some important differences.
Individuals who intervened were more likely to have had training in first aid,

life saving, self-defense, and experience in medical or police work. They were also taller and heavier (Huston et al., 1983).

Personal and Moral Values

moral values: Shared beliefs about what is good and right.

The tendency to act on one's moral views in a given situation also depends heavily on the individual's personal and moral values. Since these values are initially acquired in the home, there is usually a strong relationship between adolescents' values and those of their parents (Dudley & Dudley, 1984). Partly for the same reason, the individual's values remain remarkably stable throughout adolescence. At the same time, passage through adolescence tends to be accompanied by a decrease in the values of obedience, cheerfulness, and helpfulness associated with childhood morality, and an increase in the sense of achievement, self-respect, broadmindedness, and responsibility associated with the growth of autonomous morality. Values such as being honest and loving remain high at all ages. In a cross-cultural survey including youth aged 13 to 20 in 59 nations, excluding the Soviet Union and China, respondents were asked to pick the three things that would bring them happiness. More than two-thirds of the American teenagers (69 percent) and almost as many teenagers in other countries (60 percent) selected "love." The priority of love easily outdistanced other values like good careers, family, helping others, and freedom—reflecting the greater priority given to personal relationships among contemporary youth (*U.S. News and World Report,* 1986). However, surveys also show some differences between the sexes. In one survey of high school seniors, Bachman (1982) found that males put a greater priority on having steady work, with marriage and friendships being somewhat less important, while females put a good marriage and family life as their top priority. Also, females ranked the item on "finding meaning and purpose in my life" second, while males ranked it fifth.

As adolescents struggle with such issues as premarital sex, teen pregnancy, and the use of alcohol and drugs, they often experience considerable inconsistency between their values and behaviors. A major reason is that at this stage of development, adolescents' values are still shaped more strongly by parental influences, while their behavior tends to be determined more by peer influence (DeVaus, 1983). As a result, adolescents tend to experience a considerable degree of value-behavior inconsistency, being torn between personal and moral values acquired at home while behaving in a manner more acceptable to their peers. Another source of value conflict is the discrepancy between an adolescent's existing personal and moral values and the need to respond to new life situations, reflecting the impact of social change. For instance, the upheaval in values that occurred in the 1960s and 1970s brought an emphasis on expressive, self-fulfillment values, such as personal freedom, self-determination, and self-improvement. Some observers detect a return to traditional values among students in the 1980s, with a renewed concern over marriage, careers, and financial and material well-being (Leo, 1984). However, in his book *New Rules* (1981), Daniel Yankelovich, a seasoned observer of public attitudes, holds that the self-fulfillment values are not being surrendered; in fact, their impact has filtered down into practically all segments of society. But he sees the emphasis on self-fulfillment being modified by a new realism or countertrend toward a greater concern with relationships and commitments that bring enduring satisfaction within the larger community.

Still another source of value conflicts is the relativity of values which characterizes many aspects of American society. Perceiving all values as relativistic, many individuals are inclined to adopt the view that anyone has the right to his or her own beliefs and values over and against those realms of authority where a more definite sense of right and wrong still prevails, such as the church. At a conference of educators, participants concluded that only a small percentage of teachers and students see students as having any deep ethical sense or commitment to ethical standards. Instead, the emphasis on high academic expectations and career success together with social isolation has resulted in many students' lacking a keen sense of right and wrong or concern for others, mirroring the ethical wasteland in many segments of adult society (Naedele, 1982). The conflicts and confusion over moral values is often expressed in regard to sexual matters. When respondents in a national survey were asked what were the most important moral dilemmas they face, most responses named sexual conflicts, followed by conflicts having to do with lying, especially lying about an unwanted pregnancy and abortion (Hassett, 1981). Similarly, in the cross-cultural survey cited earlier, when teenagers were asked to name the most pressing issues they face, almost all the American teenagers (304 out of 305 of them) cited premarital sexual relations, whereas only 29 percent of the youth in other countries considered premarital sex a major problem (*U.S. News and World Report*, 1986). Apparently, many American youth are more troubled by changing moral standards and lack of standards in regard to sexual matters than popularly believed.

Not surprisingly, many youths feel the need for assistance in developing their moral values. In an extensive survey of high school and college students in 1978 and 1983, respondents were asked to what extent youth needed assistance in the development of moral values. At least three-fourths of all the groups sampled felt that the family, school, clergy, and other individuals should be involved in the development of moral values to some degree. Generally, the family was regarded as the primary influence for elementary-aged children, with the family and other individuals sharing influence among secondary school students, and other individuals more influential at the college level. All groups saw the school and clergy as having substantial influence but as being of lesser importance than the family and other individuals. College students in particular regard authorities such as the family, school, and clergy as more important earlier in life and the influence of other individuals as more important later in life (Zern, 1985).

RELIGION

religion: A system of belief, worship, code of ethics, and conduct involving belief in divine or superhuman powers.

Traditionally, churches and synagogues, along with the family, have played a major role in the teaching of moral standards and values to young people. The more important religion is to the particular family and individual, the more likely it is that religion will become a significant factor in the young person's moral development. However, with the emergence of formal thinking and greater autonomy during adolescence, even religious youths tend to go through a stage of questioning their faith, accompanied by a decrease in church attendance. During this period, youths characteristically maintain an intense interest in religious matters. According to surveys (Gallup & Poling, 1980), more than 9 out of 10 teenagers believe in God, with almost as many practicing some form of prayer or meditation. Compared to youth in the past, young people are somewhat more conservative in their religion. Almost half the Protestant teenagers and one-fifth of Catholic youth have had a religious conversion experience. At the same time, today's youth are placing more emphasis on personal rather than institutional religion. As a result, even though religious attitudes and beliefs tend to remain strong throughout the college years for many youths, there is a characteristic decline in church attendance. One study found that a smaller proportion of college seniors than first-year students attended church once a week, with a larger proportion of seniors attending church once or twice a month or not at all (McAllister, 1985).

Much depends on the particular individual and his or her religious development. We'll begin with a description of the typical developmental sequence of religious growth, including the important distinction between extrinsic and intrinsic religious orientation. We'll also discuss the importance of a relatively new phenomenon in religious development—religious cults. Finally, we'll point out some of the relationships between religion and moral behavior.

The Developmental Sequence

Mary Jo Meadow and Richard Kahoe (1984) have proposed a developmental model of religious growth that incorporates many of the established principles and findings in this field. They suggest that the individual's religious development typically begins near the extrinsic pole in their model, and progresses in a counterclockwise fashion toward an intrinsic religious orientation. (See Figure 12–4.) Not all religious persons experience the entire developmental

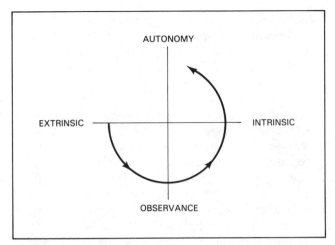

FIGURE 12–4 THE DEVELOPMENTAL SEQUENCE OF PERSONAL RELIGIOUSNESS.

Source: From Richard D. Kahoe and Mary Jo Meadow, "A Developmental Perspective on Religious Orientation Dimensions," *Journal of Religion and Health, 20* (1981):8–17. © 1981 Human Sciences Press, Inc. In M. J. Meadows and R. D. Kahoe, *Psychology of Religion* (New York: Harper & Row, 1984), p. 321.

sequence, as in other developmental models. But this is the general path they travel.

Psychologically significant religion, as opposed to empty habits, typically begins at the *extrinsic pole*. People with this orientation are inclined to use religion for their own ends, with religious beliefs and practices being motivated primarily by any number of human needs, such as insecurities, fears, and guilt. Examples would be the teenager who feels happier and more at peace with herself after she meditates, and the salesman who claims he can sell more cars after beginning the day with prayer. For some people, religion remains on a self-serving, extrinsic basis. But for many people—if not most—an extrinsic religious orientation becomes modified through affiliation with some type of organized religion.

Religiously oriented people characteristically identify with a religious institution or system, at least early in their religious development. *Religious observance* refers to a variety of religious matters, such as membership in a group, authority, beliefs, practices, ceremonies, and rituals. All of these aspects of organized religion are probably interrelated and may occur simultaneously, though some tend to occur before others. For instance, social belongingness usually occurs early, with the internalization of beliefs coming later. The shift toward institutional participation requires the member or believer to turn away from purely self-serving, extrinsic religion toward causes and tasks greater than the self. However, the shift is usually relative. That is, since most people maintain many of their extrinsic motives, religious institutions cultivate loyalty by directing attention and services to individual needs. Thus, church social

extrinsic religious orientation: The situation in which people's religious beliefs and practices are motivated primarily by any number of human needs, such as fears and guilt.

religious observance: Refers to a variety of matters, such as group membership, authority, beliefs, ceremonies, and practices that characterize organized religion.

religious conversion: An intense emotional, religious experience that results in a personal conviction or change of one's religious faith.

activities may reduce a new member's this-world loneliness while also helping to form an attachment to the church. Examples of individuals in the observance stage would be young people who join a church after attending a year-long confirmation class or those who join a cult after a religious conversion experience.

intrinsic religious orientation: The situation when people endeavor to internalize and practice their religious beliefs.

The tendency to turn away from self, which is set in motion by institutional religion, is developed further in *intrinsic religiousness*. At this stage, individuals endeavor to internalize and follow religious beliefs, that is, they *live* their religion. These are the "true believers." In contrast to individuals with an extrinsic religious orientation, those with an intrinsic orientation are more likely to have an internal sense of control over their lives, greater sense of personal responsibility, less prejudice, and engage in a variety of helping behaviors (Meadow & Kahoe, 1984). Furthermore, individuals with an intrinsic religious orientation are more likely than those with an extrinsic orientation to keep active in a religious organization during college (Woodroof, 1985). While intrinsic religion tends to grow out of the religion of observance, many people fail to achieve any great degree of intrinsic faith. In some instances, institutional leaders fail to promote intrinsic religiousness. But in other instances, individuals simply lack the personal maturity to make the transition. Most people who develop an intrinsic religious orientation also keep strong ties with organized religion, especially among mainstream religious groups. An example would be the college senior who is not only active in her own church but is also involved in campus-wide humanitarian causes and is concerned about human need in other countries.

autonomous religiousness: The most mature stage of personal religious development, characterized by greater independence of thought and concern for others than in earlier stages of religious development.

Autonomous religiousness is a step beyond an intrinsic faith and represents the most mature stage of personal religious development. Individuals who have reached this level of religious development exhibit a more advanced religious orientation characterized by greater independence of thought and practice and concern for others than those in earlier stages. Whereas most higher religions advocate intrinsic religion, they seldom promote a thoroughly autonomous faith because such independence of thought and behavior is generally against the vested interests of organized religion. Other forces also work against the development of autonomous religion. People who need the social support or sense of security of traditional religion may resist movement toward autonomous religion. Also, many individuals isolate religious experience, so that their faith remains less mature than other aspects of their development. In most cases, a strong personal inclination leads the occasional person to religious autonomy despite institutional discouragement. This might be an individual who is highly intelligent or educated, especially one with training in the reflective disciplines like philosophy or with a pronounced mystical outlook. An example would be a graduate student in philosophy, who, while struggling with existential questions in all their complexity, sees religion more as an open-ended quest than as a system of answers (Batson & Ventis, 1982).

More young people are becoming active in nontraditional religious groups. (Charles Gatewood)

These stages of religious development are not tied to specific ages but probably parallel in large measure other developmental processes, such as Kohlberg's stages of moral reasoning. Thus, the extrinsic orientation generally reflects Kohlberg's level of preconventional moral reasoning, that is, the heteronomous morality of stage 1 or the self-serving outlook of stage 2. Involvement in organized religion reflects more conventional moral reasoning, with the social aspect that accompanies religious observances associated with the interpersonal orientation of Kohlberg's stage 3 and the emphasis on doctrine and church law parallel to the emphasis on authority and fixed rules in stage 4. The development of intrinsic religion tends to parallel the attainment of the social contract orientation in stage 5 of Kohlberg's sequence. Finally, autonomous religiousness parallels Kohlberg's stage 6 of universal ethical principles. However, just as few individuals reach Kohlberg's fifth and sixth stages of postconventional moral reasoning, so relatively few individuals, much less adolescents, achieve an intrinsic religious orientation, and even fewer an autonomous religion.

heteronomous morality: The understanding of right and wrong based on external authority.

Cults

Religious cults have achieved widespread attention in recent years because they have played a more prominent role in the religious development of a substantial minority of youth than in the past. Membership in these groups is estimated to be at least 2 or 3 million people in the United States. Among the groups that receive the most public attention are the Unification Church (Moonies), the Divine Light Mission of Maharaj Ji, the Institute of Krishna Consciousness, the Children of God, and the Church of Scientology (*U.S. News and World Report,* 1982).

Religious cults vary widely in their structure and orientation. Some of these groups tend to be rather informal and loosely organized and are characterized by a disillusionment with materialistic society and a concern for helping people. But an even larger number of cults tend to be highly authoritarian and require individuals to surrender their autonomy and conform completely to the dictates of the leader. New converts are often subjected to brainwashing techniques, in which they are stripped of their previous identities, programmed with cult beliefs and practices, and made heavily dependent on the leader and the group. In some instances, individuals become converts gradually, even inadvertently, through being responsive to offers of warm, social relationships, especially during personal crises. Most groups incorporate new members through a combination of indoctrination, mind-control techniques, and an extensive system of socialization practices (Long & Hadden, 1983).

Young people may be attracted to religious cults for a variety of reasons, with such youths exhibiting one or more of the following conditions and characteristics (Swope, 1980; Dean, 1982).

indoctrination: Systematic instruction involving a set of beliefs or practices.

1. *Insecurity.* The adolescent reared in an affluent, permissive family who remains beset by uncertainties may be a prime target for the clearly defined limits and authority of cults.

2. *Inquisitiveness.* Intellectually curious youths who are questioning old beliefs and seeking meaning to their lives may be attracted to new ideas and groups, especially when approached by attractive, enthusiastic recruiters.

3. *Idealism.* Cults also exploit youthful idealism by convincing prospective members that society's ills can be solved only through the group's particular approach.

4. *Loneliness.* Young people on their own away from home who lack close friends and support groups may be especially receptive to invitations of friendship and free meals.

5. *Disillusionment.* Many converts to cults have become disillusioned with their lives, whether in reaction to heavy addiction to alcohol or drugs, sexual experimentation, or the pressures for success.

6. *Identity crises.* Severely anxious or confused youth may be attracted to cults as a way of finding their identity in a rapidly changing society that often appears to be excessively competitive and impersonal.

7. *Naiveté*. Lonely, overprotected youths who have been reared to trust others, especially their religious leaders, are easy targets for people offering friendship and guidance.

Interestingly, the majority of young people who join cults are white, single, and middle class. Membership in such groups usually brings them a sense of less confusion in their lives, often accompanied by the avoidance of alcohol, drugs, or permissive sex. Such youths may welcome the structured environment with its ready-made beliefs and friendships. At the same time, since cults usually provide a safe, regulated environment that relieves individuals of the responsibilities of making personal decisions, membership in these groups tends to undermine healthy personal and religious development.

deprogramming: The systematic attempt to reorient former cult members through an intensive teaching and counseling process.

Critics of cults claim that members have been so radically and permanently transformed that some type of deprogramming is needed to sever ties with the cult. As one ex-cult member said, "Everything that happens is exactly parallel to what a POW experiences, and for a certain amount of time, I feel I did not have my free will" (Wright, 1984). Consequently, in their attempts to reorient former cult members, many deprogramming groups feel justified in using similar tactics of kidnapping, isolation, and intensive teaching sessions, though such practices have been criticized because they violate the civil rights of former cult members. However, most deprogramming is voluntary and consists of an intensive teaching and counseling process, in which the individual is encouraged to think, ask questions, critically evaluate his or her beliefs and practices, and make decisions. Deprogramming can last from a few hours to several weeks, though the rehabilitation process may continue for several months, usually within the home of the ex-cult member. Sometimes going shopping with an ex-cult member may take an entire afternoon, mostly because the person hasn't been making personal decisions for several years (Sifford, 1983).

At the same time, studies of young people who have voluntarily defected from cults have shown that many of them willingly participated in cults. Only a minority of voluntary defectors reported that they had been duped or brainwashed. As one voluntary defector said, "I'm not angry, you know, because it was of my own volition that I was there and I could have just walked out anytime" (Wright, 1984). Voluntary defectors explained their involvement in terms of sophisticated techniques of mind control and a combination of physical and psychological deprivations. Unlike ex-cult members who were involuntarily removed from cults, these members usually saw some redeeming value in their experience. At the same time, most of them perceived their original choice, in retrospect, as a mistake (Wright, 1984). At first, many ex-cult members are furious. They want to return to the cult and get their friends out. But after counseling and rehabilitation, many of them attain a level of achievement and personal growth that is above average compared to the general population. Eventually, most of them marry and have families (Sifford, 1983).

Religion and Moral Behavior

It might seem logical to assume that religious youth have more stringent moral standards and are less likely to cheat on tests and use illicit drugs. But is there any truth to this view? Apparently so, but a lot depends on the specific behavior and circumstances. The national survey on morality cited earlier (Hassett, 1981) showed there was a significant relationship between people's religiosity and their moral behavior. As a matter of fact, the respondents' self-ratings of their religiousness turned out to be one of the single best predictors of moral standards and behavior. Moderately religious people scored significantly higher on the scale of morality than the less committed, and very religious people scored significantly higher than the moderate group. But the relationship between religiosity and morality was much stronger for institutional behavior than for personal issues. That is, religious people were much less likely to cheat on tests or income taxes, take home office supplies, or use a business phone for personal calls than other people. But they were only slightly less likely to tell little white lies, deceive their best friends, or cut into line ahead of people. At the same time, religious youths did report stricter

Youths with a personal faith are often motivated to help others. (Courtesy of MDA)

moral standards; they were much less likely to be influenced by the fear of getting caught compared to less religious people, who were more apt to say they would try to get away with what they could.

Other studies have shown varying degrees of relationships between such items as religious orientation and church attendance and moral attitudes and behavior. Using a random sample of 600 adolescents in grades 9 to 12, Hadaway, Elifson, and Petersen (1984) found a moderately strong relationship between religion and the use of drugs. The more important religion was to the individuals and the more active they were in organized religion, the lower their levels of alcohol consumption and drug use. However, the association was stronger between religion and the use of illicit drugs than for alcohol, probably because of the greater social acceptance of alcohol. Also, the importance of religion was more influential than church attendance itself on moral attitudes and behavior. At the same, other studies using this same sample found little or no relationship between religious orientation and aggressive behavior, theft, and truancy (Elifson, Petersen, & Hadaway, 1983). Another study of first-year college students by Woodroof (1985) found a marked relationship between religious orientation and involvement and premarital sexual activity. The more religious youths were and the more frequently they attended church, the less sexually active they were—including the incidence of casual sex. Also, the incidence of virginity was considerably higher among those who were very religious than among those who attended church once a week or less.

The relationship between religion and helping behavior is more complex. On the one hand, in a study using a random sample of 500 high school students Howard Bahr and Thomas Martin (1983) found a significant positive association between respondents' church attendance and their faith in people. Subjects who attended church regularly were much less likely to agree with such statements as "most people can't be trusted" or "most people don't really care what happens to the next person." However, since there is also a strong association between higher socioeconomic status and faith in people, it is not clear whether religion inclines youth to have greater faith in people or whether youths from higher socioeconomic groups are more apt to attend church and thus express greater faith in people.

Studies on the relationship between religion and helping behavior yield even more mixed results. In a review of the literature in this field, Daniel Batson and Larry Ventis (1982) found a difference in the methods used in experimental studies, especially between the use of self-ratings and behavioral measures. Studies using self-ratings of religiously and helping behavior (as in Hassett's survey cited earlier) and ratings by someone else, showed a positive, though weak, association between religious involvement and helping behavior such as coming to the aid of a stranded motorist. However, as the authors point out, such studies are more likely to be contaminated by social desirability. For instance, people who are rated as more likable and sociable tend also to be rated as more helpful, even when they are not. In contrast, none of the studies relying primarily on behavioral measures suggested that religious people are more helpful than others. Thus, more religious youth may hold more stringent moral attitudes and standards and may see themselves as more helpful and caring than others and may be so regarded by others. But when it comes to action, there is no convincing evidence that religious youth are more consistently compassionate toward others. At the same time, the further along the developmental sequence of religious growth individuals have come, especially the intrinsic and autonomous religious orientation, the more likely they are to practice their faith.

SUMMARY

The Development of Moral Reasoning

1. According to Kohlberg's theory of moral development, children and adolescents progress through a sequence of stages of moral reasoning that parallels their cognitive development.

2. Kohlberg has identified six successive stages of moral reasoning, divided into three major levels—preconventional, conventional, and post-conventional—with the highest level of moral judgment presupposing the attainment of formal thought.

3. During mid- to late adolescence, there is a marked rise in the proportion of individuals exhibiting conventional moral reasoning, with a slow but steady rise in the number of youths with stage 4 thought reflecting the general rules of laws of society.

4. Kohlberg's theory has been criticized on a number of grounds, such as that relatively few people ever attain the highest stages of moral reasoning, and that it puts undue emphasis on abstract reasoning compared to the interpersonal aspects of moral development.

Moral Behavior

5. The extent to which individuals act on their moral reasoning depends on a variety of factors, such as the existence of internalized moral standards and feelings, the immediate situation, empathetic arousal, and moral values.

6. Unethical behavior such as classroom cheating is fairly widespread among youth, though such behavior tends to be less evident among those with a strong concern for obeying their internalized moral standards.

7. Youth are more apt to engage in helping behaviors as they reach late adolescence and early adulthood, largely because of their greater sense of personal responsibility and competence in responding to human needs.

8. As adolescents struggle with such issues as premarital sex and the use of alcohol and drugs, they experience a considerable degree of value-behavior inconsistency, partly because of the tension between parental and peer influences and the relativity of values in American society.

9. As a result, the majority of young people feel youth need more guidance in the development of moral values, with authorities such as the family and school being regarded as more important in the earlier stages of development and individuals' influence more significant as people reach college age.

Religion

10. Traditionally, the churches and synagogues have played a major role in the transmission of moral standards, though many of today's youth place more emphasis on personal than institutional religion.

11. Religious development typically begins with an extrinsic religious orientation and generally progresses toward affiliation with organizational religion, with relatively fewer individuals acquiring an intrinsic religious orientation characterized by a devotion to causes and ideals beyond themselves.

12. Although religious cults have attracted a substantial minority of youth in recent years, they tend to undermine religious development because cults relieve youths of the necessity of making personal decisions.

13. Religious youths, especially those with an intrinsic orientation, tend to hold more stringent moral attitudes and standards than nonreligious youths, though they are not necessarily more likely to engage in helping behaviors than other youths.

REVIEW QUESTIONS

1. How is Kohlberg's view of moral reasoning related to Piaget's theory of cognitive development?

2. What is the basis of the sense of right and wrong in Kohlberg's preconventional, conventional, and postconventional levels of moral reasoning?

3. To what extent do adolescents attain conventional and postconventional moral reasoning?

4. What are some of the pros and cons of Kohlberg's theory of moral development?

5. What are some of the factors associated with classroom cheating and resistance to such behavior?

6. Why are adolescents more likely to engage in helping behaviors as they reach late adolescence and early adulthood?

7. What are some of the reasons youths experience value-behavior inconsistency during adolescence?

8. Explain the typical developmental sequence of religious growth.

9. What are some of the reasons youths join religious cults?

10. To what extent are religious youths more moral than nonreligious youths?

13 Delinquency

- **THE EXTENT OF DELINQUENCY**
  Age Trends
  Types of Offenses
  Sex Differences
- **CONTRIBUTING FACTORS**
  The Family
  Socioeconomic Factors
  Personality
  Psychopathology
  Drugs
- **TREATMENT AND PREVENTION OF DELINQUENCY**
  Police and Juvenile Court
  Current Approaches to Treatment
  Preventing Delinquency
- **SUMMARY**
- **REVIEW QUESTIONS**

Did you ever engage in shoplifting, petty vandalism, or underage drinking when you were a teenager? If so, you were engaging in typical adolescent behavior. Studies of self-reported delinquency have shown that more than three-fourths of Americans admit to committing one or more delinquent acts during the relatively short period of their adolescence. Most of these acts are minor infractions of the law, such as underage drinking, smoking pot, stealing, or vandalism. At the same time, youths are responsible for a significant proportion of serious offenses such as arson and burglary. Youths typically engage in such acts in small "pick-up" groups of two or three companions, usually of the same sex. Very few teenagers engage in delinquency as loners or as members of gangs. Interestingly, few teenagers see themselves as delinquents or feel they are so regarded by their friends. Only a small proportion of them—less than 10 percent—are ever caught. Even fewer are arrested and taken to juvenile court (Gold & Petronio, 1980).

THE EXTENT OF DELINQUENCY

Even so, the statistics on reported delinquency are disturbing. Of the almost 12 million people who were arrested for all criminal activities except traffic violations in 1985, a significant proportion of them were youths. Approximately 1 out of 6 of them were under 18 years of age; 1 out of 3 of them were under 21; and 1 out of 2 of them were under 25. Teenagers under 18 committed one-third of all acts of arson, burglary, theft, and motor vehicle theft. Youths under 25 accounted for two-thirds of these same crimes, plus almost one-half of violent crimes such as aggravated assault, forcible rape, and murder (*Uniform Crime Reports*, 1985).

Age Trends

juvenile delinquency: Illegal behavior committed by a youth under the legal age, which is 18 in many states.

juvenile delinquent: Generally, someone under the legal age who has engaged in illegal behavior; technically, a person convicted of such a violation in juvenile court.

Many delinquent acts in early adolescence are relatively minor offenses, such as drinking beer, smoking pot, petty thefts, and vandalism. But the frequency and seriousness of delinquent behavior rises with age. For the relatively minor offenses, such as vandalism, delinquency increases from late childhood throughout middle adolescence and then levels off somewhat, though it continues to rise until early adulthood before subsiding. More serious offenses, such as assault and robbery, accelerate from early adolescence, peak in midadolescence, and then decline (Gold & Petronio, 1980).

There are also significant differences in delinquent activities among the various age groups. Youths under 15 years of age typically engage in a great deal of nuisance behavior, such as curfew violations, disorderly conduct, and vandalism. At the same time, early adolescents commit one out of seven burglaries and thefts and one-fourth of all of the arson cases. Adolescents in the 15- to 17-year-old group commit a significant proportion of liquor law and drug abuse violations, sexual offenses, and vandalism. Youths this age also

Table 13–1   PERCENT OF ARRESTS BY CRIME INDEX AND AGE, 1985

	Under 15	15–17	18–20
Murder	1.0	7.3	13.7
Rape	5.2	9.9	12.5
Robbery	6.5	18.5	19.8
Assault	4.3	9.5	12.2
Burglary	14.2	23.8	19.5
Larceny-theft	14.2	18.5	14.6
Motor vehicle theft	9.5	28.5	19.5
Arson	26.3	14.9	10.3
Total crimes	**12.2**	**18.6**	**15.7**
Cumulative	**12.2**	**30.8**	**50.4**

Source: Adapted from data in the *Uniform Crime Reports for the United States, 1985* (Washington, DC: U.S. Department of Justice, Federal Bureau of Investigation, Government Printing Office, 1985), p. 180.

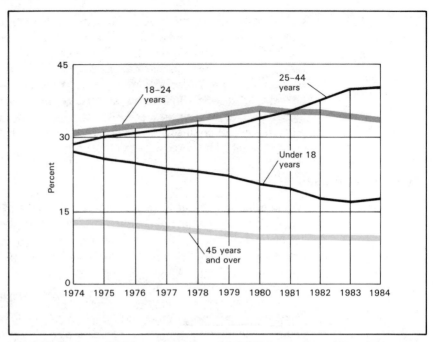

FIGURE 13–1 PERCENT OF PERSONS ARRESTED BY AGE, 1974–1984.
Source: U.S. Bureau of the Census, *Statistical Abstract of the United States, 1985,* 106th ed. (Washington, DC: U.S. Government Printing Office, 1985), p. 162.

get more involved in serious offenses, such as weapons violations, burglary, robbery, and theft. Almost 4 out of 10 motor vehicle thefts are committed by youths in this age group. Older youths in the 18- to 20-year-old group are more heavily involved in drug and liquor law violations, disorderly conduct, and sexual offenses. Youths under 21 account for over half the burglaries, robberies, and thefts, especially motor vehicle thefts. In addition, they commit over one-fifth of all homicides and non-negligent manslaughter (*Uniform Crime Reports,* 1985). (See Table 13–1.)

At the same time, 5- and 10-year trends indicate a changing pattern of arrests. There is now a decline in arrests of youths under 18 years of age, accompanied by an increase of arrests among those over 25 years of age. Authorities believe this can be accounted for mostly by the aging of the baby boomers, resulting in a smaller proportion of youths in the traditionally high-crime group of 14 to 24 years of age (*Uniform Crime Reports,* 1985). (See Figure 13–1.)

Are youths who engage in delinquency more likely to engage in crime when they become adults? Not necessarily. A lot depends on how early they begin getting involved in delinquency, how serious their offenses are, and how often they are rearrested. The earlier the age of delinquency, the more serious the offenses, the greater the number of subsequent arrests, and the more the delinquent behavior involves a delinquent peer group, the more likely the individual will commit crime as an adult. Also, the more delinquent behavior reflects an antisocial pattern, such as the disregard for the rights of others and failure to learn from experience, the greater the likelihood that criminal activity will continue into adulthood. In fact, evidence of antisocial tendencies is the single best predictor of a career in crime. In contrast, the great majority of youths who engage in occasional acts of delinquency do not go on to adult crime (Gold & Petronio, 1980; Osburn & West, 1980).

Types of Offenses

Although youths under 21 represent only about 13 percent of the general population, they commit almost half (46.5 percent) of all the serious crimes like arson, robbery, and motor vehicle thefts. When the additional percentage of offenses committed by older youths are taken into consideration, individuals under 25 commit almost two-thirds (64.7 percent) of all serious crimes (*Uniform Crime Reports,* 1985).

The biggest single category of offenses committed by young people is crimes against property. When individuals in their early twenties are included, youthful offenders account for two-thirds of all robbery, larceny, and theft and almost three-fourths of all burglary, vandalism, and motor vehicle theft. Although youths commit a smaller proportion of the crimes against people, they account for about half of aggravated assaults, forcible rapes, and murder, including non-negligent manslaughter. The second biggest category of offenses committed by young people is that of the so-called victimless crimes, including drunkenness, disorderly conduct, gambling, and liquor and drug abuse violations (*Uniform Crime Reports,* 1985).

juvenile court: An integral part of the juvenile justice system, usually for juvenile offenders referred by the police.

At least one-third of all delinquents referred to juvenile court involve status offenses. These include a variety of behaviors, such as curfew violations, running away from home, truancy, incorrigible behavior, and sexual offenses. Since many of these behaviors reflect family problems rather than criminal acts, legislation in recent years has stipulated that such problems should not be the responsibility of already overburdened police departments. Also, because of their age and vulnerability, status offenders should be kept in separate physical facilities from other delinquents and adult offenders and should be given the support services that will enable them to return to a viable home environment as soon as possible. However, the actual procedures used with status offenders vary considerably from one part of the country to another, with status offenders still being kept in detention centers with serious juvenile offenders in many places (Thornburg, 1986).

status offenses: Behaviors which are illegal primarily because they are committed by someone under the legal age, such as curfew and alcohol violations.

runaways: Youth under the legal age who leave home without parental permission or because of parental rejection.

The value of the new approach to status offenses can be seen with runaways, an increasingly important social problem. Older children and early adolescents are running away from home at earlier ages than in the past, with

Table 13–2   NUMBER OF ARRESTS BY TYPE OF CRIME, AGE, AND SEX: 1984–1985

Offense	Males under 18	Females under 18
Serious Crimes		
Larceny (theft)	255,810	95,379
Burglary (breaking or entering)	124,388	9,764
Motor vehicle theft	36,256	4,619
Robbery	26,758	1,951
Aggravated assault	28,330	5,202
Forcible rape	4,316	80
Murder	1,124	115
Arson	5,671	609
Violent crime	60,528	7,348
Property crime	422,125	110,371
Subtotal—all serious crimes	**482,653**	**117,719**
All other crimes		
Drug abuse violations	63,255	10,932
Liquor law violations	76,322	27,483
Disorderly conduct	59,944	14,291
Vandalism	81,987	7,730
Other assaults	57,280	17,259
Fraud	13,350	3,812
Stolen property	23,032	2,386
Weapons (carrying, possessing)	22,421	1,628
Forgery and counterfeiting	4,962	2,302
Embezzlement	451	195
Prostitution	708	1,638
Sex offenses (except rape and prostitution)	12,599	1,107
Offenses against family and children	1,444	804
Driving under the influence	16,089	2,420
Gambling	666	33
Drunkenness	19,112	3,697
Vagrancy	2,016	447
Curfew violations	49,258	15,985
Runaways	53,808	72,473
All other offenses (except traffic)	206,878	53,649
Total all crimes	**1,248,235**	**357,990**

Source: Uniform Crime Reports for the United States, 1985 (Washington, DC: U.S. Department of Justice, Federal Bureau of Investigation, Government Printing Office, 1985), p. 173.

a peak incidence in runaways at 14 to 16 years of age. Girls are somewhat more likely than boys to run away from home, with boys accounting for more than 4 out of 10 runaways. Also, it is estimated that there are several times as many actual runaways as those recorded in the official figures—140,000 in 1985 (*Uniform Crime Reports,* 1985). One reason so few runaways appear in the statistics is that many of them run off to the home of a nearby friend or relative. Most stay within the community and return within 4 days.

Adolescents may run away from home for a variety of reasons. In many instances, they come from an unhappy home characterized by such problems as parental conflict, faulty communication, abusive childrearing practices, or the absence of a parent through death or divorce. At school, runaways tend to be tracked as slow learners or they engage in chronic truancy (Johnson & Carter, 1980). In recent years, the use of alcohol and illicit drugs has contributed to the problem. In some cases, running away is a desperate act of asserting independence in relation to domineering parents. At the same time, many runaways are "throwaways," who are either told to leave or are victims of sexual abuse or violence who feel they have no alternative but to leave. It has been estimated that throwaways account for as many as 30 to 60 percent of runaways (Blau, 1979).

Runaways who are arrested or taken to juvenile court may feel frightened and defiant. But those who wander into the streets of the large cities may fare even worse, with drug addiction and prostitution being a common fate for runaway girls. On the other hand, youth who are fortunate enough to find help in a temporary shelter for runaways, especially those with a network of support services, may find the assistance necessary to cope with their home situation. For example, when Cheryl's parents divorced, she went to live with her mother, who remarried a year later. When Cheryl, now a 16 year old, disagreed with her stepfather about the family rules, Cheryl's mother usually sided with her new husband. Feeling betrayed, Cheryl began staying out later, spending more time away from home, and engaging in rebellious behaviors. Finally, after an argument with her mother, Cheryl ran away with her boyfriend. But 2 weeks later, after a fight with her boyfriend, Cheryl sought refuge in a shelter for runaways. There she began talking about her problems at home. With the help of the counselor, Cheryl contacted her parents and agreed to return home on the condition that all of them enter family therapy.

Sex Differences

Although males continue to commit delinquent acts more frequently than females, the gap between the sexes is narrowing. The male–female ratio of reported delinquency, which remained at 4 or 5 to 1 for many years, is now nearer 3 to 1. The male–female ratio of self-reported delinquency is even closer (Gold & Petronio, 1980). However, the ratio between the sexes varies considerably according to the types of offenses. Males are much more likely than females to commit violent crimes, such as aggravated assault, forcible rape, and murder. But there is less difference between the sexes in regard to status offenses and property crimes, with the greatest comparability occurring in larceny or theft. Female arrests actually exceed male arrests in several status offenses, notably runaways and prostitution (*Uniform Crime Reports,* 1985).

At the same time, there are some significant, though often less noticeable changes between the sexes occurring in regard to delinquency. First, females are involved in a wider range of offenses than in the past, with fewer differences between the sexes. Actually, female arrests are increasing at a faster *rate* in regard to some offenses, especially rape, motor vehicle thefts, weapons violations, embezzlement, liquor law violations, and sex offenses. Second, females are disproportionately active in delinquency in the under-15 age bracket than at older ages. Early adolescent females are especially likely to engage in aggravated assaults, larceny, and murder. As Hershel Thornburg (1986) notes, this increase in delinquency among early adolescent girls is not a chance occurrence. It probably represents: (1) an increasing proportion of girls who want to assume a more active female sex role and now see the opportunity to do so; (2) a growing number of girls in a transitional stage between the stereotyped traditional sex role and an emerging more active female role; and (3) girls who are simply unable to live up to role expectations at home, school, and among their peers. Third, while girls are somewhat less likely than boys to be arrested for delinquent acts, they are often treated more harshly than boys. Traditionally, girls are more apt to be turned over to

Females are involved in a wider range of offenses than in the past. (Ken Karp)

authorities by their parents than arrested by police. Once in the juvenile justice system, though, girls are more likely to be institutionalized, given longer terms, and institutionalized at earlier ages. Yet, compared to boys, girls tend to come from more socially disorganized families, be less severely delinquent, and less likely to become repeaters. A major reason for such discrimination is that deviant, aggressive behavior is much less acceptable among girls than among boys in our society (Rutter & Giller, 1983; Westendorp et al., 1986). Perhaps with changing sex-role expectations, females may be arrested more readily and treated more fairly in the juvenile justice system.

CONTRIBUTING FACTORS

The rapid rise in delinquency during adolescence is sometimes attributed to the onset of puberty. Yet, early developers are no more delinquent at that time than those who mature later. Also, even though girls reach puberty earlier than boys on the average, the delinquency behavior of adolescent girls increases at the same time as boys. Furthermore, the relationship between age and delinquency becomes negligible when variables other than sexual development are considered. Such findings suggest that delinquency probably has more to do with the psychological and social processes of role taking and identity formation, especially negative identity, that take place during puberty, such that delinquency is intimately associated with the cultural process of adolescence itself.

The link between delinquency and adolescence can be seen in the idea of an "optimal range of delinquent behavior." That is, delinquent behavior may be seen along a scale of deviancy, ranging from the type of behavior that evokes the admiration of one's peers to that which repels them. For example, a 15 year old who scrawls graffiti on a public wall gives little evidence of courage or skill. On the other hand, a 15 year old who throws gasoline on a stable and burns animals to death is likely to be considered disturbed or vicious, even to his or her friends. But a 15 year old who manages to steal the red flasher off a police car will likely get positive peer recognition for a brave act (Gold & Petronio, 1980).

All things considered, the causes of delinquency are so complex and interwoven, and thus poorly understood, that we will discuss them in the context of the various contributing factors to delinquency. The list of potential contributing factors to delinquency would include such things as the rapid rate of social change in American society, social and economic inequities, the mobility of families that uproots teenagers from their schools and peers, the increase in dual-career families, the importance of peer groups during adolescence, and changing sex roles. Here in this section, we'll focus on several of the major factors, such as the family, socioeconomic factors, personality, psychopathology, and drug abuse.

The Family

Youths who are arrested for delinquent behavior tend to come from families with inadequate childrearing or socialization practices, poor adolescent–parent relationships, and little family cohesiveness.

Parents of delinquent youths tend to be ill-informed or inept in regard to rearing children to become self-respecting, law-abiding individuals. Such parents tend to rely on discipline that is too harsh, too lenient, or extremely inconsistent. They're especially apt to use physical punishment without any accompanying explanation. In his work with families of aggressive and delinquent youths, Gerald Patterson (1981) identified four aspects of family life that are strongly associated with delinquency. First, there is a lack of "house rules," with no predictable family routines or expectations about what adolescents may or may not do. Second, there is a lack of parental supervision, so that parents don't know what adolescents are doing or feeling and tend not to respond to deviant behavior because they haven't seen it. Third, there is a lack of effective contingencies, or a system or rewards and punishment. Parents of delinquent youths are not consistent in their responses to unacceptable behavior. They tend to nag and shout but do not follow through. They also tend to punish undesirable behavior more than they reward desirable behavior. Finally, there are no effective strategies for dealing with family problems and crises. As a result, conflicts lead to tension and dispute but do not get resolved.

One of the single best predictors of delinquency is the lack of affectional ties between adolescents and their parents. Generally, the better young people get along with their parents the less likely they will engage in delinquent behavior. And conversely, the worse youth get along with their parents, the more likely they will eventually commit delinquent acts. Delinquents of both sexes tend to experience more parental indifference and rejection than nondelinquent youth, resulting in more mutual hostility between themselves and their parents. In a series of longitudinal studies, Stott (1982) found that the breach of affectional ties between parents and their adolescents was a major factor in more than 9 out of 10 delinquent acts. Delinquent youth usually have been threatened with rejection or have lost the support of their preferred parent. Mothers are not seen as a dependable source of affection. Feeling rejected, these youths had little or no incentive to avoid parental punishment.

Delinquent youth are also more likely to come from homes lacking in family cohesiveness. An obvious example would be delinquent youths from families broken by parental separation, divorce, or the death of a parent. But in actuality, a lot depends on the age and sex of the adolescent as well as on which parent is absent from the home. One study found that whether an adolescent engages in delinquency or not depends more on the adolescent–mother relationship among boys and girls under 14 years of age, and more on the adolescent–father relationship among older adolescents. It appears that the father's influence assumes greater importance as young people mature and take on new roles in the larger community (Gold & Petronio, 1980). Yet, the absence of the father generally has greater impact on boys than on girls, with boys from father-absent homes much more likely to engage in delinquent behavior than girls from the same homes or boys in father-present homes (Stern, Northman, & Van Slyck, 1984).

It is also important not to equate the lack of family cohesiveness with the absence of a parent through marital separation, divorce, or death. A great deal depends on the adolescent–parent relationships, childrearing practices, and emotional climate within the home. There is increasing recognition that the presence of emotional turbulence and family dissension in the home does more harm to the adolescent's emotional and psychological development than the physical separation of the parents through divorce. However, since families that eventually divorce tend to have a higher-than-average level of dissension as much as 6 years prior to separation and divorce, it's often difficult, if not impossible, to disentangle the effects of emotional turbulence from legal separation and divorce (Morrison, Gjerd, & Block, 1983). In some instances, an equally high or higher number of delinquents come from intact homes characterized by parental indifference, dissension, and mutual hostility, than from homes broken by divorce or death. In contrast, the absence of these factors in many single-parent homes sometimes results in greater family cohesiveness.

Socioeconomic Factors

The highest rates of delinquency, especially for serious offenses, are found among socioeconomically disadvantaged youth in the slum areas of the large cities (Haskell & Yablonksy, 1982). These people are characterized by low income, low education, high unemployment, and overcrowding, as well as coming from racial and ethnic minorities and single-parent families. The prevalence of delinquency among disadvantaged youth has been explained partly

---

INCLINATION

I have read somewhere that children between twelve and fourteen years of age . . . are especially apt to become murderers or incendiaries. When I recall my own adolescence (and the state of mind I was in one day) I can understand the incentive to the most dreadful crimes committed without aim or purpose, without any precise desire to harm others—done simply out of curiosity, out of an unconscious need of action. There are moments when the future looks so gloomy that one fears to look forward to it.

Leo Tolstoy, *Childhood, Boyhood, Youth* (New York: John Wiley, 1904).

Disadvantaged youth in the large cities are especially apt to engage in predatory crimes such as assault and robbery. (Marc Anderson)

on the basis of a deviant subculture. That is, because legitimate opportunities for advancement in wealth and status are lacking, it is commonly acceptable for such people to strive for achievement through other means. As a result, disadvantaged youths often turn to stealing cars, robbing homes and stores, and dealing drugs as socially acceptable means of survival. The opportunities and models for such behaviors are so widespread that youths who engage in them are sometimes referred to as "socialized delinquents." That is, they are more or less normal adolescents who behave this way mostly because they have socialized in a deviant environment.

However, the presumed link between socioeconomic deprivation and delinquency has been questioned by some sociologists in recent years. For instance, the prevalence of delinquency in the urban ghettos has been attributed to discrimination against lower-class and minority youth by police, juvenile authorities, and society generally, such that disadvantaged youth are more likely to be arrested and institutionalized than suburban youth. As a result, the gap between urban and suburban delinquency may be considerably smaller than assumed, with some authorities finding negligible socioeconomic differences (Rutter & Giller, 1983). But such studies have been criticized on several grounds, especially their small sample sizes, the failure to distinguish between serious and minor offenses, and use of self-reported delinquency (Braithwaite, 1981). In contrast, when larger representative samples are used, when the focus is on serious offenses, and when lower-class youth are distinguished from working-class and middle-class youth, significant socioeconomic differences in delinquency emerge. In fact, the higher the level of delinquency or the greater the number of offenses committed, the *greater* the socioeconomic differences in delinquency. At the same time, the pattern of delinquency varies according to the different types of offenses. Predatory crimes against people, such as aggravated assault, sexual assault, and robbery, occur three times more frequently among lower-class youth. Predatory crimes against property, such as auto theft, burglary, and vandalism, occur two times more often among the disadvantaged youth. However, there are no socioeconomic differences in three areas: *victimless offenses* such as drunkenness and disorderly conduct, *status offenses,* and *serious drug use* (Elliot & Ageton, 1980).

The greater prevalence of serious delinquency among disadvantaged urban youth should not blind us to the increasing problem of delinquency among middle-class, suburban youth. In the first place, delinquency continues to *increase* at a faster rate among youth in the suburbs than those in urban and rural areas. The 23-year trend shows suburban arrests were up 4 percent, compared to 3 percent in the cities and 1 percent in rural areas. The sharpest increases in suburban delinquency occurred in motor vehicle thefts, embezzlement, fraud, and drug abuse violations (*Uniform Crime Reports,* 1985).

The rapid rise in delinquency among suburban youth has been attributed to various factors, including influence of peer groups, breakdowns in the family, and materialistic values. Also, despite the greater material affluence in the suburbs, middle-class youth may suffer from another type of deprivation. Levine and Kozak (1979) found that a great deal of suburban delinquency occurs because of deficient socialization and inadequate parenting. Suburban youth are exposed to a strange mixture of material affluence and emotional neglect. That is, parents tend to be concerned that their young people have the necessary material means to fulfill their middle-class roles. When youths feel they lack the material objects needed to continue their middle-class lifestyle, such as, cars, clothes, pocket money, alcohol, and drugs, they may resort to stealing, with the typical suburban delinquent more willing to rob homes in his or her own neighborhood than in the past. Yet, suburban parents are not nearly as conscientious in regard to providing the emotional nurture and guidance their teenagers need. Accordingly, a great deal of suburban delinquency occurs through default, because parents don't know what their teenagers are doing and don't want to be bothered by them. As a result, parents discover only a small proportion of their teenager's offenses. Even when parents discover that their son or daughter has engaged in delinquent behavior, many parents don't do much about it. Too many parents simply ignore, scold, or punish their young people and then try to forget about it.

GANGS

Most delinquency is committed in groups, ranging from the small pick-up groups of two or three people in the suburbs to the larger, more organized drug-dealing groups and burglary rings in the inner cities.

Gangs have probably received more attention than any other type of delinquent group. W. B. Miller (1978) and his colleagues investigated gang activity in six of the larger cities: Chicago, Detroit, Los Angeles, New York, Philadelphia, and San Francisco. He found that gangs ranged in size from 10 to 15 members in cities like San Francisco to more than 30 members in the larger cities like New York. He also found changes in gang membership. Traditionally, gang members were white, from European backgrounds. But today they are mostly nonwhite, from non-European backgrounds. While there has been an increase in

female gang activity, big city gangs remain predominantly male, with females usually playing an auxiliary role. Gang members are generally in their late teens or early twenties and come from the low-income, lower socioeconomic groups in the inner cities.

Modern gangs use more sophisticated weapons, create extensive property damage, and engage in a great deal of violence against people during robberies and extortion activities. There were over 500 gang-related murders in the six cities studied in a 2-year period. At the same time, something can be done about gang violence. When former gang members were recruited to work with citizen groups to stop violence before it started, no more than seven deaths were attributed to gang violence in just 3 years in Philadelphia (Moore, 1979).

Few parents discuss the matter in a calm, concerned way that will diminish the chances of their young people engaging in such behavior again. Apparently, delinquency is regarded as an inevitable part of growing up, something to be taken lightly by parents and youth alike.

Personality

Youths who are arrested for delinquency, especially those who have committed serious offenses or become chronic delinquents, commonly exhibit certain personality characteristics. Delinquents generally have poor self-control, with a marked inability to delay gratification and a tendency to engage in impulsive behavior. They are usually antagonistic and defiant toward authorities, frequently displaying hostility and resentment. They also tend to be less honest and more hedonistic than other youth and generally lack a sense of social responsibility. Delinquent youth characteristically have poor emotional and social adjustment compared to their nondelinquent peers. Although chronic delinquents tend to have lower-than-average intelligence, most delinquents have at least average intelligence, such that intelligence itself may play less of a critical role in delinquency than other psychological and social factors (Rutter & Giller, 1983).

Delinquent youth generally have poor self-images and low self-esteem.

They're apt to describe themselves as bad, ignorant, lazy, and selfish, generally regarding themselves as undesirable people. It is sometimes suggested that the mere act of engaging in antisocial behavior may contribute to the delinquent's low self-esteem. But based on their longitudinal study of delinquents, the Rosenbergs (1978) concluded that self-esteem seems to influence delinquency more than the other way around. In some instances, the delinquents' lack of self-esteem may be readily apparent in their posture and dress as well as in their deviant behavior. At other times, they may not be fully aware of their low sense of self-worth, often displaying a cockiness and toughness admired by others. The explanation seems to be that while some delinquents may exhibit high self-esteem in their surface awareness, most of them harbor a deeper sense of inferiority or low self-esteem in the unconscious level of their psyche (Mann, 1980; Rutter & Giller, 1983).

Psychopathology

acting-out behavior: Behavior in which individuals unconsciously relieve their anxiety or unpleasant tensions through expressing them in overt behavior.

schizophrenia: A major psychological disorder characterized by social withdrawal, ego fragmentation, and distortions of thought, emotion, and behavior.

conduct disorders: Childhood and adolescent psychological disorders characterized by antisocial behavior.

Whenever you hear of teenagers shooting at passing motorists, beating up elderly people, or committing gang rapes, you may find yourself thinking "They're sick!" And in some instances, they may be. Yet, emotional disturbance varies considerably among delinquents, ranging from the milder acting-out behavior commonly seem among adolescents to the more severe psychotic disturbances like schizophrenia. One of the most common links between delinquency and psychological disturbance may be seen in the conduct disorders. These are described in the authoritative Diagnostic and Statistical Manual of Mental Disorders (Third Edition) published by the American Psychiatric Association, generally referred to as DSM-III (1980). The essential feature of these disorders is a repetitive and persistent pattern of behavior in which the rights of others and/or appropriate social rules are violated. Such behavior is considerably more serious than the ordinary mischief and pranks committed by children and adolescents. Not surprisingly, individuals with this disorder tend to come from homes with inconsistent childrearing methods, harsh discipline, rejection, absence of a parent, and/or parents who themselves are addicted to alcohol, illicit drugs, and engage in criminal activities. It isn't hard to imagine how individuals growing up in such homes acquire the characteristic personality traits associated with conduct disorders, such as poor esteem, lack of self-control, hostility toward authorities, and poor emotional and social adjustment in the home, school, and community. Conduct disorders are more common among males than females, with one exception noted below.

There are four basic types of conduct disorders, depending on whether individuals are undersocialized or socialized and aggressive or nonaggressive.

1. *Undersocialized aggressive.*
   Youth with this diagnosis lack meaningful emotional attachment with

others and appropriate feelings of guilt and remorse. They may lie, steal, set fires, break into houses, get into constant fights with others, rape, and murder, all with a characteristic callousness toward their victims. These youths are the closest parallel to the adult antisocial personality disorder.

2. *Undersocialized nonaggressive.*
   These individuals exhibit the same essential features described above, minus the aggressive component. That is, they tend to deceive and manipulate others, but they do not attack them. They are often less emotionally armored and may express the need for approval and feelings of depression. As a result, these individuals may be friendly with others in superficial ways, but they remain weak, selfish people. This particular disorder is equally common among males and females.

3. *Socialized, aggressive.*
   These youth differ markedly from the preceding two types in having normal emotional ties with others. They usually have friends and typically belong to a gang, showing loyalty toward other group members. But to the rest of the world, they behave in many ways like the undersocialized delinquents. Also, when they grow up they are apt to continue their ways in adult, antisocial, criminal activities.

Undersocialized, aggressive youth account for a large proportion of chronic delinquents. (Eugene Gordon)

4. *Socialized, nonaggressive.*
   Individuals in this category engage in socialized behavior that violates the social rules without becoming aggressive. Included here are the youth who repeatedly run away from home, skip school, shoplift, lie, deceive, and deal with illicit drugs.

The outlook for youth with a conduct disorder varies considerably. Those with the undersocialized, aggressive disorder tend to get into trouble early, usually before puberty, and become institutionalized as juveniles. In fact, they constitute a disproportionately large proportion of young offenders placed in juvenile justice centers as well as mental health centers (Westendorp et al., 1986). Also, as adults they are the most likely to be diagnosed as antisocial personalities and to be jailed for adult criminal activities. Youth with socialized aggressive conduct disorders tend to display more adequate social functioning but often persist in various illegal activities, such as fraud, into adulthood. In contrast, youth with the socialized, nonaggressive disorder often grow up and achieve reasonably good social adjustment as adults.

Drugs

The commonsense notion that drugs contribute to crime has achieved considerable support. For instance, half of all the people tested at the time of their arrest for serious crimes in New York City and the nation's capital were found to be using illegal drugs. That's two to three times as many as previously thought by government authorities. The 2-year study by the Justice Department found that 56 percent of those given urinalysis tests had been using at least one of the five illegal drugs—cocaine, PCP, heroin, amphetamines, and illegal methadone—in the previous 24 to 48 hours. Had the testing included marijuana and alcohol, the number of drug users would have been even higher. About one-third of those arrested were multiple-drug users, with cocaine being the most-used drug. The study found that 56 percent of the individuals charged with murder or manslaughter tested positive for drugs, as did 54 percent of those arrested for robbery and 41 percent arrested for sexual assault (*Philadelphia Inquirer*, 1986).

Although the high rate of delinquency is often blamed on young people's use of alcohol and illegal drugs, researchers such as Johnston, O'Malley, and Eveland (1978) point out that the association between drug use and delinquency is more complex than a simple cause and effect relationship. In the first place, many youths occasionally use alcohol or smoke pot without engaging in serious delinquency. Other youths become involved in delinquent behavior before using alcohol or illicit drugs. Then there are those who would engage in delinquent behavior whether they used drugs or not. At the same time, there are countless youths whose drug abuse problem contributes directly or indirectly to criminal activity.

One of the strongest links between drug abuse and crime, as Robert O'Brien and Sidney Cohen (1984) point out, is found among heroin users, especially unemployed heroin addicts. These drug users in particular need a vast amount of money to support their habit, usually more than they have access to by legal means. In order to support their habit, heroin addicts engage in such crimes as theft, burglary, drug dealing, forgery, and gambling. Many drug addicts commit more than one type of crime, though they tend to focus on a particular criminal activity such as drug dealing. The younger individuals become addicted to drugs, the more heavily involved in criminal activities they become later on. Furthermore, the seriousness of the criminal acts grows greater with age. Whereas muggings and purse snatchings are common among youthful offenders, burglary and armed robbery are more likely among older offenders. A study of drug-using delinquents found that the most frequent criminal activities involved property crimes, which, in descending order, included: drug sales, burglary and other types of theft, forgeries, and assaults and robberies. Yet only a fraction of their crimes resulted in an arrest: 250 crimes to one arrest. Drug sales were the least likely to be detected, with a 761 to 1 ratio, whereas acts of assault and robbery were the most likely to result in arrest with a 35 to 1 ratio.

The proportion of alcohol-addicted youth with a record of delinquency is also heavier than average compared to other youth. Although the statistics involving alcohol-related delinquency vary considerably, alcohol involvement has been found in up to one-half of all rape offenders, especially those involved in gang rapes; two-thirds of all assault offenders; and three-fourths of all murderers. Alcohol also contributes to a wide variety of other deviant behaviors, such as drunk driving, vehicular homicide, and family violence. A common explanation for the link between alcohol and violence is that alcohol releases inhibitions and intensifies certain mental states, and that the release of anger may enhance the likelihood of violence. The combination of alcohol with other drugs may result in increased violent behavior (O'Brien and Chafetz, 1982).

Almost any mood-altering drug has some potential for accentuating criminal activity. For instance, because barbiturates are sedative, hypnotic drugs, you'd think abusers of these drugs would simply fall asleep. But these drugs can also lead to argumentative, irritable behavior. Members of motorcycle gangs have been known to take "downers" before engaging in street fights, resulting in the release of aggression similar to that produced by alcohol. Drug use can also produce false feelings of bravado and ideas of suspicion and persecution. Such paranoid thoughts can lead to outbursts of violence which, in turn, can lead to criminal behavior. Ordinarily, it is usually in moderate doses that a drug leads to violence. Small doses do not generally lead to aggressiveness, and large doses are often incapacitating. At the same time, the setting in which youths take a drug, along with their mood, personality, and disposition to crime, can also alter a drug's impact on their behavior and thus accentuate the possibility of criminal activity (O'Brien & Cohen, 1984).

TREATMENT AND PREVENTION OF DELINQUENCY

For hundreds of years, youthful offenders were treated much like adult criminals, with the emphasis on handing out justice for violations of the law. Older children and adolescents were sometimes subjected to unduly harsh punishment. For instance, under the old English law, 12 and 13 year olds could be put to death for stealing small sums of money. Then with the increased understanding of human development that surfaced in the late nineteenth century, enlightened authorities instituted the juvenile court, with the aim of treating youthful offenders differently than adults. Unlike the criminal court with its philosophy of punitive justice, the juvenile court aimed at a more humane and appropriate treatment for immature offenders. Rather than fitting the punishment to the severity of the crime, as with adult defendants, judges in the juvenile court acted as wise parents dealing with youth according to the "best interests" of the child, with an eye toward rehabilitation. However, an unintended result was that juvenile courts acquired an almost absolute authority over youth, with glaring inconsistencies in the way they have been handled. For instance, a 16-year-old murderer who evokes the compassion of the judge might be released when witnesses don't show up for the hearing, whereas a 14 year old who steals a woman's purse might be institutionalized by another judge (Kaufman, 1979).

rehabilitation: Treatment that aims to restore individuals to a normal or optimum state of physical or psychological health.

The abuses of the juvenile court system received national attention in the Gault case. In 1964, Gerald Francis Gault, a 15 year old, was arrested for making an obscene phone call to a neighbor. Had he been of legal age, he might have received a small fine and a 2-month jail sentence. But being a juvenile, he was treated differently. Gault was sentenced to an institution until he reached 21 years of age. However, in the process of being convicted, Gault was denied many of his rights. Gault's parents were not notified, the neighbor who lodged the complaint never appeared in court, the judge never spoke to the complainant, and there was no record of the proceedings. Eventually, the case was taken to the U.S. Supreme Court, which in 1967 reversed the decision of the lower courts and freed Gault. The decision initiated widespread reforms in the juvenile justice system. As a result, youthful offenders now have the same civil rights accorded adult citizens, including the right to know the charges against them, the right not to incriminate themselves, the right to have a lawyer represent them, to have that lawyer cross-examine witnesses, and the right to appeal the decision to a higher court. The court decision also set in motion other reform measures, such as keeping status offenders in separate physical facilities (Prescott, 1981).

In this last section we'll take a look at the various components of the juvenile justice system, including the police, juvenile court, and correctional institutions as well as some of the current approaches to treatment and prevention.

Police and the Juvenile Court

Police are among the first authorities to come into contact with juvenile offenders, with the majority of arrests resulting from citizen complaints. Police usually have a great deal of discretionary power and may choose to deal with the juvenile offender in a number of ways: (1) they can let the offender go at the scene of the offense with only a verbal warning; (2) they can do the same thing, except at the police station, perhaps accompanied by a phone call to the individual's parents; (3) they can refer the adolescent to a social agency for help; (4) they may also choose to hold the offender in temporary custody; or (5) while in custody, they may petition for a court hearing. Police generally dispose of up to one-half of all juvenile arrests themselves (*Uniform Crime Reports*, 1985).

Once a petition has been made to the juvenile court, a given procedure is usually followed, though it may vary somewhat from one legal jurisdiction to another. Ordinarily, the youthful offender must be a given a preliminary hearing within a stated period, for example, 72 hours. Prior to the hearing

Youthful offender and police in juvenile court. (Jon Huber)

juveniles may be released to the custody of their parents or kept in a juvenile detention center, depending on the individual and the offense. At the preliminary hearing, the offender may be dismissed outright, referred to a social agency, placed on informal probation, or referred to a formal hearing. When a case warrants a formal hearing in juvenile court, there is a written petition stating the charges and verification of the youth's right to counsel at the preliminary hearing. Generally, the juvenile court hearing is conducted by the judge, with the offender, his or her parents, and others, such as the offender's lawyer, present. There may be two formal hearings: an adjudication hearing, at which the offender is convicted of the charges or not, and a disposition hearing, at which the judge determines what will be done with the offender. Between the two hearings there should be ample time for a social investigation to determine which course of action will be best for the offender.

The juvenile justice system may be compared to a giant funnel with many cracks in it (*Connecticut Juvenile Justice System*, 1978). Juvenile offenders who are arrested enter the system at the wide end of the funnel at the top. But as they progress through the system, the great majority of youths slip out through the "cracks" in the system. Most offenders are dealt with at the initial stages of the system, whether by the police, school authorities, or at the preliminary hearing. Once in the courtroom, many of the remaining offenders are found not guilty or dismissed, while others are placed on probation or supervision. Only a small minority of cases proceed through the entire juvenile justice system and are placed in a residential treatment program, and for good reason, as explained below.

Current Approaches to Treatment

Once juvenile offenders are committed by the courts, they may be placed in a variety of facilities and programs. Almost two-thirds of them (60 percent) will be put in public facilities. (See Figure 13–2.) Most of these individuals are kept in detention centers and training schools, with only a small proportion of them in smaller settings like group homes or ranches. In contrast, the majority of youth placed in private facilities stay in halfway houses, group homes, ranches, forestry camps, or farms (U.S. Bureau of the Census, 1985). The programs in these institutions tend to be based on the combined goals of correction and rehabilitation. Yet, neither of these goals is accomplished very well. Instead, institutionalized youth tend to look down on themselves and are exposed to an even more unfavorable environment with a core of serious offenders who model and reinforce negative behavior. All of this results in a poor rate of rehabilitation, such that anywhere from 60 to 75 percent of juveniles held in correctional facilities are arrested again (Rutter & Giller, 1983).

The low success rate of conventional programs has led to experimentation

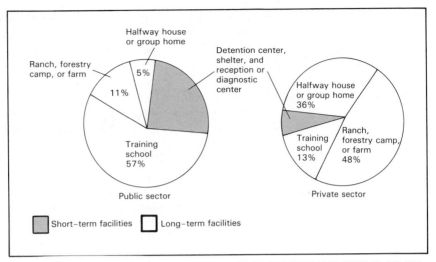

FIGURE 13–2 DISTRIBUTION OF JUVENILES IN CUSTODY BY PUBLIC OR PRI-
VATE FACILITY. Almost two-thirds (60 percent) of the 80,000 juveniles in custody are
housed in public facilities, with the remaining third in private facilities.
Sources: U.S. Law Enforcement Assistance Administration. *Children in Custody.* (Washington, DC: Gov-
ernment Printing Office, 1979), p. 6. U.S. Bureau of the Census, *The Statistical Abstract of the United
States, 1986,* 106th ed. (Washington, DC: Government Printing Office, 1985), p. 182.

probation: The suspension of a sentence of a person convicted but not yet imprisoned on condition of continued good behavior and reporting to a probation officer.

with a variety of community-based programs. Many of these programs rely on a behavioral approach to the treatment of delinquency, in which appropriate behavior is systematically rewarded and deviant behavior is either not rewarded or results in unpleasant consequences, such as a "time out" period or the temporary withdrawal of some privilege. In one program, for example, the focus is on teaching positive social skills that have a wide application in everyday life, such as problem solving, communication skills, and how to get a job. A 5-year followup study showed that the juveniles who had participated in this program while in custody were much less likely to get into trouble again than those in the control group, with a 23 percent rearrest record compared to 48 percent for other juvenile offenders (Rutter & Giller, 1983).

Authorities are also looking anew at probation, with about four juvenile offenders put on probation for every one placed in a correctional facility. Probation appears to be one of the most effective approaches to delinquency, in that individuals on probation generally have lower rearrest records than those kept in institutions (Singer & Isralowitz, 1983). At the same time, the success rate of probation depends on certain conditions. First, it is generally more effective with individuals who are arrested the first time for minor offenses, rather than those with a record of serious offenses. Second, there should be clear conditions associated with the probation, such as school attendance, work requirements, and a curfew. Third, there must be adequate supervision and assistance, with an eye toward helping the juvenile fulfill the conditions of his or her probation. However, the latter condition usually requires a reduced workload of the probation officer or case worker, which is not always possible in a more extensive application of this approach.

family therapy: Counseling for the entire family on the assumption that the disturbance of one family member reflects problems in the whole family's interactions.

Family therapy has also become an increasingly popular approach to the treatment of delinquent youth. Time and again, youthful offenders have shown improvement on an individual basis, only to relapse once they have returned home. By dealing with the home situation, family therapy increases the probability of long-term gains in the individual's behavior. Although there are over a dozen types of family therapy, as we mentioned earlier in the book, most of them assume that each family functions as a system, in such a way that each adolescent's behavior is best understood and changed within the family unit. For example, Bill is a 16 year old who has repeatedly gotten into trouble at school with his low grades, disruptive behavior in class, and skipping school. Since the school counselor discovered that Bill doesn't get along well with his parents, family therapy was recommended. In dealing with Bill and his family, the therapist soon discovered that Bill's difficulties at school were related to the family dynamics. Unlike his older sister in college and his younger brother who was the family favorite, Bill appeared to be the "black sheep" of the family who was always getting into trouble. During the next few months the therapists pointed out several factors that contributed to Bill's problems, such as his parents' harsh discipline, lack of supervision, poor communication with Bill, and a marked tendency to make Bill the scapegoat for the family's problems. In promoting more positive behavior on Bill's part, the therapist proposed the use of a family contract, a commonly used and effective method of family therapy. Bill and his parents drew up a behavioral contract regarding his desired performance at school, including attendance, study habits, and a certain grade-point average. They also agreed upon the procedures by which Bill's goals would be attained and how other family members would assist him. After 6 months Bill's school performance was no longer a problem. Although family therapy is not always this effective, it often serves as an alternative to institutionalizing youth and may also shorten the stay in cases when the individual is placed in a residential treatment program (Ault-Riche & Rosenthal, 1986).

Preventing Delinquency

All too often, our approach to the treatment and prevention of delinquency is distorted by two illusions. Either we think there's nothing we can do about it, or we believe that somewhere, somehow we'll find a cure for it. As we've already seen, there is no simple cure for the problem because delinquency is the combined product of so many factors, such as the kind of society and the times in which we live. Yet, this does not mean that we must succumb to the opposite myth that nothing can be done about delinquency. We've already seen that some strategies work better than others. Also, anything that improves the root causes of delinquency, such as social and economic inequities and family instability, indirectly helps diminish delinquency. But there are also some direct preventive steps we can take.

First, we can encourage people to identify and provide help for youth who are at high risk for delinquency. Because delinquency begins in the early school grades and chronic delinquents get into trouble sooner than nonrepeaters, the sooner we can intervene in the lives of such youth in a positive way, especially in the school, the better.

Second, intervention strategies should focus on teaching positive social skills rather than simply suppressing deviant behavior. Troubled youth need help in feeling better about themselves and acquiring the social skills and competence to find a meaningful role for themselves in society. Yet, this type of assistance is more demanding than simply controlling negative behavior.

Third, we can support the existing trend toward dealing with minor delinquency and status offenses outside the conventional juvenile justice system. Since the system's "cure" tends to worsen the problem, every effort should be made to exhaust preventive measures such as early warning and release, informal probation, counseling, and social agencies before proceeding further into the juvenile justice system.

Fourth, we can encourage better use of local police in handling juvenile offenders. Because the police are among the first to encounter delinquent youth and have great discretionary powers in handling them, there's much to be gained in training police to deal with youthful offenders in the early stages of delinquency.

Fifth, we can continue the trend toward greater use of small, community-based programs such as halfway houses and group homes. In these programs youth are more likely to be given the warmth and individual attention they need as well as integration into the community, resulting in lower rates of repeated delinquency (National Commission on Youth, 1980).

Finally, we need to realize that delinquency partly reflects the society in which we live. Consequently, anything which helps to strengthen family life, alleviate economic conditions in the urban ghettos, reduce ethnic, racial, and sexual discrimination, improve the schools, and reduce drug abuse helps, however indirectly, in the continuing struggle against delinquency.

SUMMARY

The Extent of Delinquency

1. Over three-fourths of Americans commit one or more delinquent acts during their adolescence, though relatively few of them are ever arrested for such behavior.

2. The frequency and seriousness of delinquent behavior rises with age, peaking in late adolescence and early adulthood before subsiding.

3. The biggest single category of offenses committed by youth is property crimes, with youth under 25 responsible for three-fourths of all burglaries, vandalism, and motor vehicle thefts.

4. About one-third of youthful delinquency involves status offenses, such as running away from home, with a significant proportion of runaways being rejected at home.

5. The male—female ratio of reported delinquency is about 3 to 1, with males more likely than females to commit violent crimes such as assault, rape, and murder, and females having greater comparability in status offenses and theft.

Contributing Factors

6. The causes of delinquency are complex and interwoven, and reflect the society in which we live as well as the social, cultural process of adolescence.

7. Delinquents generally come from families characterized by overly harsh or lenient discipline, lack of affectional ties between youths and their parents, and lack of family cohesiveness.

8. The highest rates of delinquency, especially for serious offenses, are found among disadvantaged youth in the inner cities, though the gap between urban and suburban delinquency continues to diminish.

9. Delinquent youths tend to lack self-esteem, self-control, and a sense of social responsibility.

10. Chronic juvenile offenders often exhibit psychopathology, especially characteristic of the undersocialized, aggressive conduct disorder.

11. Although the relationship between drug use and delinquency is complex, drug abuse is a significant factor in such crimes as theft, burglary, and drug dealing.

Treatment and Prevention of Delinquency

12. Police are among the first authorities to come into contact with juvenile offenders and dispose of up to half of all juvenile arrests themselves.

13. The juvenile justice system can be compared to a giant funnel, with the great majority of juvenile offenders slipping through the cracks in the system and only a small minority being placed in residential treatment programs.

14. Although the majority of juvenile offenders committed by the courts are placed in large public correctional facilities, the low success rate of these institutions has led to experimentation with a variety of community-based programs.

15. Practical steps in preventing delinquency include identifying high-risk youth early in their development, providing positive social skills training, diverting low-risk youth from entering the juvenile justice system, and emphasizing community-based programs.

REVIEW QUESTIONS

1. To what extent does the frequency and seriousness of delinquency increase with age?

2. Which offenses are committed most frequently by youth under 25?

3. To what extent are there sex differences in delinquency?

4. How do families contribute to delinquent behavior?

5. How important are the socioeconomic factors in delinquency?

6. Explain the four types of conduct disorder.

7. Explain the funnel model of the juvenile justice system.

8. What are the pros and cons of putting juvenile offenders on probation?

9. What role does drug use and abuse play in delinquency?

10. How do you think we can decrease the rate of serious delinquency?

14 Youth and Drugs

- **USE OF DRUGS**
  The Extent of Drug Use
  The Development of Drug Use
  Why Youths Use Drugs

- **ALCOHOL AND TOBACCO**
  The Effects of Alcohol
  Use of Alcohol
  Psychological and Social Factors
  Tobacco and Smoking

- **SELECTED ILLEGAL DRUGS**
  Marijuana
  Stimulants
  Depressants
  Other Drugs

- **PUTTING DRUGS IN PERSPECTIVE**
  Relationships with Drugs
  Treatment of Drug Abuse
  Alternatives to Drugs

- **SUMMARY**

- **REVIEW QUESTIONS**

People have been experimenting with mind-altering drugs longer than most of us realize. In ancient Greece, the Oracle of Delphi used drugs, and Homer's Cup of Helen induced sleep and provided freedom from care. And among people in the Old Testament, the mandrake root supplied hallucinogenic compounds associated with lovemaking. The ancient Assyrians sucked on opium lozenges, and the Romans ate hashish sweets 2000 years ago (Jones-Witters & Witters, 1983).

In the nineteenth century, the man who synthesized nitrous oxide, popularly known as laughing gas, held laughing gas parties in his home. Later, inhalations of nitrous oxide were sold at county fairs for a quarter, with individuals inhaling it for its pleasurable sensations and colorful fantasies. Although a few customers may have discovered final truths, most went away with little more than a laughing jag, dizziness, or an upset stomach. Students at Cambridge University held chloroform parties before its toxic effects were discovered, while American students used ether for its consciousness-expanding properties. In 1886, John Pemberton mixed coca leaves with wine in what he called a French wine coca. A year later he added caffeine from the kola nut and called the product "coca cola." It wasn't until the turn of the century that the company switched to the use of "decocainized" leaves (O'Brien & Cohen, 1984).

USE OF DRUGS

drug abuse: The chronic or excessive use of any drug, especially when this jeopardizes one's health or interferes with one's psychosocial adjustment.

illegal drugs: Over-the-counter sale and possession of such drugs are prohibited by law.

However, drug use today has become a far more serious matter, with more people using and abusing drugs than ever before. For one thing, there are more powerful drugs available through an extensive network of illegal dealers. Also, youths are beginning to use drugs at earlier ages than ever before. For instance, half of all adolescents have used alcohol and one-fourth of them have tried marijuana by the ninth grade (Johnston, O'Malley, & Bachman, 1986). The lack of judgment and emotional resiliency at this age increases the risk of harm. Then too, drug abuse is having more wide-ranging harmful effects on society—from impaired performance at school and work to increased crime and substance-related accidents and deaths. Furthermore, each year thousands of individuals are being asked to take a drug test at work, which has already become a hotly contested issue. Not surprisingly, according to a Gallup (1983) youth survey, young people say drug abuse is the biggest problem facing their generation. Many adults would agree—another Gallup poll (1986) shows that public school parents regard the use of drugs as the biggest problem facing the schools today.

We'll begin the chapter by taking a look at the use of drugs among young people. Then we'll discuss the use of alcohol and tobacco, the two most commonly used drugs among youth and adults alike. Later in the chapter, we'll examine the use of selected illegal drugs, such as marijuana and cocaine. Finally, we'll take a look at people's relationships with drugs, along with the treatment of drug abuse, and the importance of personal choice and alternatives to drugs.

The Extent of Drug Use

One of the most reliable sources of information about drug use among young people is the well-known annual high school survey and followup studies conducted by Lloyd Johnston, Patrick O'Malley, and Jerald Bachman (1986) at the University of Michigan Institute for Social Research, which is funded by the National Institute on Drug Abuse. Each year they gather extensive confidential information about drug use from a national sample of 17,000 high school seniors at both private and public schools. The data obtained from these surveys, together with the followup studies, give us a well-rounded picture of drug use among young people including high school students, college students, and other post-high school young adults up through their late twenties. Most of the data in the following section is based on the survey of the high school class of 1985, together with follow-up studies of graduates of earlier classes.

As in the past, alcohol and cigarettes are used more frequently than any illegal drug. Nearly all the high school students (92 percent) have tried alcohol by their senior year, with the great majority (66 percent) having used it in the

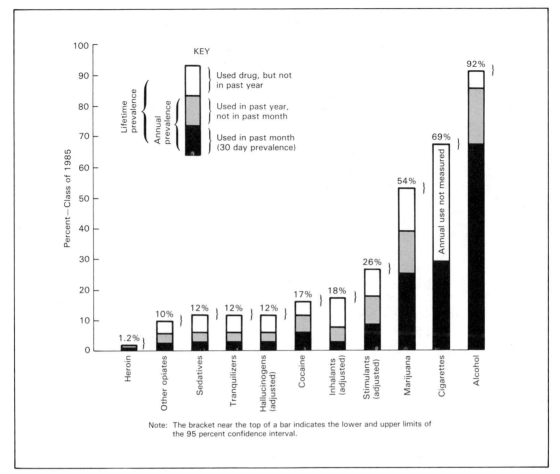

FIGURE 14—1 DRUG USE AMONG HIGH SCHOOL SENIORS.

Source: From Lloyd D. Johnston, Patrick M. O'Malley, and Jerald G. Bachman, *Drug Use among American High School Students, College Students, and other Young Adults* (Washington, DC: National Institute on Drug Abuse, U.S. Government Printing Office, 1986), p. 26.

past month. More than two-thirds of the students report having tried cigarettes at some time, with about one-third of them having smoked some during the past month. Nearly two-thirds (61 percent) of the seniors have used an illegal drug at some time during their lives. About one-fifth of them have used only marijuana. But 4 in 10 seniors have used an illegal drug other than marijuana. The most widely used illegal drugs, in the descending order of the frequency of use, are: stimulants, inhalants, cocaine, hallucinogens, sedatives, and tranquilizers. (See Figure 14–1.)

The high proportion of young people who have at least tried any illegal

Table 14—1   LIFETIME USE OF DRUGS AMONG HIGH SCHOOL SENIORS, ACCORDING TO SEX AND COLLEGE PLANS (in percents)

	Mari-juana	Inha-lants	Amyl/ Butyl Ni-trites	Hallu-cino-gens	LSD	PCP	Co-caine	Her-oin	Other Opi-ates	Stim-ulants (ad-justed)	Seda-tives	Barbit-urates	Meth-aqua-lone	Tran-quil-izers	Al-cohol	Cigar-ettes
All Seniors	54.2	15.4	7.9	10.3	7.5	4.9	17.3	1.2	10.2	26.2	11.8	9.2	6.7	11.9	92.2	68.8
Sex:																
Male	56.6	18.5	11.1	12.4	9.4	6.6	19.7	1.4	11.3	24.6	12.3	9.9	7.1	11.7	92.6	67.4
Female	51.5	12.4	4.9	8.0	5.6	3.1	14.8	0.8	9.1	27.6	11.0	8.3	6.0	11.7	91.9	69.7
College Plans:																
None or under 4 yrs	59.1	16.5	9.2	12.5	9.7	6.8	20.2	1.6	11.5	31.9	15.2	11.9	8.7	13.4	93.0	75.9
Complete 4 yrs	50.2	14.5	6.9	8.0	5.6	3.4	14.6	0.9	9.3	22.6	9.6	7.4	5.3	10.8	91.9	63.7

Source: Adapted from Lloyd D. Johnston, Patrick M. O'Malley, and Jerald G. Bachman, *Drug Use Among American High School Students, College Students, and Other Young Adults* (Washington, DC: National Institute on Drug Abuse, U.S. Government Printing Office, 1986), p. 32.

drug by their senior year in high school (61 percent) grows substantially so that by their midtwenties more than three-fourths (80 percent) of them have used an illegal drug. At the same time, the *active* use of many illegal drugs remains at levels similar to those observed among high school seniors. In fact, compared to high school students, young adults actually have lower rates of annual use of LSD, the barbiturates, and opiates other than heroin. The major exception, of course, is the use of cocaine, which continues to rise until about age 22. Although college-bound high school seniors are less likely to use most drugs than those not planning on a 4-year college degree, the differences between these two groups diminish with age, so that college students show usage rates for a number of drugs which are about average for all young adults their age. Also, college students have lower than average rates in regard to the use of LSD, barbiturates, and tranquilizers.

Males are more likely than females to use most illegal drugs at all ages. And the differences tend to be the largest at the higher frequency level. For instance, about twice as many high school males (6.9 percent) as females (2.8 percent) use marijuana on a daily basis. Among college students, three times as many males as females (4.9 percent versus 1.6 percent) use marijuana daily. The major exception to the rule that males are more frequently users of illegal drugs than females occurs for stimulant use in high school and tranquilizer use among post-high school young adults. In both cases, females report slightly higher rates of use. However, over the years the trend is toward less difference between the sexes, with the convergency of rates due mostly to the faster drop in drug use by males. (See Table 14–1.)

The use of many of the illegal drugs has generally declined since 1975, with the notable exception of cocaine. Given the growing publicity about the hazards of cocaine, one might expect a decrease in cocaine use. But lifetime prevalence and active use of cocaine has been rising with age. While about 17 percent of individuals have used cocaine by their senior year in high school, roughly 4 in 10 of them have tried cocaine by 27 years of age. Two other classes of drugs which show relatively small increases among high school seniors in the mid-1980s are PCP and opiates other than heroin.

The Development of Drug Use

One of the disconcerting findings of the annual high school surveys is the trend toward initial drug use at earlier grade levels than in the past. Most of the youths who try alcohol, cigarettes, and marijuana do so before they reach high school or the tenth grade. About half (56 percent) of the students who try alcohol and one-fourth (28 percent) of those who try marijuana have done so before high school. Furthermore, between 40 and 50 percent of the eventual users of most of the illicit drugs have tried them prior to the tenth grade. Included here are methaqualone, barbiturates, heroin, PCP, amphetamines, and tranquilizers. And a substantial minority, about one-third, of those who

Table 14–2 GRADE IN WHICH DRUG WAS FIRST USED BY HIGH SCHOOL STUDENTS, CLASS OF 1985 (in percents)

Grade in which drug was first used:	Mari-juana	Inhal-ants	Amyl/ Butyl Ni-trites	Hallu-cino-gens	LSD	PCP	Co-caine	Her-oin	Other Opi-ates	Stimu-lants (ad-justed)	Seda-tives	Barbit-urates	Meth-aqua-lone	Tran-quil-izers	Alco-hol	Cigar-ettes (daily)
6th	3.5	2.1	0.3	0.3	0.1	0.5	0.2	.02	0.4	0.7	0.8	0.6	0.3	0.6	9.7	2.6
7–8th	12.0	3.8	1.2	0.9	0.6	0.6	0.8	0.2	1.0	3.9	2.1	1.7	1.1	1.6	23.0	5.2
9th	12.5	2.4	1.1	2.6	1.8	1.2	2.2	0.2	2.2	7.5	3.0	2.2	2.0	2.8	22.8	4.9
10th	12.1	2.3	1.5	2.8	2.0	0.9	3.8	0.3	2.7	6.7	3.5	2.8	2.1	2.8	18.5	4.2
11th	8.7	3.0	2.2	2.2	1.8	1.0	5.5	0.2	2.0	4.7	1.8	1.3	1.0	2.6	11.8	3.6
12th	5.4	1.7	1.6	1.5	1.1	0.6	4.7	0.2	2.0	2.6	0.6	0.7	0.3	1.6	6.4	1.7
Never used	45.8	84.6	92.1	89.7	92.5	95.1	82.7	98.8	89.8	73.8	88.2	90.8	93.3	88.1	7.8	77.9

Source: Adapted from Lloyd D. Johnston, Patrick M. O'Malley, and Jerald G. Bachman, *Drug Use Among American High School Students, College Students, and Other Young Adults* (Washington, DC: National Institute on Drug Abuse, U.S. Government Printing Office, 1986), p. 80.

eventually try the hallucinogens, LSD, nitrites, and opiates other than heroin have tried them prior to high school. Cocaine presents a contrasting picture to nearly all other drugs in that the initiation rates are higher in the last 2 years of high school and continue up through the midtwenties. Fewer than one out of five eventual users of cocaine have initiated use of this drug prior to high school (Johnston, O'Malley, & Bachman, 1986). (See Table 14–2.)

The sequence in which young people try various drugs tends to follow a typical pattern. In a longitudinal study of high school students, Denise Kandel (1980) found that the majority of drug users begin with legal drugs before progressing to the illegal ones. Most teenagers who use a legal drug start with beer or wine. Two to three times as many beer and wine drinkers progress to hard liquor as progress to cigarettes. At the same time, more than half of those who smoke eventually use hard liquor. But few if any adolescents go straight from beer and wine to illegal drugs without first trying hard liquor or cigarettes. The usual sequence is from alcohol and tobacco to marijuana, which in turn is a crucial step towards the use of other illegal drugs, whether stimulants, sedatives, LSD, or heroin. Very few adolescents (only 2 to 3 percent in Kandel's study) go directly to illegal drugs without first trying marijuana. By contrast, over one-fourth of the marijuana users tried another illicit drug within a 6-month period.

The sequential pattern of drug use held true at all grades throughout high school and was similar for both sexes. At the same time, Kandel points out that this is not a causal progression. Many youths stop at a particular stage of drug usage and never go any further. The most likely explanation is that different factors are related to drug usage at each stage and are probably dependent on cultural as well as psychological influences. Furthermore, because each stage represents a cumulative pattern of drug use and contains fewer adolescents than the preceding one, it is important that comparisons be made among a restricted group of drug users. Youths who occasionally use marijuana especially resent being put in the same category as users of the hard drugs, and for good reasons. In fact, it is not uncommon to hear young people say, "I smoke a little marijuana, but I don't do drugs."

Why Youths Use Drugs

drug dependence: The state arising from repeated administration of a drug on a periodic or continuous basis; drug dependence is subdivided by types, such as drug dependence of the cocaine type.

Individuals may use mind-altering drugs for any number of reasons. Given the natural curiosity of the young and the availability of drugs, young people may try a drug like alcohol or marijuana mostly "to see what it's like." An even more common reason, especially for those who continue using drugs, is to enhance social interaction or "to have a good time with my friends." The various reasons people use drugs reflect several basic patterns of drug use as follows:

Experimental use includes trying one or more drugs on a short-term basis, up to 10 times per drug. Users are motivated mostly by curiosity or the desire

physical dependence:
Physiological adaptation of the body to the presence of a drug, such that the body reacts with predictable symptoms if the drug is abruptly withdrawn; sometimes known as "addiction" or "physiological addiction."

to experience new feelings, and rarely use drugs on a daily basis or to escape personal problems.

Social-recreational use occurs more among friends and acquaintances socializing with each other. Users do not usually increase the frequency or intensity of drug use, or use drugs like heroin.

Circumstantial-situational use is undertaken for a desired effect. Common examples would be students' taking stimulants to study or sedatives to go to sleep. The greatest danger is an undue reliance on drugs to the point where they become habitual.

psychological dependence: A felt need for a drug and its effects brought about by habitual use or taking a drug to avoid withdrawal symptoms once physical dependence has developed; sometimes known as "habituation."

Intensified drug use refers to a long-term pattern of daily drug use to alleviate personal problems or stressful situations. Drug use is usually integrated into the user's social and community life, and users associate mostly with other users. Behavioral change may also occur, depending on the frequency and intensity of drug use, as well as drug dependence.

Compulsive drug use generally includes frequent and intense use of drugs over a relatively long time, resulting in both physical and psychological dependence. This may include the more conventional alcohol-dependent people as well as the heroin addicts who need their daily fix to function in everyday life.

When Lloyd Johnston and Patrick O'Malley (1986) asked young people themselves why they use drugs, the responses were revealing. The reason most often given for use of *any* drug was "to have a good time with my friends." A substantial but smaller number of students indicated they used drugs "to feel good and get high." These reasons were especially characteristic of people who continued to use drugs as opposed to those who had simply tried a drug or used it occasionally. Thus, a major reason for using drugs is social-recreational. This may also help to explain the strong association between drug use and peer relationships, with those who have recently used a specific drug, such as alcohol, more likely to report that they have been with others using this drug, and that most of their friends use it. Slightly less than half (41 percent) of the seniors mentioned that they used drugs "to relax or relieve tension." The two drugs most often taken to achieve a desired effect were alcohol and marijuana, the "beer" of illegal drugs. At the same time, there has been a trend away from the social-recreational use of the amphetamines toward using them for a specific purpose, such as to lose weight or to get more energy.

Many of the self-reported reasons for using drugs clustered around coping with negative affect, such as "to deal with anger and frustration," "to get away from my problems," or "to get through the day." These reasons were mentioned by a large proportion of youth using the central nervous system depressant drugs, especially alcohol, barbiturates, and tranquilizers. Although there were few differences between the sexes in regard to the reasons for using any drug, females were slightly less likely to use drugs for social-recreational reasons, and at the high levels of frequency, somewhat more likely to mention using drugs for coping with negative affect, for self-medication, or some other functional reason. For each drug, the more frequent users gave more reasons, especially those who used a drug on a daily basis. According to the researchers, this pattern suggests that many of the heavier users, especially those who use alcohol and marijuana on a daily basis, turn to drugs for psychological coping, for example, to alleviate boredom or frustration, or to get energy.

ALCOHOL AND TOBACCO

Alcohol and nicotine (tobacco) continue to be the most commonly accepted psychoactive drugs among young people and adults, as they have been for decades. One reason is their ease of availability. Another reason is the widespread acceptance of smoking and drinking of alcoholic beverages among adults, making for less of a generation gap than in regard to other drugs. As a result, tobacco and alcohol, in that order, remain—after caffeine—the most commonly used and abused drugs among youth and adults alike. Furthermore, the two drugs tend to be used together, so that heavy drinkers are apt to be heavy smokers, and vice versa (O'Brien & Cohen, 1984).

The Effects of Alcohol

alcohol: A colorless, volatile liquid that is the intoxicating element of beer, wine, and whiskey.

When ingested, alcohol is rapidly absorbed from the stomach, small intestine, and colon and is distributed throughout all the tissues of the body. It acts as a depressant for the central nervous system, diminishing the degree of voluntary control over behavior. Taken in small to moderate amounts, alcohol acts as a social lubricant, making people feel less inhibited and more relaxed. It also slows the reflex actions and alters the individual's sense of time, memory, and reasoning ability. When taken in larger amounts, alcohol has a more marked effect on mood, perception, judgment, and bodily functions. It also has an adverse effect on the kidney and liver. In excessive amounts, alcohol may lead to coma and death, though individuals usually become unconscious before reaching a fatal dose. The amount of alcohol needed to become intoxicated varies with the individual, depending on body weight, personality, and previous experience with alcohol. (See Table 14–3.)

Use of Alcohol

Adolescents are beginning to drink at earlier ages, consume larger amounts, and remain high for longer periods of time than in previous years. In their annual survey of high school students, Johnston, O'Malley, and Bachman (1986) found that one-half of the students had already tried alcohol by the

Table 14–3 RELATIONSHIP BETWEEN BODY WEIGHT AND BLOOD ALCOHOL LEVELS

Blood Alcohol Levels (percent alcohol in blood)

Body Weight (lb)	Drinks*											
	1	2	3	4	5	6	7	8	9	10	11	12
100	.038	.075	.113	.150	.188	.225	.263	.300	.338	.375	.413	.450
120	.031	.063	.094	.125	.156	.188	.219	.250	.281	.313	.344	.375
140	.027	.054	.080	.107	.134	.161	.188	.214	.241	.268	.295	.321
160	.023	.047	.070	.094	.117	.141	.164	.188	.211	.234	.258	.281
180	.021	.042	.063	.083	.104	.125	.146	.167	.188	.208	.229	.250
200	.019	.038	.056	.075	.094	.113	.131	.150	.169	.188	.206	.225
220	.017	.034	.051	.068	.085	.102	.119	.136	.153	.170	.188	.205
240	.016	.031	.047	.063	.078	.094	.109	.125	.141	.156	.172	.188

Under .05:
driving is not seriously impaired

.05 to .10:
driving becomes increasingly dangerous; legally drunk in some states

.10 to .15:
driving is dangerous; legally drunk in most states

Over .15:
driving is very dangerous; legally drunk in any state

*One drink equals 1 ounce of 100-proof liquor or 12 ounces of beer.

Source: Patricia Jones-Witters, Ph.D., and Weldon L. Witters, Ph.D., *Drugs and Society: A Biological Perspective* (Boston, MA: Jones and Bartlett Publishers, Inc., 1983), p. 203.

time they had reached the ninth grade. Practically all of them (92 percent) had tried it by their senior year in high school. At the same time, partly because of the rising concern about alcohol abuse among teenagers, there has been a slight decline in adolescent drinking since 1980. Two-thirds (66 percent) of the seniors use alcohol each month compared to 72 percent in 1980. Daily use also declined from a peak of 6.9 percent in 1979 to a low of 4.8 percent in 1984, but has remained steady since then. Occasional heavy drinking, defined as having five or more drinks in a row, has also declined somewhat among high school students, from 41 percent in 1983 to 37 percent in 1985. However, alcohol remains a serious problem among teenagers, with nearly half of those who use alcohol reporting that they get at least moderately high when they drink, many of them getting very high on occasions. While the perceived harmfulness of alcohol has remained steady over the past few years, an increasing number of students, now including more than 4 out of 10 high school seniors (43 percent), think there is a great risk in having 5 or more drinks in a row.

Actually, the problem with alcohol gets worse after high school. Although college-bound seniors are somewhat less inclined to drink heavily than those without college plans, the situation becomes reversed during the post-high school years. Compared to all high school graduates, college graduates have an above average annual use of alcohol, with 92 percent of college students

Occasional heavy drinking, especially at weekend parties. is a big problem among college students. (Laima Druskis)

using alcohol each year compared to 89 percent of others their age. Unlike the trends among high school students, there has been little drop-off of monthly and daily use of alcohol among college students. But the most alarming trend is the increase in occasional heavy drinking among college students, most of it at parties on weekends. About 45 percent of college students drink heavily on occasion, compared to 41 percent of other young adults their age and 37 percent of high school students. Much of the increase in occasional heavy drinking has occurred among college men.

Psychological and Social Factors

Most adolescents begin drinking at home under parental supervision, especially on holidays and special occasions. The older they become the more likely they are to drink away from home and without adult supervision. By the senior year of high school most drinking is done at teenage parties with no adults present, or sitting in a car or driving around at night. Once young people are away at college, as we've seen, they are even more likely to engage in occasional heavy drinking, especially at weekend parties. At the same time, some individuals are more likely than others to engage in heavy or problem drinking. In a comparison of youths who used or abused drugs, Anthony Jurich (1985) and his colleagues found that those who abuse alcohol or other drugs tend to come from troubled families. Such families are characterized by a greater degree of mother–father conflicts, parental divorce, parental absence, poor parent–adolescent communication, and parental discipline that is either overly lenient or overly harsh. Also, the person perceived to be the strongest person in the family was inclined to use psychological crutches such as denial or drugs to cope with stress. Youths who are heavy or problem drinkers also exhibit certain personality characteristics. They tend to have low self-esteem and are impulsive, irresponsible, and rebellious. They tend to have poor relationships with their parents compared to others their age. They are also less likely to value academic achievement, and make lower grades and drop out of school in greater numbers than those without a drinking problem.

The use and abuse of alcohol is also strongly related to social and socioeconomic differences, with the incidence of moderate and heavy drinking rising with education and affluence. The number of problem drinkers with a college education is more than double the number with only an elementary school education (Harris, 1981). Drinking is also related to religious and ethnic differences. People from an Asian background and from Jewish groups, especially Orthodox Jews, tend to have the lowest proportion of problem drinkers. Conservative Protestant groups have the highest proportion of abstainers, while liberal Protestants and Catholics have the higher proportion of moderate and heavy drinkers. There's also a tendency for adolescents who attend church frequently to remain abstainers (Jessor & Jessor, 1980).

The lowest incidence of problem drinking and alcoholism generally occurs in groups that have clear attitudes and rules regarding the use of alcohol. When children are exposed to alcohol, it is usually within a strong family or religious group. In these groups, alcohol is served in very diluted form, in small quantities, usually as an integral part of a meal. Abstinence is socially acceptable. Parents set an example of abstinence or moderate drinking and regard excessive drinking or intoxication as socially unacceptable. The use of alcohol is not viewed as proof of adulthood or virility. Finally, and most important, there is widespread and usually complete agreement on the ground rules of drinking by all members of the group (Jones-Witters & Witters, 1983).

Many of the measures aimed at curbing alcohol abuse have arisen out of concern over alcohol-related accidents among youth. Although teenagers comprise only 8 percent of the nation's drivers, they account for 15 percent of all drunken drivers involved in accidents. The leading cause of death among youth in their teens and twenties continues to be accidents, especially automobile accidents, with 60 percent of all fatally injured teenage drivers having alcohol in their blood. As a result, many states now have stiff penalties for drunk driving. Although a move to create a uniform drinking age of 21 among national lawmakers failed, at least 20 states have raised their minimum drinking age to 21 since the mid-1970s. In addition, there is greater awareness that hosts and hostesses need to watch out for guests who have been drinking and make arrangements for them to get home safely without driving. Many communities are providing teenagers with interesting and constructive social opportunities, such as alcohol- and substance-free post-prom, weekend, and graduation parties. In one school in Bucks County, Pennsylvania, high school seniors were treated to an elaborate formal prom that lasted from 7:30 p.m. to 6 a.m. the next day. It included a full-course catered dinner, live entertainment, and dancing throughout the night, with pizza at 3 a.m. and a continental breakfast at 6 a.m. Such occasions provide a good atmosphere for young people to enjoy themselves without jeopardizing their health or safety (Kennedy, 1984).

Tobacco and Smoking

tobacco: Products prepared from the leaves of the tobacco plant, containing nicotine, the substance believed to cause tobacco dependence.

Most people who smoke cigarettes begin using them during adolescence. The annual survey of high school seniors (class of 1985) showed that more than two-thirds of them (69 percent) had tried cigarettes. One in eight of them was smoking daily by the time he or she had reached the ninth grade. In fact, cigarettes are used daily by a larger proportion of teenagers (20 percent) than any other drug. An estimated 12.5 percent of teens smoke half a pack of cigarettes a day. College-bound seniors are much less inclined to smoke cigarettes than those not planning to attend college. And the difference between the two groups becomes even more pronounced with age. That is, the proportion of students who smoke daily drops off in college, whereas it increases among other young adults the same age, with 14 percent of college students smoking daily compared to 26 percent among other young adults. Compared to college students, almost twice as many post-high school young adults smoke half a pack a day, 18.5 percent versus 9.4 percent. Cigarette smoking is also higher among females at all age levels, with an ever greater difference between the sexes in college than in high school—17.5 percent of college women smoke compared to just 10 percent of college men.

Fortunately, many young people have become more aware of the dangers of smoking in the past decade. The proportion of teenagers who believe that smoking one or more packs of cigarettes a day entails great harm for the user has risen from 51 percent in 1975 to 67 percent in 1985. Greater realization of the health hazards of smoking, as well as other uses of tobacco, has led to a decrease in smokers at all age levels. Smoking among high school students dropped significantly between 1977 and 1981, with slight flucutations in tobacco use since then. At the same time, these gains should not blind us to the fact that nearly a million teenagers begin smoking each year, and that about one-third of them are not yet convinced that cigarette smoking is a health hazard.

Young people smoke for a variety of reasons, many of them unclear even to themselves. Although some teenagers initially try cigarettes because of curiosity, others do so out of defiance, because it's something they are "not supposed to do." Then too, many youth feel it is sophisticated to smoke, which is not surprising in light of all the seductive advertising in newspapers and magazines. Smoking is also associated with sociability—something to do after a meal, talking to friends, and at social gatherings. Teenagers are also much more likely to smoke if at least one parent smokes, especially the same-sex parent. There is also some evidence that teenage smokers are more extroverted, happy-go-lucky, and frank, but less agreeable than nonsmokers their age (Smith, 1980).

Smoking is a difficult habit to break. Even conservative estimates imply that half or more of those who quit smoking eventually resume the habit. In addition to the psychological dependence on the smoking habit, nicotine is considered to be physically addictive by many authorities. As a result, smokers build a tolerance to nicotine and need to smoke a larger number of cigarettes or ones with a higher nicotine content to get the same effect. The average smoker smokes 20 to 30 cigarettes a day—one about every 30 to 40 minutes. And since the biological half-life of nicotine in humans is approximately 20 to 30 minutes, habitual smokers keep their systems primed with nicotine during most of their waking hours. Although nicotine has a variety of effects on the body, it is primarily a stimulant which increases the respiratory rate, heart rate, and blood pressure. Withdrawal from habitual smoking, and therefore nicotine, produces a variety of symptoms, including nervousness, headaches, dizziness, fatigue, insomnia, sweating, cramps, tremors, and heart palpitations (Jones-Witters & Witters, 1983).

LOWER TAR AND NICOTINE

Although studies have shown that there is no such thing as a "safe" cigarette, those who are not yet able to quit smoking would be well advised to switch to brands with the lowest possible tar and nicotine (T/N) content.

In an American Cancer Society study conducted from 1960 to 1972, the average mortality of low T/N smokers was 16 percent lower than that of high T/N smokers. The comparable figure for lung cancer mortality was 26 percent. Moreover, low T/N smokers found it easier to quit smoking altogether than high T/N smokers.

However, it is important to remember that besides tar and nicotine, cigarette smoke contains a host of other poisonous gases such as hydrogen cyanide, volatile aromatic hydrocarbons, and especially carbon monoxide—possibly a critical factor in coronary heart disease and fetal growth retardation, among other things. While some hazards are reduced slightly by cigarette filters, certain filtered brands have been found to actually deliver more carbon monoxide than those without filters.

Adapted from 1985 American Cancer Society, *Cancer Facts & Figures* (New York: American Cancer Society, 1985), p. 19.

Cigarette smoking has become a major health hazard in our society. The American Cancer Society estimates that cigarette smoking is responsible for 85 percent of the lung cancers in men and 75 percent of the lung cancers in women. The cancer death rates for male smokers is twice that of nonsmokers, with the rate for female smokers being two-thirds greater than for nonsmokers. The higher cancer rates for men reflect that in the past more men than women smoked, and smoked more heavily. But now that this practice has become reversed, the gap in mortality rates between male and female smokers is narrowing each successive year. Smoking is also implicated in cancers of the mouth, pharynx, larynx, esophagus, pancreas, and bladder. Furthermore, smoking causes about 30 percent of all cancer deaths, is a major cause of heart disease, and is linked to conditions ranging from colds and gastric ulcers to chronic bronchitis and emphysema. A little-known fact is that the combined use of tobacco and alcohol—and the two usually go together—multiplies the carcinogenic effects of both. Although the switch to cigarettes with lower levels of tar and nicotine in recent years has slightly lowered the health risk of smoking, there is "no safe cigarette"—as the American Cancer Society keeps reminding us (Cancer Facts and Figures, 1985).

SELECTED ILLEGAL DRUGS

The widespread abuse of alcohol and tobacco is an apt reminder that legal drugs are still the worst killers. In fact, national drug abuse statistics show that more Americans die or suffer medical emergencies from using prescription drugs improperly than from the use of all illegal drugs combined (Ostrow, 1982). At the same time, this should not blind us to the growing menace of illegal drugs. There are now a wide variety of powerful drugs that are manufactured, distributed, and consumed apart from medical and legal supervision and without age restrictions, which multiplies the potential for drug abuse. In this section, we'll describe some of the effects and patterns of use of the most commonly used illegal drugs, including marijuana, the stimulants, the depressants, inhalants, hallucinogens, and heroin. But keep in mind that people who use drugs tend to use more than one of them at a time, without fully understanding the interactive effects of the drugs they are using. People use more than one drug at a time for a variety of reasons, such as to enhance the effect of one drug, to counteract the undesirable effects of another drug, or to achieve a less expensive high by combining one drug with a less expensive drug. In any event, multiple drug or polydrug use is becoming more and more popular, with single drug users being in the minority (O'Brien & Cohen, 1984).

Marijuana

marijuana: A derivative of the Cannabis plant which is usually smoked for its relaxing and disinhibiting effects.

Marijuana is made from the common hemp plant *Cannibis sativa*. This plant was once widely grown in the American colonies as a source of fiber for clothing and rope, and to a lesser extent for medicinal uses, and is now grown in most parts of the world. The potency of the drug varies with the content of THC (tetrahydrocannabinol), the main active ingredient of the Cannabis plant. The most powerful form of the drug, made from the resin of the female plant, is called hashish. The least potent form, marijuana, is made from the leaves, stems, and flowering tops, and is commonly grown in Mexico and the United States. Although marijuana is often classified as a hallucinogen, it is considerably milder than the other drugs in this class such as LSD, and is used much more widely, so that it is generally discussed separately.

Marijuana is usually inhaled through smoking, though it can also be ingested. The common physical effects are a slight rise in pulse rate, lowered blood pressure, dilation of the pupils with a reddening of the eyes, dryness of the mouth and throat, and increased frequency of urination. Marijuana has an adverse effect on concentration, reading, comprehension, thinking, and short-term memory, so that habitual marijuana use may lower school performance. Marijuana also impairs vision and retards the reaction time and performance of mental and motor faculties. Although such changes are usually slight at low-dosage levels, they become more marked at higher levels. Over three-fourths of those who use marijuana say they sometimes drive while they are high, with marijuana users overrepresented in highway accidents. Driving under the combined influence of marijuana and alcohol is especially dangerous (Jones-Witters & Witters, 1983).

The psychological effects of marijuana, especially at lower dosage levels,

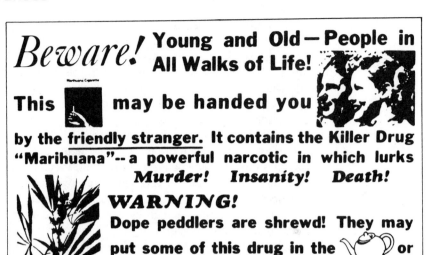

FIGURE 14–2 ANTIMARIJUANA POSTER DISTRIBUTED BY THE FEDERAL BUREAU OF NARCOTICS IN THE LATE 1930s.

Source: In Patricia Jones-Witters, Ph.D., and Weldon L. Witters, Ph.D., *Drugs and Society* (Boston, MA: Jones and Bartlett Publishers, Inc., 1983), p. 116.

depend largely on the personality of the user, his or her expectations of the drug, the circumstances in which it is used, and the user's past experience with the drug. Some common psychological effects are a feeling of relaxation, inner satisfaction, a free flow of ideas and imagination, exhilaration, and an altered sense of time and space—minutes may seem like hours and near objects appear distant. Yet much depends on the individual. New users may report little or no effect or overwhelming effects, depending on their makeup and moods. An anxious person may become more anxious or a depressed person more depressed. Marijuana is generally pleasurable at low levels, though unpleasant at very high levels. In short-term occasional use, marijuana may act as an aphrodisiac by releasing the central nervous system inhibitions on behavior. It also increases dilation of the blood vessels in the genitals and delays ejaculation in the male, though decreasing the sperm count. However, high doses over a long period of time are associated with a lowered libido and impotence.

Because marijuana cigarettes contain strong tars and some of the chemicals in marijuana are carcinogenic, chronic use of marijuana may be damaging to the lungs. However, since smokers in the United States are not exposed to as high a level of tars because marijuana is smoked less than tobacco, there may be a longer interval of time for the pathological effects of marijuana to show up. Frequent use of marijuana may lead to psychological dependence,

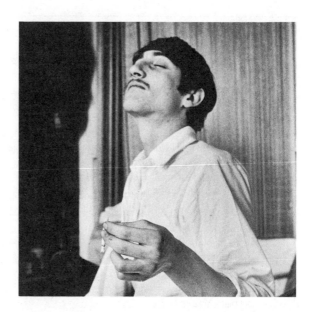

Marijuana remains the most widely used illegal drug in the United States. (Charles Gatewood)

with only mild physical addiction if at all. However, abrupt withdrawal from habitual use may produce irritability, sleep disturbance, appetite and weight loss, sweating, and gastrointestinal disturbances.

Although marijuana use among teenagers has declined somewhat in the past few years, it remains the most widely used illegal drug in the United States. According to the annual high school survey by Lloyd Johnston (1986) and his colleagues, about one-fourth (28 percent) of teenagers have tried marijuana by the time they reach the ninth grade, and more than half of them (54 percent) by their senior year. About 1 in 20 seniors (4.9 percent) use marijuana on a daily basis, down considerably from double that number in the late 1970s. College-bound seniors are less inclined to use marijuana than their counterparts not planning to attend college. And the difference between these two groups increases throughout young adulthood, with the ratio of daily marijuana use among college students compared to others their age being 3.1 versus 4.6. Nevertheless, the proportion of individuals who try marijuana rises with age, with three-fourths of individuals in the 18 to 25 age group having used it at some time. Although occasional use of marijuana is only slightly higher among males than females, males are about twice as likely to use it on a daily basis in high school, with the differences between the sexes becoming less evident during young adulthood.

About half the marijuana users in high school get high for 1 to 2 hours, with one-third of them staying high for 3 to 6 hours. But there has been a downward trend in the degree and duration of highs in recent years, with only two-thirds of the seniors saying they get either "moderately" or "very" high in the

THE COFFEE CANTATA

Johann Sebastian Bach (1685–1750) wrote the Coffee Cantata around 1732, when the new drink began to leave the male society of coffee houses in Germany and invade private homes, where ladies began to consume it. In the cantata, the father is angry because his daughter is a coffee addict. He threatens not to provide her with a husband unless she promises to stop drinking it. Here are a few excerpts from their dialogue:

FATHER: O wicked child! Ungrateful daughter, why will you not respect my wishes and cease this coffee drinking?

DAUGHTER: Dear Father, be not so unkind; I love my cup of coffee at least three times a day, and if this pleasure you deny me, what else on earth is there to live for?

DAUGHTER: [continues in solo aria]: Far beyond all other pleasures, rarer than jewels or treasures, sweeter than grape from the vine. Yes! Yes! Greatest of pleasures! Coffee, coffee, how I love its flavor, and if you would win my favor, yes! Yes! let me have coffee, let me have my coffee strong.

FATHER: Well, pretty daughter, you must choose. If sense of duty you have none, then I must try another way. My patience is well nigh exhausted! Now listen! From your dress allowance I will take one half. Your next birth-day should soon be here; no present will you get from me.

DAUGHTER: . . . how cruel! But I will forgive you and consolation find in coffee . . .

* * *

FATHER: Now, hearken to my last word. If coffee you must have, then a husband you shall not have.

DAUGHTER: O father! O horror! Not a husband?

FATHER: I swear it, and I mean it too.

DAUGHTER: O harsh decree! O cruel choice, between a husband and my joy. I'll strive no more; my coffee I surrender.

FATHER: At last you have regained your senses. [The daughter sings a melancholy aria of resignation.]

TENOR: And now, behold the happy father as forth he goes in search of a husband, rich and handsome, for his daughter. But the crafty little maiden has quite made up her mind, that, ere she gives consent to marriage, her lover must make a solemn promise that she may have her coffee whenever and wherever she pleases.

From Andrew Weil, M.D., and Winifred Rosen, *Chocolate to Morphine* (Boston: Houghton Mifflin Company, 1983), appendix.

mid-1980s compared to three-fourths who reported this in the mid-1970s. Also, the proportion of teenagers who stay high for 3 hours or more has declined from about a half to a little over a third in 1985. Apparently, fewer high school students are now using marijuana; students are using it less frequently, and are smoking fewer joints at a time.

The major reason marijuana use has declined among youth is the growing recognition of its harmful effects. Between 1975 and 1985, the proportion of high school seniors who felt that the regular use of marijuana is harmful almost doubled, from 43 percent to 70 percent. The increased awareness of the potential harmfulness of marijuana is probably due to the dramatic short-term changes in mood, behavior, and self-control associated with the drug in addition to the long-term physiological impact. At the same time, there has been a trend toward decriminalizing marijuana, with the majority of states making the possession of marijuana a misdemeanor. Some states distinguish between private and public possession, reserving the more serious penalties for the latter (Jones-Witters & Witters, 1983).

Stimulants

stimulants: A group of drugs that stimulate the activity of the central nervous system, such as cocaine and the amphetamines.

amphetamines: A widely used group of stimulants that reduce fatigue and help maintain a high level of efficiency for short periods of time.

cocaine: A stimulant derived from coca leaves, which, though legally classified as a narcotic drug, induces intense euphoric effects for a short period of time.

freebasing: Conversion of the stable salt form of an alkaloid, like cocaine, into the less chemically stable but more potent form "freed" of the ionic salt; using a drug in this form.

These are substances that act on the central nervous system, causing the person who takes them to feel more lively. He or she may become more talkative, restless, and be unable to sleep. With the exception of the caffeine-type stimulants (coffee, tea, chocolate, and cola), the most commonly used stimulants are the amphetamines and cocaine.

About one in four high school seniors (26 percent) have used *amphetamines,* with an even larger number of them having used one of the over-the-counter drugs such as diet pills and stay-awake pills. Overall use of these substances is generally higher among females than males, mostly because of their use for dieting. About 4 in 10 (42 percent) high school females have used diet pills, and 1 in 9 of them (11 percent) in the past month. Although regular use of amphetamines is not generally regarded as physically addictive, fatigue and mental depression often follow disuse of the drug, so that individuals may develop a psychological dependence on them. The general use of amphetamines has declined in recent years among all high school and college students as well as young adults. A major exception is the increased use of stay-awake pills among students, especially in college (Johnston, O'Malley, & Bachman, 1986).

Cocaine, another powerful stimulant, is similar to the amphetamines in reducing fatigue and inducing a euphoric state, though the effects of cocaine last only a short period of time. Furthermore, the initial use of cocaine generally occurs at older ages than the other illegal drugs. Only one in six teenagers (17 percent) has tried cocaine by his or her senior year in high school. But lifetime, annual, and current use (that is, in the past 30 days) continue to increase with age. As a result, 4 in 10 youths have tried cocaine by 27 years of age. About 1 in 5 of them have used it in the past year, and 1 in 10 the past month. (See Figure 14-3.) The drug tends to be used more frequently among youth in urban areas in the Northeast and Western regions of the United States, and more among males than females.

When taking cocaine, individuals may snort it or smoke it. About 44 percent of those in the 12- to 17-year-old group who have tried cocaine report "freebasing," or smoking cocaine, as opposed to snorting it. About one-third of regular cocaine users of all ages say they have freebased (Nesbit, 1986). Cocaine is generally used in the evening with others present, frequently at a party, but more often with just one or two people. The most common reasons given for using cocaine are "to see what it's like," "to get high," and "to have

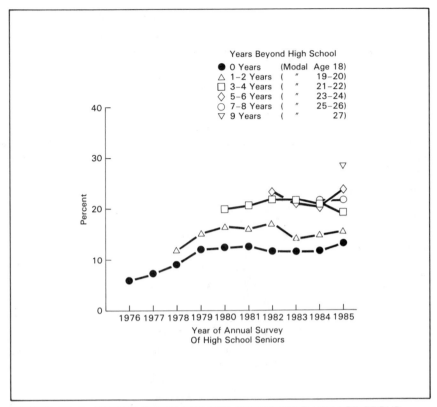

FIGURE 14–3   TRENDS IN COCAINE USE AMONG YOUTH BY AGE GROUP.

Source: Lloyd D. Johnston, Patrick M. O'Malley, and Jerald G. Bachman, *Drug Use among American High School Students, College Students, and Other Young Adults* (Washington, DC: National Institute on Drug Abuse, Government Printing Office, 1986), p. 159.

a good time with my friends." Compared to nonusers, individuals who use cocaine are more likely to use other illegal drugs, especially marijuana, to smoke cigarettes, and to drink heavily. It's not uncommon to use cocaine concurrently with marijuana or alcohol or both. Cocaine is not regarded as being physically addictive, but psychological dependence can become quite strong in some cases. Although less than 5 percent of the high school seniors report they have tried to quit cocaine and are unable to do so, cocaine use becomes heavier in the next 3 or 4 years following high school, suggesting that psychological dependency either develops rather slowly or with relatively low frequency among moderate and light users. In 1985, more than three-fourths of the high school seniors associate great risk with cocaine and disapprove of experimenting with it. But, as we've seen, older youths try it in greater numbers, such that the popularity of cocaine has increased dramatically in recent years. However, the tragic drug-related deaths of Len Bias and Don Rogers, two nationally known athletes, in quick succession in the early

Using cocaine. (Marc Anderson)

crack: A concentrated form of cocaine that is more potent and more likely to lead to drug dependence than other forms of cocaine, primarily because it is inhaled.

part of 1986, accentuated the growing danger of cocaine use among young adults (Nicholi, 1986).

"Crack" is a more concentrated form of cocaine that comes in small, soap-like pellets or "rocks" that are smoked in pipes. Although cheaper than the usual forms of cocaine, the effects of crack are also shorter, such that crack users need more of the drug to maintain their habit. Crack is also more addictive, mostly because it is more potent and inhaled into the lungs, where it is absorbed into the bloodstream in about 10 seconds. Thus, crack not only produces a higher "high" but a deeper "crash," leaving the user desperate for more. At the same time, inhalations of this powerful drug can lead to a brain seizure, cardiac arrest, and paranoid psychosis.

Although crack was used initially by youths in the urban areas of the large coastal cities such as New York, Miami, and San Diego, it is now consumed by a wide variety of people, including professional athletes, middle-class business managers, and people in the entertainment world. A survey of 485 drug hotlines found that the use of crack had not only jumped dramatically but had spread to at least 25 states and 16 major cities. About one-third of those calling the national hotline (800-COCAINE) in May 1986 were crack users, whereas no callers had even mentioned the drug a few months earlier. Crack users also become addicted much more rapidly than they do with regular cocaine. People are reporting full-blown addictions in 6 months or less, compared to a year or more for cocaine powder. At the same time, the danger of crack, along with increased media coverage of the drug problem, has aroused a groundswell of concern among law-enforcement agencies, urban residents, and people nationwide that may eventually help in the deterrence of drug abuse (Doyle, 1986).

Depressants

depressants: A diverse group of drugs that reduce the activity of the central nervous system.

barbiturates: Any of a group of barbituric acid derivatives used in medicine as sedatives.

methaqualone: A sedative hypnotic drug that produces effects similar to those of the barbiturates.

The depressants are a diverse group of drugs that diminish the activity of the central nervous system, producing anything from mild sedation to a coma, depending on the dosage. (See Table 14-4.) The most common depressants are the barbiturates, methaqualone, and the minor tranquilizers. Along with alcohol and the stimulants, the depressants are among the most used and abused drugs in our society.

About one in eight teenagers (12 percent) has taken a sedative, such as one of the barbiturates or methaqualone, by his or her senior year in high school. Although the barbiturates or "sleeping pills" are commonly prescribed to reduce anxiety and promote sleep, adolescents often use the short-acting barbiturates to lower their tension and inhibitions, to take the edge off a pep-pill habit, or along with alcohol—an especially dangerous habit because of the additive effect of the barbiturates when combined with other drugs. Methaqualone, a sedative hypnotic that produces effects similar to those of the barbiturates, enjoyed a rapid rise in popularity during the 1970s as a recreational drug. Youths who take methaqualone tend to get very high for longer periods of time than with many of the other drugs. The most probable explanation is

Table 14—4   HOW THE BODY AND MIND CAN REACT TO BARBITURATES AND OTHER DEPRESSANTS

	BODY	MIND
Low dose ↑ ↓ High dose	Drowsiness Trouble with coordination Slurred speech Dizziness Staggering Double vision Sleep Depressed breathing Coma (unconscious and cannot be awakened) Depressed blood pressure Death	Decreased anxiety, relaxation Decreased ability to reason and solve problems Difficulty in judging distance and time Amnesia Damage to brain

Source: Patricia Jones-Witters, Ph.D., and Weldon L. Witters, Ph.D., *Drugs and Society* (Boston, MA: Jones and Bartlett Publishers, Inc., 1983), p. 175.

sedatives: A group of drugs that depress the activity of the central nervous system, including the barbiturates and minor tranquilizers.

the youthful users' claim that methaqualone enhances sexual desire. Both the barbiturates and methaqualone have a high potential for physical and psychological dependence. Fortunately, the use of these depressants among teenagers has declined by half since the mid-1970s (Johnston, O'Malley, & Bachman, 1986).

tranquilizers: A group of drugs which produce sedative and antianxiety effects.

The minor tranquilizers, such as those known by the trade names of Valium and Librium, are another class of depressants that alleviate anxiety. Although tranquilizers are widely used by adults, they are also used by a significant number of teenagers, with one out of eight (12 percent) of them having used a tranquilizer by his or her senior year in high school. The majority of users first try the drug in early adolescence, and have used it only several times in their lives. Although these drugs are somewhat safer than the barbiturates in regard to overdosing and physical dependence, they are habit-forming and do have serious effects. Valium combined with alcohol has killed people (Walker, 1982). Use of the tranquilizers peaked in the late 1970s and has since declined among young people (Johnston, O'Malley, & Bachman, 1986).

Other Drugs

inhalants: Any group of substances inhaled to alter one's subjective state, such as plastic (model) cement; sometimes designated as "abused volatile substances."

There are a variety of other drugs that are used mostly on occasions by a substantial minority of youth. These include the inhalants, hallucinogens, heroin, and the other opiates. Use of these drugs is generally higher among noncollege-bound teenagers and young adults generally than college students. Also, the lifetime prevalence of these drugs is two and one-half times higher among males than females.

During high school, about one in five teenagers (18 percent) use inhalants—substances whose fumes are inhaled to make the user feel good or high. The two classes of inhalants used most frequently by adolescents are the amyl and butyl nitrites, which go by the street names of "poppers" or "snappers" and such brand names as Locker Room and Rush. Butyl nitrite got its reputation as an aphrodisiac by its apparent effect if used

hallucinogens: Drugs that produce sensations such as distortions of time, space, sound, color, and other bizarre effects.

during sexual relations near orgasm. However, afterwards users may have a tremendous headache and feel nauseous for a while. As a result, most of those who try this inhalant don't continue to use it. The inhalants tend to be used more commonly by younger adolescents who have usually discontinued their use by late high school, often failing to report their earlier experience with the drug. Although most teenagers experiment with inhalants in transition to other drugs, some of them become chronic users. The inhalants do not lead to physical addiction, but they can have adverse reactions and also long-term carcinogenic effects that may not appear for 10 to 30 years (Jones-Witters & Witters, 1983).

About one in eight teenagers (12 percent) have tried one of the hallucino-

LSD: One of the psychedelic drugs, taken mostly for its consciousness-expanding properties.

PCP: One of the psychedelic drugs that has widely varying effects on the central nervous system, including symptoms resembling those of an acute schizophrenic disturbance.

gens during high school. This diverse group of mind-altering drugs, including LSD and PCP, is variously known as the psychedelic (consciousness-expanding) or the psychotomimetic (psychosis-mimicking) group. Taken in small amounts these drugs bring about mild feelings of euphoria. Larger amounts produce more marked reactions, ranging from horror to ecstasy, from minor distortions of body image to loss of ego boundaries, and from intensification of color and depth to hallucinations. LSD is the most widely used hallucinogen, with the majority of users taking it only occasionally to get very high. Although LSD does not cause physical dependence, it does have moderate potential for psychological dependence and abuse. Most former users of LSD have discontinued the drug because of their fear of physical damage, psychological harm, or an emotionally upsetting experience with the drug. PCP, which is often present in other drugs sold on the streets such as mescaline, is regarded with fear by many people because it sometimes leads to intense feelings of paranoia, unpredictable violent behavior, and a complete loss of reality testing. The actual use of PCP is probably more extensive than indicated in the high school surveys because many PCP users do not report their use of the drug (Johnston, O'Malley, & Bachman, 1986).

heroin: A drug derived from opium taken mostly for its psychological effects.

opiates: A group of drugs that are derivatives of, or are pharmacologically related to, products from the opium poppy, and are used medically primarily for the relief of pain.

morphine: An opiate that produces sedative effects through depressing the nervous system; used medically primarily for the relief of pain.

Heroin is generally regarded as the most infrequently used illegal drug, with only 1.2 percent of high school seniors admitting that they have used it. A much larger proportion of adolescents (10 percent) have used one of the other opiates such as codeine. Drugs such as heroin, codeine, and morphine are classified as opiates or narcotics, synonymous terms for drugs derived from opium, which is made from the juice of the poppy flower. Codeine, a common ingredient in cough medicines, is used by teenagers to get a high, often with beer or wine. Morphine, a standard painkiller, is sometimes obtained illicitly through fake prescriptions or drugstore robberies. When heroin was extracted from morphine and proved effective in combating opium and morphine addiction, it was received with such high hopes that it was named from the word "hero"—heroin. Yet, it has turned out to be the most physically addictive and destructive of all the drugs, accounting for more drug-related deaths than any other drug except alcohol (Cross & Kleinhesselink, 1980).

Heroin use occurs in two stages. In the early stages it is taken mostly for its psychological effects, which include an early surge of feelings of warmth and peace likened to a prolonged orgasm. Users feel relaxed and drowsy, with an easing of pain, fears, and worries. They also experience reduced hunger and sex drives for 3 to 4 hours. However, after becoming addicted, which may occur after using the drug daily for only several weeks, users enter a second stage of heroin use in which they take it mostly to avoid the unpleasant withdrawal symptoms, such as vomiting, diarrhea, convulsions, sweating, and twitching muscles. Largely because it is only available illegally and is expensive and addictive, heroin is the least used illegal drug, and is more common among noncollege youth and males in the ghettos of the large cities.

PUTTING DRUGS IN PERSPECTIVE

The discussion of young people's use of drugs is especially difficult because of the lack of a clear and consistent approach by parents, educators, and youth alike. Some people equate the use of any illegal drug with drug abuse. As a result, what passes for drug education becomes a thinly disguised attempt to scare young people away from disapproved drugs by exaggerating their dangers. Yet, the lectures, pamphlets, and filmstrips that adopt this approach often stimulate curiosity, making the prohibited substances even more attractive. In contrast, other people distinguish between the moderate recreational use of drugs and an abusive, pathological use of any drugs, including legal drugs. Drug education based on this approach tends to emphasize the importance of understanding the various types of drugs, including their potential for danger, and the value of individuals making a personal decision about the use or disuse of drugs. More in line with this latter approach, in this section we'll take a look at people's overall relationships with drugs, the treatment of drug abuse, and some of the alternative to drugs.

Relationships with Drugs

pharmacology: The science that deals with the properties of drugs and their effects on living organisms.

set: Psychological make-up or behavior of a drug user: if a good result is expected, it is more likely to occur.

setting: The circumstances in which a drug is taken.

The effect of a drug is not simply a matter of what drug a person chooses to consume. It also depends on the relationships individuals form with a given drug. Many factors determine people's relationships with drugs. The drug itself is obviously a crucial factor, with the whole field of pharmacology devoted to discovering the makeup and effects of drugs. But the overall effect a drug produces also depends on many other factors, such as the person's physical and psychological makeup, his or her needs and traits, and past experiences with the drugs. The person's psychological "set" or expectations of a given drug also contribute to the psychological effect of a drug. It is common knowledge that teenagers who expect to get high on a couple of beers will begin to act that way regardless of the amount of alcohol they have consumed. Another factor that modifies the effects of a drug is the social "setting," or total environment, in which a drug is used. A few drinks of an alcoholic beverage may produce sexual arousal in one situation, whereas the same amount of alcohol consumed by the same person may lead to drowsiness and fatigue in another set of circumstances. The set and setting are thought to be of even greater importance in the use of the psychedelic and hallucinogenic drugs than other drugs (Jones-Witters & Witters, 1983).

In their book *Chocolate to Morphine,* Andrew Weil, M.D., and Winifred Rosen (1983) provide some guidelines for establishing good relationships with drugs as follows:

1. *Recognition that the substance you're using is a drug and being aware of what it does to your body.* The people who develop the worst

relationships with drugs have little understanding of the drugs they're using. They think that coffee is just another beverage or that cigarettes are harmless because they are sold legally, despite the warning on the package. People need to realize that *all* drugs have the potential to cause trouble unless they control their use. A necessary first step is to understand the nature and the effects of the substances being used. (See Figure 14-4.)

2. *Experience of a useful effect of the drug over time.* People who use drugs regularly often find that their early experiences with them were

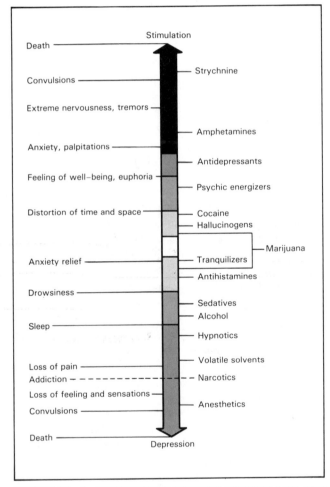

FIGURE 14—4 COMPARISON OF DRUG EFFECTS AND ACTIONS.

Source: By R. W. Earle, Ph.D.—University of California, Irvine—in Patricia Jones-Witters, Ph.D., and Weldon L. Witters, Ph.D., *Drugs and Society* (Boston, MA: Jones and Bartlett Publishers, Inc., 1983), p. 71. Used by permission of artist and author.

their best. The effects seem to diminish the more often a drug is used. People in bad relationships with drugs tend to use them very heavily but get the least out of them. Frequency is the most important factor in determining whether the effect of a drug will last over time. When the experience people like from a drug begins to fade, that is a sign that they are using too much. Those who ignore the warning and increase their consumption of a drug risk sliding into a worse relationship with the drug.

3. *Ease of separation from use of the drug.* People who are in a good relationship with drugs can take them or leave them. By contrast, one of the striking features of a bad relationship with drugs is dependence on a drug—psychological or physical. When people have become dependent on a drug—whether it's caffeine, nicotine, alcohol, or some illegal drug—the drug controls them more than they control the drug. Thus, drug dependence is a major factor in determining drug abuse.

4. *Freedom from adverse effects on health and behavior.* People vary in their susceptibility to the adverse effects of drugs; some people can smoke cigarettes all their lives and never develop cancer of the lung, mouth, or throat; others cannot. Some people can take a drink or two before dinner without it disrupting their family lives or work, while other people's drinking contributes directly to their troubles at home or on the job. Using drugs in ways that produce adverse effects on one's health and behavior is another major characteristic of drug abuse.

Weil and Rosen point out that bad relationships with drugs begin with an ignorance of the nature of the substance and the loss of the desired effect with an increasing frequency of use, gradually leading to greater difficulty in leaving the drug alone, or drug dependence, with the eventual impairment of one's health and social functioning. The endless news stories about the drug-related problems of various athletes and celebrities should be ample proof that anyone can get into bad relationships with drugs. Yet, there is a tendency for each of us to feel "it can't happen to me." But nobody is immune from drug abuse. For instance, a prominent physician experimented with cocaine, but later gave it up. Yet he remained addicted to tobacco smoking. When told that his heart arrhythmia was aggravated by his heavy cigar smoking, up to 20 cigars a day, he stopped smoking for 14 months. But he said the torture of not being able to smoke "was beyond human power to bear," and he resumed smoking, though on a somewhat more moderate basis. Eventually, he developed cancer of the jaw and mouth that was also attributed to smoking. However, despite 33 operations for cancer and the construction of an artificial jaw, this physician continued to smoke until his death at the age of 83. His repeated efforts to stop smoking and the suffering he endured make him a tragic example of tobacco addiction. His name was Sigmund Freud (Altrocchi, 1980; Jones, 1953).

Treatment of Drug Abuse

Freud's tobacco addiction serves as a reminder that the abuse of legal substances such as alcohol, tobacco, and prescription drugs are a continuing part of the drug problem. Thus, the problem of drug abuse must include the excessive and chronic use of *any* drug. We've already indicated two important criteria of drug abuse, namely the existence of drug dependence, whether physical or psychological, and the adverse effects of drugs on the user's health and social functioning. There is also a fairly general agreement about a third criterion of drug abuse. That is, when individuals orient their existence around the drug experience, whether as a way of attaining a desirable high or coping with stress, they are abusing drugs. Individuals who fulfill any of these conditions, or a combination of them, are in need of help.

There is a wide variety of programs available for treating drug abuse. Many of these programs were begun in the golden years of the 1970s when the government and other agencies invested billions of dollars in dealing with the widespread use of heroin and other drugs by American service personnel in Vietnam. The various treatment programs differ in regard to a number of issues, such as whether they are voluntary or involuntary, selective or nonselective, inpatient or outpatient, medical or nonmedical, and whether the goal is abstinence or controlled use of the substance.

A major difference in the approach to drug abuse is whether individuals receive treatment in a residential or outpatient program. Most of these programs assume that drug abuse is a symptom of underlying psychological problems and aim at a change in lifestyle as well as abstinence. There are several hundred residential treatment programs, including medically oriented, hospital-based programs and therapeutic communities staffed largely by former addicts. Hospital treatment programs typically last 1 or 2 months, though some last up to a year. Therapeutic communities tend to require longer periods, from about 9 to 18 months. But both types of programs have a relatively high failure rate. Individuals may not be ready for the complete change in lifestyle demanded by the program. Or they become bored and frustrated, with most dropouts leaving in the first 30 days. A large proportion of people who enter these programs drop out and reenter the program at a later date, with relapse being the rule rather than the exception (Jones-Witters & Witters, 1983). In the outpatient programs, individuals visit the treatment facility at certain intervals and participate in a variety of programs including individual and family counseling, encounter groups, educational and career programs, training in life skills, and work assignments in the community. Yet, overall effectiveness in outpatient drug-free programs has also been poor, with some studies suggesting a success rate in the range of 25 percent (Gressard, 1979).

A comprehensive approach to treatment suggested by Turanski (1982) would include components for youths at different stages of drug use and abuse. A preventive component, especially for teenagers whose parents abuse drugs, would include individual, group, and family therapy, as well as the availability of educational and career counseling and support groups. A second component would focus on teens at-risk who are currently recreational users of drugs but are in danger of future escalation of drug use. This would include the resources mentioned above plus workshops in life-skills training and alternative activities to drug use. A third component would provide intensive and long-term residential treatment for the more severe drug abusers. A final component would include aftercare and support for youth and their families during the transition back to the community, crisis intervention if needed, and guidance in developing new activities and peer relationships. In reviewing the National Institute of Drug Abuse findings in regard to treatment and prevention programs, Craig Thorne and Richard DeBlassie (1985) suggest the need for greater use of family therapy, community and educational programs, the use of behavioral strategies, and acceptance of the view that addiction is something the individual can and has to take responsibility for in order to achieve lasting improvement.

Alternatives to Drugs

The notion of personal accountability for drug abuse brings us to the importance of personal choice in the use or disuse of drugs. In the continuing battle against drug abuse, it isn't enough to regulate the *supply* side of drugs, that is, making certain illegal drugs are harder to get and attaching stiffer penalties to their use, as important as this may be. We must also focus on the *demand* for psychoactive drugs.

A crucial issue in the demand for drugs involves the individual's decision whether to use drugs or not as well as the needs and motives that are being met by drugs. The marked increase in perceived harmfulness of various drugs during the same period in which daily use of many illegal drugs dropped in half among high school students suggests that there is more of a rational decision-making component in drug use than often assumed (Johnston, O'Malley, & Bachman, 1986). Earlier attempts to dissuade young people from using drugs through exaggerating the dangers of drugs proved highly ineffective during the early 1970s. More recent approaches in drug education have attempted to inform young people of the facts and consequences of various drugs and encourage them to make their own decisions about the use or disuse of drugs. The importance of personal choice is apparent in a study of drug use and abuse among several hundred high school students (Scott, 1978). When asked why they continued to use drugs, users frequently cited the "payoff" of drugs. When asked what might make them stop using or abusing drugs, the most common answer was "my decision." When former drug users and abusers were asked why they had stopped, a large percentage of them indicated there was "no payoff" from drugs or they no longer had a

Youths engaged in socially constructive activities have less need for drugs. (Laima Druskis)

"need for drugs." And the most frequently given reason for staying off drugs was "willpower" and "self-determination," all of which underlines the importance of individuals' making a personal choice in regard to drugs.

It is equally important to assist young people in recognizing their needs and motives for taking drugs and helping them to find more constructive ways of satisfying these needs. (See Table 14-5.) Although this approach is appropriate at any age, it is best begun during early adolescence before the critical stage of drug use begins. In discussing the alternatives to drugs, Andrew Weil and Winifred Rosen (1983) point out that a major reason people take psychoactive drugs is to vary their normal experience or consciousness, and that this is a normal need that exists in everyone to some extent. Infants rock themselves into a blissful state, and many children discover that whirling or spinning is a powerful technique to change their awareness. There are an endless number of ways that adults vary their experiences, including listening to music, fasting, dancing, hang-gliding, falling in love, and jumping into cold water after taking a hot sauna. The possible variations of consciousness range from states associated with the lowered input into the brain, such as in relaxation and meditative states, to states associated with an increased input into the brain, such as in experiences with psychedelic drugs, schizophrenic states, and mystical rapture. Furthermore, individuals vary in their desire for novel experience. For instance, people who score high on measures of sensation-seeking

sensation seeking: The degree to which a person is willing to try new experiences and take risks.

Table 14—5 ALTERNATIVES TO DRUGS

Experience	Corresponding motive	Drugs used*	Possible alternatives
Physical	Desire for physical well-being, relaxation, desire for more energy.	Alcohol, tranquilizers, stimulants, marijuana.	Aerobic exercise, dance, sports, swimming, outdoor work, yoga.
Sensory	Desire to magnify sensory experience: sound, taste, touch; need for sensual, sexual stimulation.	Hallucinogens, marijuana, alcohol.	Sensory awareness training, experiencing sensory beauty of nature, sky diving, sauna, massage, sexual intercourse.
Emotional	Relief from anxiety or psychological pain, desire for emotional insight.	Any, especially narcotics, alcohol, barbiturates, tranquilizers, cocaine.	Individual counseling, group therapy, instruction in self-improvement.
Interpersonal	To gain peer acceptance, to communicate with others, assert independence.	Any, especially alcohol, marijuana, cocaine.	Participation in growth groups, social skills training.
Social	To promote social change, to improve environmental conditions, to help others.	Marijuana, cocaine, psychedelics.	Support social causes, help in social service projects, such as tutoring disadvantaged students.
Political	To promote political change.	Marijuana, cocaine, psychedelics.	Participate in community action groups, lobby for partisan and nonpartisan causes.
Intellectual	To escape boredom, intellectual curiosity, to gain understanding.	Stimulants, sometimes psychedelics.	Intellectual stimulation through reading, discussion, creative games and problem solving, hypnosis.
Creative-aesthetic	To improve creative performance, enhance enjoyment of arts, such as music, dancing.	Marijuana, stimulants, psychedelics.	Seek instruction in producing and/or appreciating art, music, drama, and creative hobbies.
Philosophical	To find a meaningful life, enrich one's values, clarify one's personal identity.	Psychedelics, marijuana, stimulants.	Take part in discussions, seminars, courses on ethics, literature, philosophy, values.
Spiritual-mystical	To develop spiritual insights, higher levels of consciousness.	Psychedelics, marijuana.	Seek methods of religious, spiritual development; study world religions, meditation, yoga.

*The possible combinations of drugs in any category depend largely on individual use and greatly exceed the drugs listed.

Source: Adapted from A. Y. Cohen, *Alternatives to Drug Abuse: Steps toward Prevention* (Washington, DC: National Institute on Drug Abuse, No. 14.; Department of Health, Education, and Welfare, 1973), in Patricia Jones-Witters, Ph.D. and Weldon L. Witters, Ph.D. *Drugs and Society* (Boston, MA: Jones and Bartlett Publishers, Inc., 1983), pp. 379.–380.

are more likely to experiment with drugs, engage in a variety of sexual practices, and seek excitement in risky sports. Those who score low in the sensation-seeking motive may prefer more familiar and peaceful patterns of behavior, such as taking a walk or playing chess. Morton Zuckerman (1983), who has investigated this phenomenon extensively, thinks that a person's sensation-seeking tendencies may be partly determined by biological factors. People who score high on sensation-seeking have low blood levels of monoamine oxidase (MAO), an enzyme that regulates the concentration of certain neurotransmitters in the brain. In any case, the sensation-seeking motive tends to peak during the college years and grows weaker with age, somewhat parallel to the patterns of drug use.

In finding alternatives to drugs, people need to realize that drugs alone do not "contain" highs. The possibilities for such states exist within the human nervous system. Drugs mostly serve to trigger such states or provide an excuse to notice them. Even then, drugs do not make people high automatically. They must learn how to interpret the physical effects of drugs as the occasion for their highs, as we discussed earlier in regard to the importance of set and setting. As Weil and Rosen (1983) point out, one of the curious frustrations of the human condition is that we cannot feel high whenever we want to. It seems necessary to work for highs or resort to tools outside of ourselves to achieve them. One of the major advantages of psychoactive drugs is that they can make people feel temporarily different faster and more powerfully than other means. But they also have the disadvantages of reinforcing the notion that desirable states of consciousness come from outside ourselves, thereby making people feel more inadequate and incomplete without drugs. Drug-induced states also lose their effectiveness with time, leading to increased use and all the dangers that come with drug dependence.

In contrast, people need to realize that there are other, more constructive ways to achieve desirable states of consciousness, to have a good time with their friends, or to manage stress. But most of these ways require an investment of time, effort, and the acquisition of skills—all of which may appear initially as disadvantages. A major advantage of such an approach is that it helps us to realize we have the capacities within ourselves to feel good, and can do so without huge expenditures of time or money for special equipment or jeopardizing our health. It's also important for people to continue experimenting with new means of achieving desirable states of consciousness as they mature and change. This may be especially difficult for people who have abused drugs, and who have grown accustomed to the physical sensations of drugs and consider them necessary components of the experience—not to mention that drugs can produce the experience more directly and easily. People who have relied on drugs to make their lives interesting or tolerable may find it hard to imagine life without drugs. But if they are determined enough, they can eliminate drugs from their lives and never miss them. Even more fortunate are those who realize earlier in their development that it isn't necessary to use drugs to achieve desirable states of consciousness or to manage stress.

SUMMARY

Use of Drugs

1. Although people have been experimenting with mind-altering drugs for centuries, drug use and abuse have become more serious problems today, especially among youth.

2. Practically all adolescents have tried alcohol and nearly two-thirds of them have used an illegal drug by the time they have reached their senior year in high school.

3. Adolescents are starting to use drugs at earlier ages than ever before, beginning with the legal drugs such as alcohol and cigarettes before progressing to the illegal drugs such as marijuana.

4. The most common reasons mentioned for using drugs are to have a good time with friends, to get high, and to relieve tension.

Alcohol and Tobacco

5. Alcohol and tobacco continue to be the most commonly used and abused substances, mostly because of their availability and widespread acceptance.

6. Adolescents are beginning to drink alcohol at earlier ages, consume larger amounts, and remain high for longer periods of time than in previous years.

7. Occasional heavy drinking, mostly at weekend parties, has become a serious problem among young people.

8. Many of the measures aimed at curbing alcohol abuse have arisen out of concern over alcohol-related accidents, with over half of the fatally injured teenage drivers having alcohol in their blood.

9. Cigarettes are used daily by a larger proportion of teenagers than any other drug, though increased awareness of the health risk posed by smoking has led to a decline in cigarette smoking in recent years.

Selected Illegal Drugs

10. Marijuana remains the most widely used illegal drug, with three-fourths of young people having tried it by their midtwenties.

11. The most commonly used stimulants, apart from the caffeine-type substances, are the amphetamines and cocaine, with increasing use and abuse of cocaine becoming a more serious problem among older youth in recent years.

12. Although use of depressants such as the barbiturates, methaqualone, and tranquilizers has declined in the past decade, about one in eight adolescents have used them at one time or another.

13. A variety of other illegal drugs, such as the inhalants, hallucinogens, and the various opiates, are used by a substantial minority of youth, mostly on an occasional basis.

14. Although heroin is the most infrequently used illegal drug, it is one of the most physically addictive and destructive of all the drugs.

Putting Drugs in Perspective

15. The effects of a drug depend on a number of factors in addition to the drug itself, including the individual's expectations, the setting in which the drug is taken, the physical and psychological makeup of the individual, and his or her past experience with the drug.

16. Bad relationships with drugs begin with an ignorance of what the drug does to one's body and mind and the loss of effect with an increasing frequency of use, gradually leading to greater psychological and/or physical dependence on the drug, with the eventual impairment of one's health and social functioning.

17. There are a wide variety of approaches to treating drug abuse, but their overall effectiveness and success rate continue to be rather low.

18. A major approach to preventing drug abuse is to inform young people of the facts and consequences of drugs and the importance of making a personal choice in regard to the use or disuse of drugs.

19. Equally important is teaching adolescents how to recognize their needs and motives for taking drugs and how to satisfy these needs without recourse to drugs.

REVIEW QUESTIONS

1. Describe the typical sequence in which teenagers begin using the various legal and illegal drugs.

2. What are the most common reasons given for using drugs?

3. To what extent has alcohol abuse become a serious problem among young people?

4. Why is cigarette smoking such a health risk among youth?

5. What are some of the reasons marijuana continues to be the most widely used illegal drug?

6. Select one specific drug and describe the typical effects it has on users as well as its potential for psychological and physical dependence.

7. Explain some of the factors that determine the effects of any drug on a given person.

8. What makes for a bad relationship with a drug?

9. What do you think is the best approach to helping a young person overcome his or her abuse of alcohol?

10. What are some constructive alternatives to using drugs?

15 Psychological Disorders

- **PSYCHOLOGICAL DISTURBANCE IN ADOLESCENCE**
- **ANXIETY AND EATING DISORDERS**
  Anxiety Reactions
  Phobias
  Obsessive-Compulsive Disorders
  Anorexia Nervosa
  Bulimia
- **DEPRESSION AND SUICIDE**
  Manifestations of Depression
  The Range of Depression
  Adolescent Suicide
  Contributing Factors
  The Prevention of Suicide
- **SCHIZOPHRENIA**
  Adolescent Schizophrenia
  Contributing Factors
  Outlook
- **THERAPY**
- **SUMMARY**
- **REVIEW QUESTIONS**

Carol, a high school student, takes exactly the same route to and from school every day. Recently, while walking home with another girl, she tried walking down a different street. Midway down the unfamiliar block, Carol felt dizzy and her heart began pounding. She broke out in a cold sweat. She quickly excused herself, ran back to the corner, and went home the usual way.

Ed is a freshman at college who has been doing poorly in school. Each time he sits down to study he finds himself worrying and daydreaming. He suffers from feelings of worthlessness and makes excuses to keep from going out with his friends. Ed is also having trouble getting to sleep at night and has begun drinking several beers before bedtime. He sleeps until noon but still feels tired all the time.

Amy is an eccentric teenager who has few friends. Most of the time at home she keeps to herself in her room. For the past several months, Amy has complained that students are talking about her and calling her names behind her back. She has also begun hearing voices that accuse her of engaging in homosexual acts. Recently, after she began staying up most nights and refused to attend school, she was taken to a local mental health center for an evaluation.

PSYCHOLOGICAL DISTURBANCE IN ADOLESCENCE

agoraphobia: An irrational fear of open spaces or public places.

depression: An emotional state of dejection, with feelings of worthlessness, sadness, and pessimism about the future.

psychological disorders: Conditions characterized by painful symptoms of personal distress and significant impairment in one or more areas of functioning.

externalizers: Individuals whose psychological conflicts are manifested in external behavioral problems such as aggression.

internalizers: Individuals whose psychological conflicts are manifested primarily in internal symptoms like depression.

All three of these youths are experiencing significant psychological disturbances. Carol is suffering from agoraphobia, one of the anxiety disorders. Ed exhibits several of the classic symptoms of depression, a common disorder that is often masked behind other problems, especially alcohol and drug abuse. Amy has been diagnosed as suffering from schizophrenia, a severe psychological disorder that is more likely to make its initial appearance during adolescence or early adulthood than at any other period of life. We'll discuss these and other disorders in this chapter. But first, we'll take a look at the incidence of psychological disturbances at adolescence.

Admissions of children to clinics begin to increase gradually at 6 or 7 years of age and reach a peak at 14 or 15 years of age. The increase in admissions at 6 or 7 is probably related to the child's beginning school. Problems that have been ignored or put up with at home may no longer be tolerated in the classroom. Then too, starting school itself is often stressful and may precipitate psychological problems. The peak in psychological disturbances at midadolescence probably reflects the normal developmental problems of this age coupled with the inherent stress of adolescence. The fact that severe psychological disorders peak at this age in other countries such as Britain, Japan, and the USSR suggests that there is a link between adolescence and psychological disturbance (Sarason & Sarason, 1984).

The extent to which psychological disturbances are recognized as such during adolescence depends partly on the forms in which they are manifested. Thomas Achenbach (1978) makes a useful distinction between adolescents who are externalizers, and those who are internalizers: The classification depends on the extent to which a child or adolescent's problems are manifested primarily in conflicts with the outside world or in intrapsychic conflicts. Externalizers are more likely to have problems with aggression, delinquency, and sex, whereas internalizers tend to have problems with depression, anxiety, obsessions, phobias, and somatic complaints. Achenbach suggests that the difference between externalizers and internalizers is partly the product of adolescents' socialization. That is, externalizers generally have parents who have more overt problems and who exhibit little concern about their adolescents. Consequently, these youths have learned to express their aggressive impulses overtly, resulting in more behavior problems in school and with the police. On the other hand, internalizers tend to come from more stable homes, with parents who have fewer overt problems and show more concern about their adolescents. As a result, these youths characteristically react to stress with inner conflicts. The fact that a greater proportion of boys are classified as externalizers and are more likely to express their difficulties in outward behavior may help to explain why about twice as many boys as girls are referred for psychological help during adolescence (Taube & Barrett, 1985). The same distinction may also help to explain the preponderance of

Quiet, introverted youths may have trouble expressing their problems to others. (Ken Karp)

lower-class youth referred for help. In one study, Achenbach and Edelbrock (1981) found that lower-class parents reported a greater incidence of problems and disturbances among their children and adolescents than did middle-class parents, with many of these problems involving undercontrolled, externalizing behavior.

Experienced clinicians have long realized that it is more difficult to distinguish between normal and disturbed behavior at adolescence than at any other period of life. Because adolescence is an intense period of rapid physical, cognitive, and emotional growth, symptoms that might suggest abnormality in adults often turn out to be considerably less serious in adolescents. Thus, clinicians are often reluctant to apply the more serious diagnostic labels to adolescents, lest this should make the client's situation worse. Weiner (1980) reports that a disproportionately high percentage of adolescents are classified in the less serious categories, such as the transient situational disorder, now known as an adjustment disorder (DSM-III, 1980). This category includes relatively minor disturbances of adjustment brought on by developmental and situational factors such as stress, rather than deep or long-standing psychological conflicts.

adjustment disorders: Maladaptive reactions to a particular stressful situation that result in impaired functioning and symptoms beyond a normal response to such stress; can be expected to decrease when the stress passes.

At the same time, Weiner reports that youths diagnosed with the less serious labels are just as likely to receive psychotherapy and remain in treatment as long as those with more serious diagnoses. Nor do they easily outgrow their problems. A 10-year followup study showed that youths originally diagnosed with the milder situational or adjustment disorders were just as likely to reappear in therapy and for just as many subsequent periods of psychological help as those with more serious disorders. Finally, of the youths who returned for help, an almost equal number of those with the milder diagnoses (11.2 percent) were subsequently diagnosed as schizophrenic, as compared to 14 percent of those with the personality and anxiety disorders. All this suggests that youths who are sufficiently disturbed to warrant treatment are often experiencing more than just growing pains.

The close link between adolescent development and disturbance presents two opposing dangers to adults. One danger is to overinterpret signs of disturbance at adolescence, thereby undermining the youth's self-confidence and creating self-fulfilling expectancies among adolescents and adults alike. The other danger is to dismiss serious problems at adolescence, thereby avoiding the needed treatment and allowing problems to get out of hand. In order to minimize these dangers, parents and other concerned adults might assess the troubled adolescent's behavior in light of the following questions:

- Has the problem lasted beyond the expected age?

- Is it frequently displayed?

- Does such behavior resist ordinary efforts to change it?

- How seriously does it interfere with the adolescent's relationships with adults and peers?

- Does the behavior interfere with school work?

- If such behavior continues, will it interfere with adult adjustment?

ANXIETY AND EATING DISORDERS

anxiety: A vague unpleasant feeling of apprehension; warning of impending danger.

anxiety disorders: Psychological disorders characterized by symptoms of inappropriate or excessive anxiety or by attempts to escape from such anxiety.

In mild to moderate doses anxiety may help mobilize young people to meet some threat or emergency. But when anxiety becomes excessive and disproportionate to the situation at hand, it has a disruptive effect on behavior. In the anxiety disorders, the anxiety may be experienced in several ways. In a generalized anxiety reaction, the anxiety itself becomes the predominant disturbance. Either the person is anxious all the time or suffers from periodic anxiety or panic attacks. However, in a phobic disorder, the anxiety is evoked by some dreaded object or situation, such as riding in an elevator. In the obsessive-compulsive disorder, anxiety occurs if the person does *not* engage in some thought or behavior, such as repeatedly washing one's hands, which otherwise is senseless and embarrassing. In still other mild, related disturbances, such as the eating disorders, the individual's anxiety centers around his or her eating habits and physical appearance.

Anxiety Reactions

anxiety reaction: Acute or chronic feelings of apprehension accompanied by disabling symptoms such as dizziness and nausea.

When anxiety becomes intense or disabling, especially in the absence of any specific danger, the individual may have an anxiety disorder. In an acute anxiety attack, individuals feel agitated, restless, and apprehensive in relation to others; these feelings are often accompanied by physical symptoms such as lightheadedness, nausea, and sleeping difficulties. Individuals themselves are frequently baffled by what is happening. Yet, upon closer examination, the heightened anxiety often reflects disturbances in their development, whether in their feelings about themselves, their newly awakened sexual impulses, or their close relationships with parents or peers.

For example, a 15-year-old girl became noticeably more fearful after her family's move to a new city. Her grades in school began to suffer, she had trouble sleeping, and she could not seem to make new friends. A concerned counselor discovered that the girl's anxiety reflected a disturbance in the home—namely, repeated threats of a marital separation, as well as the girl's conflicting feelings about becoming more independent of her parents. Another contributing factor was an intense rivalry with an older sister. The family's move simply precipitated the anxiety attack, which then provided the opportunity for clarifying some of the girl's conflicts.

Youths may also experience chronic anxiety, which is often the result of unresolved acute anxiety. In this case, undue anxiety may erupt over some aspect of the individual's life, whether grades, entrance into college, sex, or close relationships with peers. But the lack of a crisis often leaves the anxiety untreated.

Phobias

phobias: Persistent and irrational fears of a specific object or activity, accompanied by a compelling desire to avoid it.

A phobia is a persistent and irrational fear of a specific object or activity, accompanied by a compelling desire to avoid it. Most of us experience an irrational avoidance of selected objects, such as snakes and spiders, but this generally has no significant impact on our lives. By contrast, when the avoidance behavior becomes a significant source of distress to the individual and interferes with his or her everyday behavior, the diagnosis of a phobic disorder is warranted (DSM-III, 1980). Simple phobias generally involve a circumscribed stimulus, such as the fear of heights or the fear of closed places. Social phobias involve an irrational fear of and a compelling desire to avoid situations in which the individual may be scrutinized by others, such as the fear of speaking in public.

school phobia: A persistent and irrational fear of school, accompanied by a compelling desire to avoid it.

Although *school phobia* is more common among children, it is more serious when it occurs at adolescence. School phobias usually reflect a dread of some aspect of school, such as an authoritarian teacher or sports. According to a dynamic view, a school phobia may reflect an internal conflict involving separation anxiety in relation to overprotective parents. In many instances, the father may fail to counteract the mother's possessiveness. Sometimes, a school phobia reflects negative experiences at school, perhaps in relation to a critical teacher or excessive competitiveness among peers. In many instances, anxiety is directed toward specific aspects of the school environment—for example, test anxiety or entrance into college. School phobia among adolescents is often just one aspect of chronic maladjustment with a long history of problems. Chronic phobics tend to break off contact with their friends and mope around the house without accomplishing anything. Consequently, school phobia is usually a more serious condition among adolescents than in children, a difficult condition to treat, and perhaps a prelude to poor adult adjustment in work-related situations (Weiner, 1980).

Agoraphobia is typically the most severe phobic reaction and the one for which people most often seek treatment. Agoraphobia is a cluster of different fears, all of which evoke intense anxiety about panicking in unfamiliar places. Situations commonly avoided include being in crowds, such as a crowded store, in elevators or tunnels, or on public transportation or bridges. This type of phobia tends to occur in the late teens or early twenties, though it can occur later in life. It is also much more prevalent among young women than men. During the outbreak of this phobia, individuals are often housebound. If they do go out, great care is taken to avoid certain situations such as being in an elevator or crossing a bridge.

Obsessive-Compulsive Disorders

obsessive-compulsive disorder: A preoccupation with unacceptable ideas which appear without external stimuli, accompanied by the urge to engage in repetitive activity to ward off their imaginary threats.

The essential features of this disorder are recurrent obsessions or compulsions, or, as is usually the case, both. Young people afflicted with this disorder may find themselves unable to put certain unacceptable ideas out of their heads. Typically, these obsessions involve thoughts of lust and violence, partly because of their association with the individual's anxiety and guilt. Then too, the individuals may have the urge, almost against their wills, to engage in repetitive activities to ward off the imaginary threats. Most of these compulsions fall into two categories and involve checking rituals, such as making certain the door is locked, and cleaning rituals, such as changing clothes several times a day to insure that they are spotless. Common examples would be the teenager who bathes compulsively, habitually sets fires, or engages in ritualistic acts to maintain neatness. In one case, an individual washed her hands over 500 times a day despite the raw skin and painful sores that resulted. She had a strong fear of being contaminated by germs and felt she could temporarily alleviate her fear only by washing her hands (Davison & Neale, 1982).

MULTIPLE PERSONALITY

Multiple personality is one of the dissociative disorders in which the individual alternates between two or more distinct personalities. Or there may be a dominant personality with a variety of subordinate ones.

The disorder is most often diagnosed among late adolescents and young adults and is especially common among females. The majority have been victims of physical or sexual abuse in childhood. It appears that when faced with overwhelming stress, such children escape emotionally by imagining separate personalities to deal with the suffering (Coons, 1980).

Most cases of multiple personality have certain features in common. First, there are marked differences in the memories and mannerisms of each personality. Also, each personality tends to emerge in response to a particular situation. Then too, each personality takes on a particular role or emotional experience for others. That is, one personality may take on a controlling role; another one may play a more rebellious role. In many cases, the dominant personality is unaware of the activities of the other personalities, which may account for the "lost time" reported by people suffering from multiple personality (Putnam, 1982).

During treatment, therapists help the person to integrate the different personalities, sometimes through the use of hypnosis. But it is usually a long and difficult process. In the famous case of Sybil, for example, it took 11 years to integrate her 16 personalities into a stable, coherent personality (Schreiber, 1974).

Anorexia Nervosa

Since feeding is such a crucial part of development among infants and children, it should not be surprising that the eating disorders often mirror emotional problems. Unlike obesity, which is common among adolescents and adults alike, anorexia nervosa and bulimia are two related eating disorders which characteristically occur at adolescence.

anorexia nervosa: An eating disorder characterized by a severe loss of appetite and weight.

The essential features of *anorexia* are a fear of becoming fat or obese, accompanied by a disturbance in body image and a refusal to maintain normal weight. A loss of 25 percent of body weight, along with other physical signs such as the suspension of menstrual periods, is usually sufficient for a diagnosis of anorexia. The weight loss is usually accomplished by a reduction in total food intake, especially foods high in carbohydrates and fats, self-induced vomiting, use of laxatives or diuretics, and sometimes extensive exercise (DSM-III, 1980).

Anorexia tends to occur during early and midadolescence, commonly at 14 or 15 years of age. It is 9 to 10 times more frequent among girls than among boys, and more common in all-girl families. Parents of anorexics tend to be older than average and from the higher socioeconomic groups. The girl is characteristically overweight for her age and height. She is intelligent and makes good grades. But she is also conscientious and perfectionistic. Often, she begins restricting her food intake after being teased about her weight,

AN ANOREXIC GIRL

Nancy, the oldest of three girls in an affluent family, began exhibiting anorexic symptoms as she approached her sixteenth birthday. Although she was only slightly overweight for her age, Nancy suffered from an intense fear of getting fat. Her fears, coupled with a desire to become socially attractive, led her to experiment with ever stricter diets.

Nancy blamed much of her troubles on her father, a well-known trial lawyer, who constantly threatened her when she wouldn't eat. At the dinner table he would start out reasoning calmly with her, but inevitably ended up yelling at her. During this same period Nancy also had numerous fights with her boyfriend, mostly because of his attempts to get her to eat. But the more her father and boyfriend pressured her to eat, the more she resisted. As a result, her weight dropped from 140 pounds to 88 pounds in less than a year.

Nancy exhibited many of the characteristic symptoms of anorexics. She was depressed, easily upset, frequently lost her temper, lost interest in her boyfriend, and ceased having her monthly periods. She also suffered from low self-esteem. Despite being one of the better students at school, she intensified her study efforts, acknowledging that even her best efforts "were never good enough."

When Nancy finally agreed to see a psychiatrist, she discovered that her poor eating habits had a lot to do with her low self-esteem and were an overreaction to strict, overcontrolling parents. After a joint consultation with Nancy's parents led them to rely more on negotiation than threats, she gained greater freedom in her daily affairs and began feeling better about herself. Later, during a 2-week stay in the hospital, she was told that she would be allowed to attend her boyfriend's senior prom if she resumed sensible eating habits, which she did, and continued doing so.

Now 21, Nancy has no trouble with her eating habits or weight control, but admits she has not fully gotten over the fear of getting fat.

either by peers or adults. Then she becomes meticulous about counting calories, what she eats, and how she eats it. When parents put pressure on the girl to eat, she may react by hiding food or throwing it away. Sometimes the girl will eat and later vomit the food. The denial of food together with a reluctance to reduce activity leads to severe physical problems.

Anorexic girls generally have emotional conflicts from growing up in strict families. Typically, they have been "model" children and are quiet and obedient. Yet, they lack a firm sense of personal identity and autonomy. They also suffer from a disturbed body image, so that they do not realize they are getting dangerously thin, even when they see themselves in the mirror. The eating habits of these girls have been so regulated by their parents that they have not learned to interpret the inner cues signaling the need for food. Instead, they have an obsessional need to control their lives primarily through their eating habits, often engaging in elaborate rituals to ensure that they will not eat too much (Bruch, 1978).

Since most anorexics deny the illness and are uninterested in or resistant to treatment, they are notoriously difficult to treat. In some cases, the course of anorexia may continue in an unremitting way, leading to starvation and death. However, the mortality rate has been greatly reduced in recent years through greater recognition and prompt treatment of the illness. More commonly, anorexia consists of a single episode with a full recovery. Salvador Minuchin (1978/1981) and others point out that the long-term success usually necessitates family therapy, because the anorexic symptoms are seen as an expression of the patient's attempted solution to the family dysfunction. Minuchin and his colleagues report a high success rate for such patients after long-term followups.

Bulimia

bulimia: An eating disorder characterized by excessive overeating or uncontrolled binge eating followed by self-induced vomiting.

Bulimia is closely related to but distinct from anorexia. Whereas the aim of the anorexic is to lose weight, the bulimic tries to eat without gaining weight. The essential features of this illness are episodic eating sprees or binges, accompanied by an awareness that the eating pattern is abnormal, a fear of not being able to stop eating voluntarily, and the depressed mood and self-disparaging thoughts that follow the eating binges.

Unlike anorexia, which occurs in early and midadolescence, bulimia typically occurs in late adolescence, most frequently at 18 years of age. However, like anorexia, bulimia is more common among girls, especially those in the middle and upper socioeconomic groups. Although the estimated frequency of the illness varies considerably, about one in five college-aged women are involved in bulimic behavior (Muuss, 1986).

Bulimia is usually triggered by an unhappy experience, whether in school, work, or close relationships. It also increases greatly at exam time. The eating patterns may vary considerably. Some girls binge and purge occasionally, while with others eating becomes a continuing obsession. The food is usually gobbled down quite rapidly, with little chewing. Once eating has begun, there is a feeling of loss of control or inability to stop eating. For instance, Maria, a 19-year-old bulimic, felt an eating spree coming on gradually, but felt helpless to resist. During a short period of time she consumed a gallon of ice cream, a dozen doughnuts, and several boxes of brownies. Afterwards she felt despondent and self-critical. Such binges are generally terminated by abdominal pain, sleep, social interruption, or induced vomiting. The vomiting decreases the physical pain, thereby allowing either continued eating or termination of the binge. Although the eating may be pleasurable, it is usually followed by disparaging self-criticism and depression (Muuss, 1986).

Individuals with bulimia show great concern about their weight and appearance and make repeated attempts to control their weight through dieting, vomiting, and the use of diuretics. Frequent weight fluctuations due to alternating binges and fasts are common. Although bulimia is not usually incapacitating, except in the few individuals who spend their entire day in binge eating and self-induced vomiting, without treatment the illness tends to get progressively worse. Also, since these individuals feel their life is dominated by conflicts about eating, therapy is usually indicated.

DEPRESSION AND SUICIDE

Depression rivals the anxiety disorders as the most widespread psychological disturbance among adolescents and adults alike. (Robins et al., 1984). Although fluctuations in mood are normal in everyone, including feelings of sadness at times, when such feelings persist and become the dominant mood, clinicians ordinarily speak of depression. Even then, the term *depression* covers a wide range of disorders, from the relatively mild feelings of sadness to the more severe melancholy associated with a major depression.

Even though fewer than 10 percent of adolescents between 10 and 17 years of age are diagnosed as being primarily depressed, as many as half of them display some of the symptoms of depression, such as mood disturbances, self-deprecation, crying spells, and suicidal thoughts. As adolescents mature, they tend to manifest depression in more adult terms, such that the incidence of depression increases among older youth (Weiner, 1980). It has

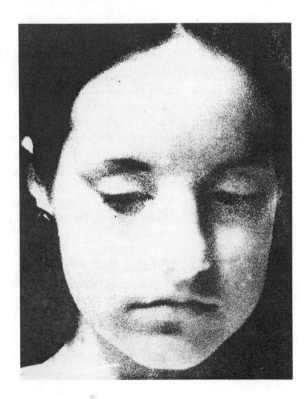

Half of all adolescents show some symptoms of depression at one time or another. (Camerique)

been estimated that at any one time one-fourth of all college students are suffering from symptoms of depression. About half of these students will require some type of professional help. Furthermore, more than three-fourths of all college students suffer some type of depressive symptoms at one time or another during each year. Depressed students tend to experience difficulties in sleeping, loss of appetite, long periods of sadness, diminished pleasure in everyday activities, feelings of worthlessness, and in some cases suicidal thoughts (Beck & Young, 1978).

From mid- to late adolescence and up, depression tends to be much more common among females than males. One explanation is that females are closer to their emotions and can admit illness more readily than males do. But females are also twice as likely to seek help for depression, which further distorts the figures (Robins et al., 1984). Furthermore, it is now thought that the actual prevalence of depression among males may be closer to that of females, except that depression among males is often masked as alcoholism and drug addiction. Evidence for this can be seen in a study among the Amish people in Pennsylvania which showed that depression is equally common among men and women, largely because the Amish prohibit the use of alcohol and drugs (Egeland & Hostetter, 1983).

Manifestations of Depression

A major difficulty in recognizing depression in adolescents is the frequency with which they mask their feelings and express their depression in other symptoms. For one thing, adolescents find it difficult to admit to themselves or others self-critical attitudes or doubts about their being competent. Then too, they are inclined more toward doing things than thinking about them, so their feelings of depression frequently get expressed in substitute symptoms or behavioral equivalents of depression. Some common signs of masked depression are boredom, a preoccupation with bodily complaints, constant fatigue, and excessive complaining. Adolescents often express their depression through a lack of interest in school, difficulties in concentrating, poor performance, and dropping out of school. Depression is one of the most frequent reasons for dropping out of school, especially in college.

The masking of depression often consists of efforts to ward off depression. One common form is restlessness, which may consist of keeping busy or searching for new and stimulating activities. What may initially appear as a lively enthusiasm for life on closer inspection often turns out to be a desperate effort to keep busy so that the individual will not have sufficient time to think. Adolescents may also attempt to escape depression by a flight towards or away from people. Adolescents may exhibit a need for constant companionship or the search for more interesting friends. Or their attempts may take the form of avoiding people, spending more time by themselves, or pursuing solitary activities. Still another way of warding off depression, especially among older youth, is alcohol and drug abuse. Youth who are unable to realize a satisfying degree of excitement and companionship through more adaptive pursuits may turn to sex and drugs as a way of combatting depression. In some adolescents, depression may be expressed primarily through indirect appeals for help, usually in some type of problem behavior, such as attempted suicide, running away, stealing, truancy, and other acts of defiance or delinquency (Weiner, 1980).

The Range of Depression

It is important to distinguish between the mild, short-lived symptoms of depression that so often accompany adolescence and the more severe types of depression that have more serious consequences and usually require treatment.

Adolescents often experience the more transitory forms of depression characterized by feelings of emptiness, uncertainty about one's personal identity, and depersonalization. Such experience is associated with the loss of childhood and the transition into adolescence. Youths with this type of depression are not so preoccupied with inward hostility, as many depressed adults are, as much as with the lack of feelings of esteem and control over their lives. They are bothered by how to evaluate and express these feelings. This is by far the most common and benign type of depression among youth and is usually outgrown as adolescents acquire a firmer self-identity, autonomy, and social skills.

major depression: A prominent and relatively persistent state of depression manifested in poor appetite, insomnia, restlessness or apathy, fatigue, feelings of worthlessness, and recurrent thoughts of death.

Another type of depression is more serious and usually requires treatment. Adolescents suffering from a major depression have a long history of problems, repeated experiences of self-defeating behavior, and difficulties relating to others. A major depression during adolescence is usually related to a great deal of family stress, lack of family cohesion, and little support for coping with stress and loss. This type of depression is more marked and persistent than the characteristic adolescent depression and is more likely to lead to suicide or suicide attempts.

bipolar disorder: An affective or mood disorder involving periods of extreme elation and depression; sometimes known as manic-depression.

Another type of depression which has received greater attention among youth in recent years is that in which individuals may experience an alternation of elated and depressive moods. Popularly known as manic-depression, this condition is now labeled *bipolar disorder*. This disorder generally appears in the form of a manic episode, in which the individual exhibits such symptoms as an expansive mood, increased social activity, talkativeness, sleeplessness, and reckless behavior. In due time, the initial manic episode may be followed by a period of normal activity and eventually a depressed episode, and then another normal period. Actually, the sequence of moods varies somewhat between the different types of this disorder. In addition to the manic episodes, there are other characteristics that distinguish bipolar disorder from a major depression. The bipolar disorder is much less common than

CASE HISTORY OF DONNA

Donna, a 19-year-old college student, experienced the most serious type of depression, which eventually led to a suicide attempt. Donna felt that she had never been a happy person, placing much of the blame on her unhappy childhood. Her parents fought constantly before getting a divorce when Donna was 6 years old. She continued to live with her mother, but felt abandoned by her father, who had moved to California and telephoned only on her birthdays.

When Donna's mother remarried, Donna felt even more left out because of her mother's attention to her new husband and their preoccupation with the stepfather's three younger children. Her mother recalled that Donna had always been a somewhat sad person. Since childhood, Donna had remained shy and introverted, making few friends. Tall, thin, and not too attractive, Donna also suffered from a poor body image and an inferiority complex. Although she had a good mind, Donna made only mediocre grades, mostly because of her preoccupation with her inner feelings and problems.

Donna's problems intensified when she went away to college. Much of the stress came from her conflicts with an incompatible roommate, demanding studies, and a part-time job. The final blow came when Donna received a note from her boyfriend ending their tenuous relationship. A call to her mother brought only an admonition, "Grow up! Losing a boyfriend isn't the end of the world." Feeling desperate and lonely, Donna took an overdose of sleeping pills. Fortunately, she was discovered in time when her roommate returned from a date later that evening.

After a brief stay in the university hospital and several months in outpatient therapy, Donna decided to return to a less stressful environment in her home town. With added support and encouragement from her parents, she took an apartment and enrolled part-time in a community college near her home. Donna feels she is a stronger person now and has a more positive attitude towards herself and her future.

depression and tends to be more equally distributed among both sexes. Unlike a major depression, which may occur at any age, bipolar disorder usually appears before age 30. It is also more likely to run in families (Hirschfield & Cross, 1982). The use of lithium carbonate is often helpful in controlling the mood swings in this disorder. However, it is difficult to find the correct dosage and to get patients to take their medication continuously. Individuals may stop taking their medication, mainly because it takes away their feelings of well-being when they are in a mildly elated state. Yet, if they do not resume the medication, they may experience another manic or depressive attack as great as that experienced before they began taking medication (Walker, 1982).

Adolescent Suicide

suicide: The act of taking one's life intentionally.

Depression is also a major risk factor in suicide, with at least half of all the suicides in the United States being committed by people suffering from depression (Greenberg, 1982). Suicide is relatively rare among children and early adolescents. But as youths gain more self-identity and autonomy, from midadolescence on, both the potential for suicide and the number of reported suicides rise rapidly. Furthermore, the suicide rate among young people has almost tripled in the past 30 years. There are now about 12 reported suicides per 100,000 youths in the 15 to 24 age bracket in the United States, with the suicide rate among males about four times that of females (U.S. Bureau of the Census, 1985). (See Figure 15–1.) For youths 15 to 24 years of age, suicide is now the second leading cause of death among males, after accidents, and the third leading cause of death among females, after accidents and cancer (Robbins, Angell, & Kumar, 1981).

It is widely recognized that reported suicides are greatly outnumbered by unreported suicides, attempted suicides, and other types of self-destructive

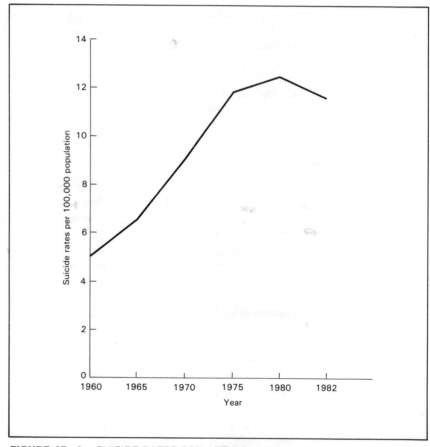

FIGURE 15–1.   SUICIDE RATES FOR 15 TO 24 YEAR OLDS.

Source: National Center for Health Statistics, in Carl A. Taube and Sally A. Barrett, eds., *Mental Health, United States, 1985* (Washington, D.C.: U.S. Department of Health and Human Services, DHHS Publ. No. [ADM] 85-1378; Government Printing Office, 1985), p. 150.

behavior. For instance, David Phillips (1981) found that automobile fatalities in California increased 31 percent 3 days after a suicide received wide publicity. Many of these accidents may have been unconsciously caused by people intent on suicide. Most were single-car accidents involving people close to the age of the suicide victim. The victims also lingered fewer days before dying than those in other car accidents.

Although adolescents are less likely than adults to take their lives, they are at least as likely to attempt suicide. Whereas there are about 6 to 10 suicide attempts for each suicide in the general population, the ratio of attempted to actual suicides among high school and college students ranges as high as 50 to 1 (Weiner, 1980; Goleman, 1986). And while suicide threats are several times more common among young women than men, young men are four times more likely to actually commit suicide. A major reason is that men tend to use more swift and violent means, such as a gun or hanging. By contrast, women are much more likely to use pills or some form of poisoning, which often permits intervention (U.S. Bureau of the Census, 1985). As a result, many of these unsuccessful suicide attempts are regarded as a cry for help. It may be that women are more able than men to cry out for help with their emotional needs, such that twice as many women students as men seek help from college counselors. Yet the higher risk is the silent and isolated college man who does not turn to anyone for help (Goleman, 1986).

Contributing Factors

In some instances, suicide represents an impulsive act in response to a stressful situation. The most common precipitators of suicide and attempted suicide are the breakup of a love relationship, disruptions at home, and doing poorly in school (Gispert et al., 1985). The tragedy is that such problems are often transitory, whereas the solution—suicide—is permanent. In other instances, suicide is the last in a long series of frustrations and bouts with self-defeat. Many suicidal youths have a long history of problems; they experience an escalation of their problems at adolescence, as well as social isolation and a sense of hopelessness and helplessness. In still other instances, suicide may be precipitated by severe emotional disturbance. One study of patients hospitalized after attempting suicide found that four out of five of them were suffering from depression. Curiously, severely depressed youths are more likely to take their lives as things are looking up for them. When they are most depressed, they may not have sufficient energy to take their lives. Usually, it's when they start to feel better and get their energy back that they act on their underlying hopelessness and commit suicide (Greenberg, 1982).

Young people who commit suicide often feel overwhelmed by a combination of problems. In one study of teenage suicide, 505 suicide attempters seen in the emergency room of a large hospital were compared with the same number of individuals matched for age and sex who had been treated in the

Suicidal behavior sometimes results in single-car accidents. (Irene Springer)

emergency room but not for suicide. Among the attempters, girls outnumbered boys three to one. Compared to the controls, the attempters had more psychological problems, greater history of alcohol and drug abuse, and more prior psychological treatment. Furthermore, their families had more psychological problems, more history of suicide, and more parental absence through death or divorce (Garfinkel, Froese, & Hood, 1982). Two recurring themes in the lives of suicidal youth are the loss of love, often the loss of a parent at an early age, and a sense of hostility, guilt, and self-blame associated with this loss. Consequently, such individuals remain especially vulnerable to the loss of love in later life, with broken love relationships being a major precipitating factor in suicides. A study of high school and college students who had entertained serious suicidal thoughts showed that such individuals were more likely than their classmates to view their parents as having many conflicts with each other, to have poor relationships with their parents, to see at least one of their parents as angry or depressed most of the time, and to perceive themselves as having a drinking or drug abuse problem (Wright, 1985).

The Prevention of Suicide

The prevention of suicide among young people has received greater attention in recent years. One approach has been to make it more difficult to engage in self-destructive behavior, such as having tighter control over the prescription

WARNING SIGNALS

Here are some warning signs of suicide:

- Expression of suicidal thoughts or a preoccupation with death
- Prior suicidal attempt
- Depression over a broken love relationship
- Despondency over one's situation at school or work

- Giving away prized possessions
- Abuse of alcohol or drugs
- Marked personality changes
- Change in eating habits
- Change in sleeping habits
- Sense of hopelessness

of sedatives and taking protective measures against people driving while they are intoxicated. Another approach is to increase campus and community awareness and the resources for dealing with suicide. Many communities have established suicide hotlines that are available 24 hours a day, though these are more apt to be used by low-risk individuals. Even then, it is important to know how to respond to a suicide threat. In the first place, the old myth that those who threaten suicide seldom carry it out has been disproven. Instead, it is now believed that most suicidal individuals express some warning signs, such as an indirect or direct expression of suicidal feelings or sense of hopelessness. Then too, allowing the suicidal person to talk about his or her thoughts does not necessarily give such ideas greater force. Instead, providing these people with an opportunity to talk about their suicidal thoughts may help them to overcome such wishes and know where to turn for help. But it's important not to analyze their motives or show anger. It's usually better to express warmth and concern, without overreacting to the threats of suicide. Friends and helpers also need to distinguish between dealing with the immediate situation and helping to get the person to professional help for the longer-standing problems. One of the more successful suicide prevention programs is in the public schools in Fairfax County, Virginia, just outside Washington D.C. Every high school and intermediate school teacher is trained in the causes and symptoms of suicide. The program includes seminars for students and meetings for parents to discuss adolescents' stress-related problems with mental health professionals. Special counselors are also available at the school to work with troubled teens (*On Campus*, 1985).

SCHIZOPHRENIA

Schizophrenia, one of the most severe psychological disorders, is especially apt to make its initial appearance during adolescence or early adulthood. Although schizophrenia is relatively rare, affecting about 1 in 100 people in the general population, it is so disabling that as many as half of the mental hospital beds in the United States are occupied by people with this disorder (U.S. Department of Health and Human Services, 1981). Schizophrenia literally translated means "split mind," but the split is not between a Dr. Jekyll and Mr. Hyde type of personality; rather, it's a splitting of the different psychic functions of thought, feeling, and behavior such that the person is out of touch with social reality. Even though the singular term "schizophrenia" is commonly used for the sake of convenience, it actually refers to a group of related disorders which exhibit a wide diversity of symptoms. The essential features of these disorders consist of disordered thinking, including delusions; disturbed perception, including hallucinations such as hearing voices; blunted or inappropriate emotions; and social withdrawal.

Adolescent Schizophrenia

The risk of schizophrenia rises rapidly after age 15 and peaks in late adolescence and early adulthood. Males are more at risk before 25 years of age, females after 25 (Lewine, 1981). The initial onset of schizophrenia is more difficult to detect among adolescents than adults. Only about 30 to 40 percent of adolescent schizophrenics manifest the clear symptoms associated with this disorder. The rest initially show a mixed picture in which the characteristic schizophrenic symptoms, such as incoherent thinking and inappropriate emotions, are peripheral to other complaints, especially depression and behavior problems. Like depression, schizophrenia may be masked by other symptoms. Instead of appearing schizophrenic, the adolescent may exhibit behavior problems such as fighting and stealing, especially common among males, or may act depressed, as is often the case with females (Weiner, 1980).

The onset of this illness may occur abruptly, with marked changes in personality appearing in a matter of days or weeks. Or it may occur in a gradual, insidious deterioration in functioning over a period of years, usually an unfavorable sign. During the initial phase of the illness, individuals tend to have difficulty communicating with others, become socially withdrawn, and neglect personal hygiene, school work, or their jobs. The active phase of the disorder is often precipitated by intense psychological stress, such as rejection in love or the death of a parent. During this period, individuals display the characteristic symptoms of the disorder, such as delusions, hearing voices, and other bizarre behavior (DSM-III, 1980).

Contributing Factors

Despite extensive research and treatment, the causes of schizophrenia are not fully understood. However, it is generally thought that this disorder results from an interaction of the individual's inherited predisposition to the illness and environmental stress. So far, the evidence strongly suggests that there is a genetic predisposition to schizophrenia. The 100 to 1 odds of anyone becoming schizophrenic rise to 10 to 1 among those with a parent or sibling who is schizophrenic and 50:50 among those with an identical twin diagnosed schizophrenic. Furthermore, it seems that an identical twin of a schizophrenic individual is about as likely to develop this disorder whether reared apart or with the schizophrenic twin. Also, children of schizophrenic parents who are adopted by nonschizophrenic parents have an increased risk of getting this disorder, while children of nonschizophrenic parents who are adopted by someone who becomes schizophrenic are unlikely to catch the illness. Thus, there is a genetic predisposition to this illness, though the predisposition by itself is not usually sufficient for the development of schizophrenia (Nicol & Gottesman, 1983).

Although there is no environmental factor that will invariably produce schizophrenia in people who are not related to a schizophrenic, environmental factors often play a significant role in triggering the development of schizophrenia. For many years, authorities emphasized the role of overprotective but rejecting mothers and weak, ineffectual fathers in the development of the disorder. More recently, the focus of attention has been on deviant family patterns, especially faulty communication. In one study, deviant communication by parents, such as mixed messages, proved to be an accurate predictor of whether their adolescents would be diagnosed as schizophrenic 5 years

CASE HISTORY OF SANDY

Sandy was a shy, lonely, but otherwise normal child until she suffered an acute schizophrenic breakdown at age 17. She was the last born of three children and felt inferior in comparison with her older brother and sister. Her sister, the oldest, was especially attractive and outstanding in school. Sandy, however, was rather plain in appearance and made only average grades in school. She had no real friends outside the family. She attended church with her family, but felt uneasy there because of her guilt about sex. Sandy said that she masturbated under the covers for fear that God would see her.

Although she had planned on attending college away from home, she was apprehensive about being separated from her family. Late in the spring of her senior year in high school, Sandy became obsessed with some strange ideas. She felt that her mother was out to destroy her, and that only her father, who was God, could save her. She begged him to take her to work with him, but he refused, not clearly understanding her reasons. Then one day, Sandy ran out of the house screaming incoherently. At first, neighbors thought she was on drugs, even though Sandy had only smoked marijuana occasionally.

After a quick consultation with the family's clergyman, Sandy was taken to a local hospital, where she received excellent treatment in the psychiatric unit. Within several weeks she was acting normally and resumed life at home. She spent the next year at a local college, living at home. Then she went away to college, where she made several good friends and better-than-average grades.

Sandy made a good recovery because she had supportive parents, got prompt treatment, made a favorable response, and was fortunate enough to be treated by psychiatrists who had a positive attitude about the possibilities for recovery for such individuals.

later (Doane et al., 1981). In hopes of identifying the psychosocial triggers of schizophrenia, some researchers have begun following the "high-risk" children born to a schizophrenic parent and comparing the experiences of individuals who become schizophrenic with those who do not. In this way, eventually we may be able to detect the early warning signs and the circumstances that worsen an individual's chances of contracting this disorder. So far, it appears that children who eventually become schizophrenic are more likely to have a mother whose schizophrenia was severe and chronic and/or a mother who underwent complications at the child's birth; further, it is more likely that children who become schizophrenic have been separated from their parents and have created disturbances in school (Watt et al., 1984).

Outlook

There is a considerable difference of opinion about the outlook for individuals who have suffered an acute schizophrenic episode. For a long while clinicians adhered to the principle of thirds, that is, about one-third of the individuals who have experienced a schizophrenic episode make a good recovery, another one-third make a partial recovery with occasional relapses, and still another third remain chronic schizophrenics. However, with improved methods of treatment, including the use of powerful antipsychotic drugs, more favorable attitudes toward those afflicted with the disorder, and more sophisticated research strategies, a larger proportion of schizophrenic individuals are making at least a partial recovery. For instance, an extensive followup study of 1000 people who suffered a schizophrenic episode by Manfred Bleuler (1978) showed that one-fourth of them resumed normal functioning, one-half to two-thirds of them alternated between normal life and a recurrence of the active phase of the illness, and only about 1 in 10 of them remained schizophrenic for the rest of their lives. Actually, studies by Arieti (1981) and others have shown that how well an individual recovers from a schizophrenic episode depends on a variety of influences, especially the following factors:

1. *Premorbid adjustment.* The more adequately the individual functioned prior to the illness, the better the outcome.

2. *Triggering event.* If the illness is triggered by a specific event, such as the death of a parent, the possibility of recovery is more favorable.

3. *Sudden onset.* The more quickly the disorder develops, the more favorable the outcome.

4. *Age of onset.* The later in life the first illness develops, the better.

5. *Affective behavior.* The presence of conscious anxiety and other emotions, including depression, are favorable signs. A state of hopelessness not accompanied by depression is a poor sign.

6. *Content of delusions and hallucinations.* The more the delusions involve feelings of guilt and responsibility, the better the outlook. Conversely, the more the delusions and hallucinations blame others and exonerate the individual, the more severe the illness.

7. *Type of schizophrenia.* Individuals with the diagnosis of paranoid schizophrenia have the best chance of recovery; those diagnosed undifferentiated schizophrenic the worst.

8. *Response to the illness and treatment.* The more insight individuals have as to what makes them ill and the more cooperative they are with their therapists, the better their chances of recovery.

9. *Family support.* The more understanding and supportive the families of these individuals are, the better their chances for a good recovery.

Mark Vonnegut, son of writer Kurt Vonnegut, was an excellent student in college. While in his twenties and living in an experimental community with other youths, Mark experienced a combination of stressful events such as his parents splitting up and his girlfriend running off with another man, all of which contributed to his acute schizophrenic disorder. Yet, with prompt medical treatment Mark made a good recovery and later entered medical school (Vonnegut, 1975; Gorman, 1984). Mark's experience points up some of the predictive factors mentioned here: the better the person's prior adjustment, the more quickly the disorder develops; and the better the patient responds to the treatment, the more favorable the chances of recovery.

THERAPY

therapy: Refers to the overall treatment of psychological problems, and may include counseling, medication, and hospitalization.

antisocial personality: An individual who violates social norms without evidence of guilt, sometimes known as a sociopath or psychopath.

Adolescents are often referred for help because a parent, teacher, or social worker notices that "something is wrong." Since adolescents spend much of their waking day at school, it should come as no surprise that anywhere from one-third to one-half of all adolescents referred for help have school-related problems, such as learning difficulties or behavior problems (Weiner, 1980). Among older youth 18 to 24 years of age, the primary presenting problems vary somewhat according to sex. For males, the four most common psychological disorders are: alcohol abuse/dependence; drug abuse/dependence; phobia; and antisocial personality. The most frequent disorders among females are: phobia; drug abuse/dependence; major depressive episode; and alcohol abuse/dependence. Troubled youths may be seen in a variety of therapeutic settings. The majority of them will be treated in one of the publicly funded community mental health centers, with a much smaller number seeing a mental health specialist, such as a psychiatrist or psychologist in private practice. Youths with the more serious disorders, such as schizophrenia, are

psychotherapy: A form of therapy in which a trained therapist performs activities that will facilitate a change in the client's attitudes and behaviors, either in a one-to-one or group setting.

initially treated in an institutional setting, with individuals up to 24 years of age representing about one-fourth of all admissions to public and private psychiatric hospitals (Taube & Barrett, 1985).

Individual psychotherapy with adolescents is generally more difficult than with children or adults. For one thing, adolescents are still in a process of clarifying their identity and autonomy and may be uncommunicative, skeptical, and uncooperative in therapy. Then too, many of them have been referred by their parents, and the therapist must establish a relationship of trust and cooperation before significant progress can be made. Also, adolescents' behavior is often unpredictable, with youthful clients feeling overwhelmed by their problems in one session and feeling little or no need for help the next. Furthermore, acting-out behavior, such as running away or vandalism, often complicates the treatment program. All things considered, it is usually best for therapists to focus on the adolescent's present situation rather than the achievement of deep insight. It is also important for the therapist not to pay undue attention to the adolescent's problem behaviors, but to look beyond the symptoms to the unfulfilled needs being expressed. In this way, the therapist can assist the adolescent to articulate more positive goals and to acquire the necessary skills to achieve them. Although it is not realistic to ask adolescents to make a long-term commitment to therapy, it appears that the greater number of sessions attended, at least up to 26 or so, the more specific goals are attained (Richmond, 1978; Howard et al., 1986).

group therapy: All those forms of therapy in which a leader meets with a group of clients, including preexisting groups such as families and those consisting of strangers.

There is a growing realization that troubled youths may make faster progress when treated in groups. Among the major reasons are that use of the group process permits the adolescent's identification with others with similar problems and encourages conformity to group norms; in addition, there is mutual help and support among group members. Another advantage is that the therapist is able to observe directly the adolescent's behavior with others. Group therapy may take place in outpatient mental health centers or, as is frequently the case, as an integral part of a residential treatment program. Family therapy is especially valuable for observing the adolescent's behavior within the family group and for eliciting parental involvement. However, a major difficulty in explaining this type of therapy is that there are at least 14 different schools of family therapy, which vary in their theoretical approaches, range of techniques, and procedures used (Gurman, 1983). Some family therapists see the parents for an initial session and then the adolescent with his or her parents in the remaining sessions. Other therapists prefer to see all family members together from beginning to end. Still other family therapists may see only the parents and the adolescent who is the identified patient.

A common feature of the various family therapies is the assumption that families operate as systems, such that the adolescent's problem behaviors are best understood and treated as an integral part of the family unit. Thus, many problem behaviors, such as delinquency, running away, and drug abuse, may reflect larger problems within the family, especially difficulties separating from

Volunteers answering a hotline for troubled youths. (Leo Rutigliano/Decisions Center, Inc.)

the family. In many instances, especially with adolescents who exhibit severe problems, family therapy may serve as an alternative to hospitalization or help to decrease the length of stay in a residential treatment program (Ault-Riche & Rosenthal, 1986). At the same time, a number of studies remind us that family therapy is not always or automatically more effective than individual therapy. In one comparative study, many of the adolescents' problems were unrelated to family functioning, though some problems and types of families, such as black families in this study, were especially appropriate for family therapy (Pardeck et al., 1983). In one case, a 12-year-old delinquent male living in a home with a disturbed and turbulent marriage made positive behavioral changes following therapy despite the lack of improvement in the parental marriage (Kanner, 1984).

There is also a wide range of community services available to troubled youth, including hotlines, group homes, halfway homes, shelters, and peer programs. Hundreds of hotlines have been established in the last decade or so, especially in large cities. Some hotlines are more general in scope; others specialize in a particular type of problem like drug abuse or suicide prevention. Halfway houses provide support for the young person's transition from the hospital or residential treatment program back to the community. Shelters and group homes provide a similar type of support, often for those who have not been institutionalized. Frequently, these agencies are aimed at particular types of troubled persons, such as runaways, drug abusers, or unwed mothers. Group homes are usually staffed by a professional or paraprofessional with a high involvement of young people. The aim is to provide positive adult models, while encouraging individual responsibility and self-governance. Peer programs, another innovative type of help, are springing up everywhere. Although peer programs are staffed primarily by young people—frequently those who have successfully overcome a problem similar to that being dealt with—they are closely supervised and supported by professionals in most cases. Peer programs have been established for a wide variety of problems, including smoking, alcohol and drug abuse, sexual problems, and depression. In some cases, peer programs can change behavior just as effectively, or more so, than conventional therapies (Johnson, Graham, & Hansen, 1981).

SUMMARY

Psychological Disturbance in Adolescence

1. Clinic admissions tend to reach a peak at midadolescence, probably because of the developmental problems of this age coupled with the inherent stress of adolescence.

2. A greater proportion of boys than girls are referred for psychological help during adolescence, partly because the boys' characteristic problems with aggression and undercontrolled behavior are more likely to come to the attention of authorities.

3. It is more difficult to distinguish between normal and disturbed behavior at adolescence than at any other period of life, with parents and concerned adults risking overinterpreting signs of disturbance among adolescents or prematurely dismissing them until problems get out of hand.

Anxiety and Eating Disorders

4. In an anxiety disorder, the person may suffer from intense anxiety most of the time or from periodic panic attacks that are disproportionate to the situational threat.

5. Phobias are excessive, unrealistic fears, with school phobia being more serious when it occurs at adolescence than in childhood.

6. Adolescents with the obsessive-compulsive disorder are unable to put certain unacceptable ideas out of their heads, and have the urge to engage in repetitive acts to ward off the imaginary threats.

7. Anorexia, an eating disorder commonly found among adolescent girls, is characterized by a disturbance in body image and eating habits that results in a loss of normal body weight.

8. Bulimia, which is common among older adolescent girls, involves episodic eating binges, accompanied by a fear of not being able to stop eating voluntarily, and eventually followed by a depressive mood.

Depression and Suicide

9. The symptoms of depression are common among youth, though often masked in a variety of ways, such as in boredom, excessive complaining, and dropping out of school.

10. Depression appears to be more common among females because they are more likely to seek help for their depression whereas males are more likely to mask their depression in alcohol and drug abuse.

11. There are many types of depression, ranging from the mild, short-lived depression that accompanies adolescence to the more severe types of depression that require professional treatment.

12. Suicide is now the second leading cause of death among males and the third leading cause of death among females in the 15 to 24 age bracket.

13. Young people who attempt or commit suicide often feel overwhelmed by a combination of problems, including the loss of love relationships, depression, and family stress.

14. A major approach to the prevention of suicide is greater campus and community awareness of the problem and more resources for dealing with it.

Schizophrenia

15. Schizophrenia, the most frequently occurring and most disabling of the psychoses, is characterized by severe disturbances in thought or perception, blunted or inappropriate emotions, and social withdrawal.

16. The risk of schizophrenia rises rapidly after age 15 and peaks in late adolescence and early adulthood.

17. Schizophrenia usually results from an interaction of the individual's inherited predisposition to the illness and environmental stress.

18. How well an individual recovers from a schizophrenic episode depends on a variety of factors, such as the prior adjustment to the illness, the age of onset, and how rapidly the disorder develops.

Therapy

19. Troubled youth may be seen in a variety of therapeutic settings, including mental health centers, professionals in private practice, and psychiatric hospitals.

20. Individual therapy with adolescents is usually more difficult than with adults and usually focuses on improving practical behaviors rather than the achievement of deep insight.

21. There is a growing realization that troubled youth may often make faster progress when seen in groups, especially in family therapy.

22. Troubled youth also have a wide range of community services available to them, including hotlines, group homes, halfway houses, shelters, and peer programs.

REVIEW QUESTIONS

1. What is an anxiety reaction?

2. What are some of the conditions associated with school phobia?

3. Explain the eating disorder of anorexia.

4. How is bulimia like and unlike anorexia?

5. What are some symptoms of masked depression in adolescents?

6. To what extent is depression more common in females than males?

7. What are some of the psychological factors that contribute to attempted and actual suicides?

8. What are the symptoms of adolescent schizophrenia?

9. What type of adolescent would be at high risk for developing schizophrenia?

10. What are some of the possible advantages and disadvantages of family therapy?

16 Transition to Adulthood

- **LEAVING HOME**
  Peaceful and Stormy Departures
  A Family Matter
  Leaving for College
  Returning to the Nest
- **TAKING HOLD IN THE ADULT WORLD**
  Autonomous Decision Making
  Preparing for Economic Independence
  Establishing Close Relationships
- **FROM YOUTH TO ADULTHOOD**
  Continuity
  Change
  Individual Differences
- **SUMMARY**
- **REVIEW QUESTIONS**

To determine the developmental tasks college students are currently working on, Bruce Roscoe and Karen Peterson (1984) administered a variety of inventories to 485 undergraduates at a midwestern university. They found that while the students regarded themselves as early adults, most of them were simultaneously working on tasks from several developmental stages. More than three-fourths of the students were engaged in the developmental tasks of adolescence, including achieving mature relationships with peers, acquiring emotional independence from parents, learning socially responsible behavior, and acquiring a set of values. At the same time, almost as many of them were still working on developmental tasks from even earlier periods, such as developing attitudes toward self, toward social groups, and concepts for everyday living. Only about half the students were engaged in the typical developmental tasks of early adulthood, such as taking on civic responsibility, finding a congenial social group, getting started in a career, and selecting a mate. Such findings suggest that people of this age are characteristically in *transition* to adulthood, rather than having fully arrived.

LEAVING HOME

leaving home: Refers to the psychosocial separation from one's family of origin, and includes the achievement of emotional autonomy as well as moving out of the family home.

In their longitudinal study of adult development, Daniel Levinson (1978) and his colleagues made a similar discovery, namely, that the transition to adulthood is more complex and strung out than popularly understood. College students, like others their age, are usually in the initial stages of adulthood. Levinson has called this period the "early adult transition," which lasts roughly from the late teens to the early twenties. During this period individuals face two major developmental tasks: leaving the preadult world and initiating early adulthood. We'll take a look at the first developmental task in this section and the second one in the next section.

A major part of leaving the preadult world is separating from one's family of origin—popularly known as "leaving home." The external aspects of this transition involve moving out of the family home, becoming less dependent financially, and taking on new roles and responsibilities. Its internal aspects involve increasing differentiation of self from parents and reducing emotional dependence on parental authority and support. Although these two aspects of leaving home are related, it is the inner psychosocial transition that is the most crucial.

early adult transition: Levinson's term for the transitional stage between adolescence and adulthood; the initial stage of adulthood.

Peaceful and Stormy Departures

It's not uncommon to hear teenagers tell their friends, "I've got to get away from home. My parents are driving me crazy." But they seldom act on these feelings before 17 or 18 years of age. Most of them lack the financial resources to set out on their own, and generally remain at home at least until graduation from high school. After 18, young people are more inclined to act on their convictions and begin pulling up their roots in earnest. Graduation from high school usually provides the occasion for their departure, with college, military service, jobs, and short-term trips serving as the acceptable vehicles for getting away from home. Attitudes toward parents tend to soften somewhat, with 18 to 22 year olds less likely to blame their parents for their problems, seeing them, instead, as less important in their lives (Gould, 1978).

In many instances, leaving home is a gradual, relatively peaceful affair. For instance, Kim had always gotten along well with her parents throughout high school. As a teenager, she had spent time away from home at summer camps and felt comfortable looking out for herself. During college Kim returned home regularly during holidays and part of each summer, partly to see her friends back home. However, by the time she had entered medical school at the same university where she had attended college, Kim felt more at home among her friends at school and visited home less often. Kim kept in touch with her parents regularly by phone and letters, but more as an adult who enjoyed sharing her experiences than as an adolescent in need of direction. When Kim's parents disagreed with her on occasion, they respected their daughter's right to live her life in her own way. Gradually, they not only made the adjustment to her grownup status but felt proud of having a daughter who was her "own person."

In some instances, the departure from home is a stormy one. John, who constantly fought with his father about school grades and the use of the family car, hastily moved away after graduating from high school. He worked at a service station for a year while attending a nearby community college part-time. Even when he was refused financial aid from the state because of his father's income, John preferred his independence to financial support from his father. Debbie, a 20 year old with a clerk's job, frequently argued with her mother over taking her own apartment. She was told that she should not leave home until she married. Her friends were not surprised when she suddenly moved in with her boyfriend, whom she had known for only 4 months. In Debbie's case, an abrupt physical departure simply added physical distance to psychological alienation and transferred emotional and financial dependence from her parents to her boyfriend. By contrast, John's stormy departure may have provided the needed opportunity for assuming responsibility and growth.

A Family Matter

How peaceful or stormy the departure from home is depends as much on the parents as on young people themselves. A study of over 500 parents showed that parental concern over their children aged 16 to 35 was strongest when the family was in the midst of launching young people from the home and the oldest was between 21 and 25 years of age. Parents were more likely to be distressed at this stage, especially those parents who were anxious and pessimistic about their kids. Greater parental distress was seen among parents who relied less on other family members and more on professional helpers. On the positive side, parents with a strong sense of personal mastery were more likely to seek professional help with their family problems at this stage (Menaghan, 1978).

In his book *Leaving Home,* Jay Haley (1980) points out that maladjustment at this age often reflects problems in separating from home as well as in the dynamics of the family itself. Separation troubles can take many forms, including drug addiction, delinquent behavior, emotional disturbance, suicide, failure at school or work, or paralyzing apathy. In many instances, troubled young people have difficulty leaving home because their parents are unwilling to "let go." Such parents may complain of a young person's problems at school or with drugs, while deriving an unconscious satisfaction from knowing that they are still needed as parents. Physical presence in the home is not necessary for the expression of dependence. A young person might be in a mental institution or be a member of a religious cult. The constant thread is the youth's refusal to succeed by becoming independent. It is as if the young people sacrifice their own personal growth for the sake of preserving family unity. Ironically, such problems often occur in "model" families with overprotective parents.

The healthier the family, the more parents desire, and in most cases actively promote, the disengagement of their young. For one thing, the presence of young people in the home is frequently an added source of aggravation at a time when marital satisfaction is at its lowest point in marriage. Many parents of older adolescents also express agreement with the statement "I regret the mistakes I made in raising my children" (Gould, 1978). In some instances, parents may be too eager to get rid of their children. For example, one father took over his son's room as a study as soon as the son left for college. One young woman who was married and quickly divorced expressed disappointment that "her room" was no longer available when she wanted to return home in the period following her divorce. Wiser parents may actively promote disengagement, but in a more gradual and constructive way. That is, parents may expect their young people to earn part of the money for their education, to help with chores while at home, and yet be reassured that they have a home base while they are starting out on their own.

Leaving for College

Young people understandably experience some conflict over leaving home—wanting to leave on the one hand, yet apprehensive about leaving on the other—mostly because of their uncertainty over the future. Leaving for college helps to tip the balance toward greater independence, though not without a certain degree of inner anguish. As a result, incoming college freshmen tend to experience a great deal of anxiety about leaving home, often coupled with anger toward their parents. Girls are even more likely than boys to report resentment toward their parents, perhaps because of the overprotective attitude toward girls. But as students adjust to college, they come to feel closer to their parents. The most likely explanation is that anxiety and anger toward one's parents are a normal part of the emotional disengagement of leaving for college. Only when such feelings persist long after the transition to college do they seem to indicate a deeper problem (Goleman, 1980).

Going away to college usually improves the relationship between young people and their parents. This was brought out in a study by Kenneth and Anna Sullivan (1980) which included 242 male students from 12 high schools in Pennsylvania, New Jersey, New York, and Massachusetts. The Sullivans compared two groups: those who attended residential colleges and those who remained at home and commuted to school. Students completed questionnaires during high school and again during college, as did their parents. Questions were asked about affection and communication between parents

Going away to college generally improves the communication between young people and their parents. (Laima Druskis)

and students and the students' independence. The level of the young man's independence was measured by whether he felt his family encouraged him to make his own decisions or criticized him and tried to dictate things like hairstyle, and the degree to which he made decisions without their help. Results showed marked differences between the two groups, with those who had gone away to college reporting more affection and better communication with their parents than those living at home.

Leaving home, in the more inclusive sense of making a successful psycho-social transition to adulthood, appears to be an essential part of one's ego-identity development as well as adjustment to college. In one study involving 132 students in various stages of college, Steven Anderson and William Fleming (1986) administered a variety of instruments measuring the students' home-leaving strategies, ego-identity, and college adjustment. The extent to which students had left home was measured in terms of four related variables; economic independence, living apart from one's parents, perception of personal control, and emotional attachment to parents. The results showed that all four aspects of leaving home were highly predictive of the students' ego-identity and college adjustment. Compared with other students, individuals with strong ego-identity were more likely to have attained economic independence, their own living quarters, greater personal control over their lives, and positive feelings and attachment to their parents. Results for college adjustment were similar, with higher scores for college adjustment associated with all four measures.

Returning to the Nest

When Brad graduated from college he went home to live with his family a while before he found a job and lived on his own. Similarly, Marie, a 23-year-old graduate student living on her own, decided to move back home with her parents temporarily to "regroup," as she put it. Brad and Marie are typical of a trend in the 1980s. Like birds flying back to the nest, more young people are moving in with their parents, sometimes after years of absence. The number of individuals 18 years or older living at home increased from 13.7 million in 1970 to 19.1 million by 1982. The increase in "nesters," as some social service workers call them, is partly due to there being more people this age in our society. But much of the increase is due to economic pressures in recent years, such as inflation, a tight job market, and the high cost of housing. As a result, about 6 out of 10 men and 5 out of 10 women 18 to 24 years old still live with their parents. (See Table 16–1.)

Returning home is often a mixed blessing that neither young people nor their parents may prefer. People like Brad and Marie may feel that they are failures because they expected to be independent by their twenties. They may have to swallow some of their pride and give up some personal freedom for the safety of the nest. Parents who anticipated their children would be fully

Table 16–1 **LIVING ARRANGEMENTS OF PERSONS 15 YEARS OLD AND OVER BY AGE AND SEX**

		Percent living:			
Age and sex	Total (1,000)	alone	with spouse	with parents/ relatives	with non- relatives
Male					
15–19 years	9,222	0.7	1.1	95.7	2.5
20–24 years	10,055	6.8	21.2	57.6	14.4
25–44 years	35,517	10.5	66.3	14.5	8.7
Female					
15–19 years	9,103	0.7	5.3	90.8	3.2
20–24 years	10,411	4.9	34.6	49.0	11.5
25–44 years	36,639	6.9	67.5	21.3	4.3

Source: From U.S. Bureau of the Census, *Statistical Abstract of the United States, 1987*, 107th ed. (Washington, DC: 1986), p. 46.

launched by 21 or 22 may be annoyed to have their son or daughter back home. Although many nesters pay room and board, making such arrangements is awkward and the payments are usually small. Parents often complain that the payments rarely cover the added expenses. Also, parents and their young should be careful to avoid falling back into the relationship they had before the young person first moved out. Parents walk a very fine line. They want to be supportive of their youth, but they don't want to be overprotective. Nevertheless, parents may find themselves asking "What time did you get in last night?"—forgetting that their son or daughter is older now. Young people, too, must maintain their independence and not forget they are adults. They must be careful not to let their temporary living arrangements undermine their self-confidence and taking charge of their lives. One 25-year-old man sought help from a counselor, admitting he was ashamed to go out on a date because he had difficulty telling people he was still living at home. Whenever he got to that question—your place or mine—he was embarrassed to say, "I'm living with my parents" (Shapiro & Lounsberry, 1983).

Given the large number of potential problems, young people should avoid returning home when possible. It might be better to live in an inexpensive one-room apartment or move in with a friend. Yet, in some cases, it may be wiser for young people to return home to temporarily regroup before setting out on their own. When this is the case, they should not be embarrassed or confuse their temporary living arrangements with progress in the more important developmental tasks of this period, such as clarifying their ego-identity and assuming the responsibilities of adulthood as soon as they are appropriate. Furthermore, young people and their parents may find this a satisfying time for strengthening their ties before a son or daughter leaves home again, usually for good. It's been said that the two most important things parents can give their young are roots and wings. Healthy people need both.

TAKING HOLD IN THE ADULT WORLD

Just as leaving home means more than simply moving out of the family house, so becoming an adult involves more than getting a job and finding a place of one's own, as important as these accomplishments are. Taking hold in the adult world—the second major developmental task in the transition to early adulthood—has to do with a new level of psychosocial engagement with the world. This involves trying out new roles, assuming more adult responsibilities, and finding an optimal fit between one's self and society. The major components of this developmental task are: learning to make decisions in an autonomous way, preparing for economic independence, and forming satisfying close relationships with one's peers, especially one's special partner.

Making decisions on one's own is an important part of growing up. (Courtesy NCR)

Autonomous Decision Making

This involves young people making decisions by and for themselves. That is, they must not only make their own decisions with a minimum of assistance from others, but they must also do so without an undue need for approval from their parents and other adults. Admittedly, this is a difficult goal for anyone, much less someone at this stage of life. One reason is that most young people have had only limited experience in making important life decisions. Even when they are given the opportunity to make an important decision, such as whether to attend college and, if so, which college, they usually decide jointly with their parents who often share the major burden of the expenses. Another reason has to do with the inherent difficulty of making good decisions, which requires experience and good judgment as well as vigorous information processing, weighing alternatives, and making a final choice. Then too, young people may become overwhelmed by the vast array

FREER TO CHOOSE

According to a survey of 1500 Americans carried out by Yankelovich, Skelly, and White, the majority of Americans feel they have more freedom of choice in how to live their lives than their parents did. Most feel that their parents' lives were hemmed in by all kinds of social, educational, and economic constraints which they themselves have escaped.

The survey found that 73 percent of Americans feel they have more freedom of choice than their parents did, 17 percent feel they have the same level of choice as their parents, and 8 percent believe they have fewer choices. Only 2 percent were not sure.

Furthermore, 80 percent feel confident they will be able to carry out these choices and live the way they truly want to. Individuals said things like "I can do what I want with my life; they couldn't." "My parents would say the same things I say, but they haven't done what I've done."

Americans who perceive their parents this way do not necessarily feel they are better people than their parents. But they do believe they have more options in the important areas of education, work, sex, marriage, family, travel, friends, possessions, where to live, and how to live. Would you agree?

Daniel Yankelovich, *New Rules* (New York: Random House, 1981).

of options in regard to schools and careers, especially if they haven't clarified their goals. Lack of a sense of direction makes decison making all the more agonizing at this stage of life.

Autonomous decision making is especially crucial at this stage when so many important life decisions are being made. Probably three of the most important decisions of one's life are made during this period: deciding whether to attend college and, if so, which one; choosing a career; and selecting a marriage partner. In explaining the importance of decisions during early adulthood, Daniel Levinson (1978) has observed that one of the great paradoxes of development is that we are required to make crucial choices before we have the judgment and self-understanding to choose wisely. Yet if such choices are put off until we feel ready, the delay may produce even greater costs. This paradox is especially true in the choice of a career or starting a family.

A major problem for people in their twenties is the terrifying feeling that their choices are irrevocable. When young people decide against graduate school or against having children, or they forego travel abroad, they often have the feeling that they will have to live with that choice forever. Yet this is generally a false fear, since change is not only possible but often inevitable. In fact, one of the distinctive requirements of this period is keeping in balance two coexisting tasks: to explore options and to create a stable structure for one's life. On the one hand, young people need to explore their options, expand their horizons, and put off firmer commitments until their options are clearer. Yet they must also create an initial adult life structure giving them roots, direction, and continuity. Those who "hang loose" too long may find themselves drifting from one thing to another, while those who make rigid commitments may needlessly shut off personal and professional growth. One of the most important competencies of early adulthood is to learn to make basic life choices one's self, with the realization that these may well need to be modified, without feeling one has betrayed one's "dreams" (Levinson, 1978).

Preparing for Economic Independence

Most individuals at this stage of life are still engaged in *preparing* for their careers and economic independence, rather than having fully established themselves in these areas of life. Thus, this tends to be a time of great concern, change, and economic instability. Most students are working in part-time jobs that are not related to their career goals. Compared to young people in earlier eras, today's students express greater concern about choosing fields with good opportunities and choosing courses which will be useful to them in their jobs as well as expressing concern about their grades in school and getting appropriate work experience (Freeman, 1982). Even after graduation, early adulthood continues to be a time of change and experimentation. The typical worker in his or her twenties tends to hold several brief jobs before settling into a more promising position (Crittenden, 1980).

Young people's attitudes toward finances play an important role in their lives, as was shown in a national survey on the subject reported by Carin Rubenstein (1981). When people were asked how important money was, respondents of all ages ranked money third, just after love relationships and work. Actually, people have become more money conscious in recent years, mostly because of the increased economic pressures in society. And people in the 18- to 25-year-old group are the most money conscious of all. Individuals this age worry more about money and are more dissatisfied with their financial situation than those in their thirties and forties. About one-third of those under 30 are postponing going back to school, getting a place of their own to live, and buying a house. Yet, many of them are willing to make financial sacrifices to accomplish these goals, and they expect to accomplish these things in the near future. Furthermore, the younger respondents are the most fervent believers in the American dream. That is, they believe that hard work, brains or talent, ambition, education, patience, and frugality (in that order) pay off. At the same time, some young people express a certain degree of cynicism, insisting that greed, having the right connections, and luck are necessary for success. But those who believe in the value of hard work are happier with their jobs, financial situations, and personal growth and are more optimistic about their future.

Young women may experience added difficulties in achieving economic independence. Traditionally, the majority of women have sought economic stability primarily through marriage. Today, however, more young women are getting a college education, working before marriage, marrying later, and

Combining career and family responsibilities is not easy. (Ken Karp)

combining marriage and a career. A major issue for women is economic parity with men, with women on the average earning only two-thirds as much as men with the same education and job. Women who choose a traditionally feminine job in the lower-paying service sector are at an even greater disadvantage (Lewin, 1984). Fortunately, as we noted in an earlier chapter, an increasing proportion of women are choosing careers and jobs in a wider spectrum of fields.

Married women face additional challenges, such as combining the responsibilities of marriage and work. Unfortunately, as we discussed in an earlier chapter, women are still assuming a disproportionate amount of responsibility for house chores and childrearing. Yet, women who work outside the home tend to feel greater self-esteem and share power more equally with their husbands than other women. Furthermore, the more satisfied couples are with their joint income, the happier they are with their marriage relationship. Couples generally experience more problems in regard to how they spend their money. More than three-fourths of couples married 10 years or more favor pooling their financial resources, though women are somewhat less enthusiastic about this than their husbands. A lot depends on both partners' attitudes toward their relationship. More than 8 out of 10 husbands and wives

voluntary marriage: The view adopted by marriage partners who assume that they will stay married only as long as they remain "in love."

who believe marriage is a lifetime relationship favor pooling, whereas only 5 out 10 spouses who subscribe to the idea of "voluntary marriage" like the idea. Couples living together generally feel even less favorable toward pooling their money and property. Actually, couples who keep their money separate have fewer fights over money. But they also have weaker ties, making it easier for them to dissolve their relationships. The longer an unmarried couple lives together, the more likely they are to pool their resources (Blumstein & Schwartz, 1983).

Establishing Close Relationships

People at this stage of life are characteristically engaged in substituting friends for family. Young people are looking for two kinds of relationships with their peers: a network of social ties, including congenial friends of both sexes, and a close emotional attachment to one special person, whether friend, lover, or spouse. Erikson (1980) holds that achieving closeness with one's peers, especially a partner of the other sex, is the major developmental task of this period. The high priority of intimate relationships also has been confirmed in numerous surveys, with over 9 out of 10 people this age regarding close relationships as very important in their lives (Rubenstein, 1983). At the same time, Erikson points out that the need for intimacy is rarely fully satisfied, with the result that young people are especially prone to feelings of loneliness. In fact, people in the 18- to 25-year-old group suffer more from loneliness than any other age group. Having loosened the emotional ties with their families, people this age are acutely aware of the discrepancy between their search for intimacy and their failure to find it (Rubenstein, Shaver, & Peplau, 1982).

According to a survey on friendship, over two-thirds of young adults report having between one and five close friends along with a network of casual and work friends. Most individuals have little difficulty distinguishing between close and casual friends, with the quality of the relationship being more important than the frequency of association. The two most important qualities in a friend are loyalty and the ability to keep confidences, followed by warmth, affection, and supportiveness. Friends are so important to people this age that in a crisis most of them say they would turn to their close friends rather than to their families. Same-sex friendships tend to be more common than opposite-sex friendships, though individuals of both sexes are increasingly likely to report having both types of friendships. At the same time, men and women feel that opposite-sex friendships are different, mostly because of the potential for sexual involvement. Almost half of those with opposite-sex friends have had sex with one of their friends at one time or another (Parlee et al., 1979).

Sexual involvement among friends raises the question about the similarities and differences between friendships and love relationships. A study comparing friendships and love relationships was conducted among a variety of young adults, including college students and community members, both single and married. The results showed that friendships and love relationships share many similarities, such as acceptance, trust, and mutual assistance. Yet, two additional clusters of characteristics make love relationships unique, namely the "caring" cluster and the "passion" cluster. Thus, love relationships generally exhibit greater exclusiveness and sexual intimacy. But lovers also experience greater ambivalence, conflict, distress, and mutual criticism, so that while love relationships are more satisfying they are also more frustrating than friendships. In contrast, friendships tend to have greater mutual acceptance and stability, especially among same-sex friends (Davis, 1985).

The search for a romantic partner is especially crucial at this stage of development, as shown in a survey of love and romance. When the respondents were asked what they looked for in their romantic relationships, over half of them indicated "love" followed closely by "companionship." Other qualities such as romance, sex, and financial security were ranked much further down the list. Most individuals this age regard romance as the initial stage of love. Typical descriptions of romance included walking on a moonlit beach, kissing in public, and making love all weekend. In contrast, love was defined more in terms of the enduring qualities of the relationship, such as compatibility, friendship, and devotion. Two-thirds of the respondents feel that love may last a lifetime. Yet, they speak from limited experience. Half of them have been in love for less than 5 years. And among the married couples, one in three feel that their marriages may end in divorce. One of the most disturbing aspects of close relationships is the uneven emotional involvement between the partners. Half of the couples feel that they have a love relation-

Young couple in Central Park, New York. (United Nations/photo by L. Barns)

ship in which one partner loves more than the other. Furthermore, women, who are more often on the giving end of such relationships, tend to be more demanding, more critical, and less happy in their love relationships. One explanation is that women have higher expectations for intimacy and thus react more negatively to the reality of close relationships. Women also say that they want more verbal responsiveness from their partners, the type they have with their women friends. They complain that men, especially the strong silent types, communicate too little in their love relationships (Rubenstein, 1983).

Individuals and couples alike tend to have different priorities in regard to the importance of close relationships, as was brought out in a study by Philip Blumstein and Pepper Schwartz (1983) involving cohabiting couples as well as married couples. In about half of the married couples, one partner is relationship-centered while the other is more work-centered. In many instances the husband devotes most of his energies to the job and the wife to the needs of the relationship. Yet, this is not always the case. Actually, there is a tendency for the partner most inclined to be understanding and compassionate, regardless of sex, to put the relationship first. In a few instances, the roles were reversed, with men putting the relationship first. In one-fourth of the married couples and slightly more of the cohabiting couples, both partners were relationship-centered. Generally, these were the happiest couples of all. Cou-

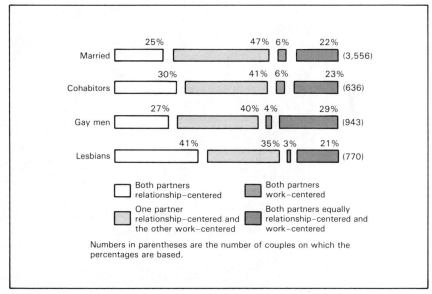

FIGURE 16–1 RELATIONSHIP-CENTERED COUPLES, WORK-CENTERED COUPLES, AND MIXED COUPLES.

Source: From *American Couples* by Philip Blumstein, Ph.D., and Pepper Schwartz, Ph.D. (New York: William Morrow and Company, 1983), copyright © 1983 by Philip Blumstein and Pepper Schwartz. By permission of William Morrow and Company, 1983, p. 171.

ples in which one partner was relationship-centered were somewhat less happy in their marriage. Couples in which both partners were work-centered were the least happy and committed to their marriages. (See Figure 16–1.) Now that work demands so much of a person's energies and there are more dual-career couples, there is an increased risk that the marriage relationship may easily become a secondary aspect of the individual's and couple's life.

FROM YOUTH TO ADULTHOOD

As youths in their late teens and early twenties take hold in the adult world, they increasingly view themselves as adults. They tend to dress and act somewhat differently than they did at younger ages. Also, they may relate to others in a more mature way. But in the process of growing up, do young people undergo significant change or growth in their personalities? Or do they remain essentially the same kind of people they were in their youth? Although longitudinal studies do not support either extreme, they do provide some evidence for both views. That is, youth exhibit both continuity and change in their personalities as they move into adulthood.

Continuity

continuity: The state or quality of being continuous or connected, as in the continuity of personality traits.

Apparently, there is considerable continuity in personality development between adolescence and adulthood. In one study, Jack Block (1981) has observed several hundred individuals for more than 20 years, beginning in junior high school, then again in their late teens, again in their thirties, and later when these subjects were middle-aged. Data was gathered not only from the subjects themselves but also from their parents, teachers, and spouses. The results have shown a striking pattern of stability. Block has found significant correlations between the subjects' ratings of themselves in high school and ratings of themselves in their thirties on virtually every one of the 90 rating scales. Thus, individuals who were cheerful and outgoing as teenagers continue to exhibit these same qualities as young adults. And individuals who were bothered by mood fluctuations when they were in high school were still experiencing mood swings in midlife. Similar support for the continuity of personality comes from another longitudinal study by Bachman, O'Malley, and Johnston (1978). Their data was collected from a national sample of over 2000 boys in the tenth grade and again at varied periods throughout the next 8 years. Again, the dominant picture was one of stability in various areas of development. For instance, in regard to achievement, students who had been successful in high school also tended to be successful in their careers and maintained high achievement aspirations throughout adulthood. By contrast, those who had not done well in high school experienced less success in their careers and entertained lower achievement aspirations.

Cheerful and outgoing youth tend to show these same qualities as adults.
(Laima Druskis)

introversion-extroversion: A trait dimension in which people who describe themselves as quiet and introspective are designated as introverts and those who see themselves as active and sociable are extroverts.

In another longitudinal study, Paul Costa and Robert McCrae (1980) discovered that some personal traits are generally more stable than others. They found the highest degree of stability on measures of introversion-extroversion, which assesses gregariousness, warmth, and assertiveness. Thus, individuals who were expressive and outgoing in their teens were apt to remain that way as adults. The researchers found almost as much consistency in some specific traits such as anxiety, impulsiveness, hostility, and depression. Although activity levels generally decreased for everyone across time, they also remained quite stable within individuals, with those who were active in their youth remaining more active and energetic than their peers as adults.

Change

Costa and McCrae (1980) also observed certain changes in personality in the course of development. They frequently found that with increasing age there were slight drops in activity levels, excitement-seeking behavior, hostility, and impulsiveness. But they attributed these changes to the mellowing that comes with age and experience. Thus, the relative differences between individuals remained intact, such that an impulsive 17 year old may be a bit less impulsive by the time he is 30 years old, but he is still likely to be more impulsive than his peers. Similarly, men and women tend to achieve greater feelings of self-esteem, self-control, and personal mastery as they grow older, mostly because of the maturing effect of life experiences (Brim & Kagan, 1980).

Daniel Levinson (1978) and his colleagues found that many of the individuals in their study changed significantly during their transition to adulthood. One reason is that young people make so many crucial life choices at this stage. Also, most of the people in their study moved out of the family home during this period, thereby facilitating a certain degree of personal growth as well. Probably the most important reason for the change during this period has to do with the need for adapting to the extensive changes in a complex, rapidly changing society. Compared to people in the past, individuals now have a greater number of personal options and are being encouraged to forge their own way, rather than to follow in their parents' footsteps. Consequently, most of the people in Levinson's study were already moving toward lives that were quite different from those of their parents, a shift that increased throughout adulthood. Creative individuals such as artists and novelists exhibited the greatest break with their past, with professionals such as doctors and lawyers the next greatest. Factory workers and executives were somewhat less likely to make significant changes in their lives. Few individuals strongly rejected their families, but most held only tenuous ties with their family's ethnic and religious traditions.

Individual Differences

age-graded influences: Events and experiences that commonly occur at a given stage of life, such as puberty.

non-age-graded influences: Events and experiences that are unique to the individual rather than associated with a given stage of life, such as a divorce.

The degree to which people change with age varies even more widely among individuals as they move into adulthood. One reason is the increased importance of non-age-graded influences compared to age-graded influences (Baltes, 1979). Up through adolescence, much of our development is associated with changes that generally occur at a given age, such as puberty and high school graduation. By contrast, adult development depends more on events and influences that may occur at any age, such as the decision to change careers. Thus, the person who drops out of college permanently at 20, delays marriage until 30, and changes careers at 45 will have a different development and outlook than someone who graduates from college at 22, marries at 25, and remains in the same career throughout middle age.

How much people change also depends on the different priorities they assign to stability or change in their lives. Thus, people with traditional values and outlook tend to exhibit a high degree of stability in their lives *unless* something happens to them. The events that would be likely to change deeply ingrained patterns usually have to be pretty dramatic ones, such as an unwanted divorce, the death of a child, or being taken hostage in a crisis situation. In these cases, individuals may undergo marked changes in their outlook and personality. In contrast, those who put a greater value on personal growth may continue changing to a greater extent throughout their lives. Orville Brim

---

RELIVING THOSE GOLDEN DAYS

Have you ever attended a school reunion? Do you wonder what your former classmates are doing now? You may be interested to know that going back to school reunions gives people an opportunity to reflect on their lives and to make comparisons about themselves and others.

In order to find out who attends reunions and why, Douglas Lamb and Glenn Reeder (1986) conducted a survey among the graduates of a high school in a middle-class community on the east cost. The school held a 5-day reunion which everyone who ever graduated from the school was invited to attend. From this roster, the researchers selected 1500 graduates, including 800 returnees and 700 nonreturnees.

People who returned were similar to those who didn't, with the two groups being almost identical in regard to age, sex, marital status, education, and how happy they were at the present time. However, there was one important difference. The returnees rated themselves happier and more popular while in high school than did the nonreturnees.

When asked why they wanted to attend the reunion, the majority of graduates indicated it was to renew old friendships. In some cases, the likely presence of one special person was the main reason for attending. Other reasons were to reminisce, to relive the fun and feelings they experienced in high school, and to look for changes, usually in the form of comparisons between themselves and others.

The older grads were especially interested in renewing old friendships. They spent most of their time with friends and little time meeting new acquaintances. By contrast, the younger grads were more interested in seeing how "others have changed" and in having them see "the ways I have changed." Many of them were interested in comparing their progress with others in their graduating class. At the same time, some younger grads had a sense of cynicism and expected that people would brag about themselves and their jobs. About one in four people in all age groups admitted making a special effort to look good, such as going on a diet, starting an exercise program or purchasing special clothes. One woman even rescheduled eye surgery to avoid wearing her thick glasses to the reunion.

When asked how they felt about the reunion, the majority of returnees were glad they attended. Three-fourths of them indicated that since the reunion they had found themselves thinking about "how I was then" and "how I am now" and were generally pleased with the changes in their lives.

From Douglas H. Lamb and Glenn D. Reeder, "Reliving Golden Days," *Psychology Today* (June 1986):22–30. Reprinted from *Psychology Today* Magazine, © 1986 The American Psychological Association.

(1980) holds that people are dynamic organisms who continue striving to master their environment, such that they naturally change and grow, *unless* they get stuck. Brim points out that we are now in the midst of a revolution in human development away from the traditional pattern of continuity toward greater change and growth throughout the lifespan. Medical technologies, such as plastic surgery and organ transplants, techniques of behavior modification, and the encouragement of change provided by thousands of support groups are all part of this trend. Furthermore, perhaps more than ever before, people are deliberately trying to change and improve themselves.

What about yourself? Was your adolescence any indication of what you are like now? If you were a bit shy and had only a few friends, are you still this way? Or, have you changed a great deal? Perhaps you feel more at ease with others and now have a wider circle of friends. What about the classmates you have known over the years? You may find yourself making such observations if you decide to attend a class reunion. See the boxed item on "Reliving those golden days."

It is also encouraging to remember that there is a tendency for humans to change toward adaptation and health. That is, adolescents and young adults tend to grow toward adaptation rather than away from it, and characteristics that interfere with personal and social adjustment are the most likely to change. For instance, at least one study has shown that adolescents who change the most in their transition to adulthood are those who were characterized as brittle, passive, and negativistic, traits that are valued neither by society nor by the youths themselves. Intellectually bright, productive youths tended to change the least, mostly because they had already achieved a positive adjustment (Block, 1981). Realizing this, parents and teachers as well as young people themselves might adopt a more positive attitude toward youth, especially troubled youth, thereby helping them to make the desired changes in their lives as they move into adulthood.

SUMMARY

Leaving Home

1. We began the chapter by noting that the transition to adulthood is a complex and lengthy process, usually lasting from the late teens into the twenties.

2. A major developmental task of this period is leaving home, which involves a psychosocial transition as well as moving out of the family house.

3. Whether young people make a peaceful or stormy departure from home depends partly on the parents and the dynamics of the family itself, with some parents having difficulty letting go of their young.

4. Going away to college generally promotes the emotional disengagement and autonomy associated with leaving home.

5. The tendency for young people to move back with their families, if only temporarily, presents an opportunity for strengthening family ties as well as a potential problem in attaining autonomy.

Taking Hold in the Adult World

6. A second major developmental task in the transition to adulthood has to do with taking hold in the adult world.

7. Young people must learn to make important life decisions on their own without an undue need of approval from their parents or other adults.

8. Individuals at this stage of life are generally engaged in preparing for economic independence rather than having achieved it, and are in school or the early stages of their career.

9. Young people are also searching for satisfying relationships with their peers, including a close relationship with one special person, whether lover or spouse.

10. Couples in which both partners are relationship-centered tend to be happier than those in which one person is more emotionally involved in the relationship.

From Youth to Adulthood

11. During the transition to adulthood, individuals exhibit both continuity and change.

12. There is considerable stability in personal development, especially in regard to such qualities as introversion-extroversion, anxiety, impulsiveness, hostility, and depression.

13. At the same time, individuals also exhibit varying degrees of personality change because of many factors, including the mellowing that comes with age and experience, the crucial decisions being made in early adulthood, the adaptation to social change, and the increased emphasis on personal growth in contemporary society.

14. There are wide differences among individuals in regard to the degree of personal change during their transition to adulthood, largely because of the greater influence of non-age-graded influences in adulthood and the different priorities individuals place on stability and change in their lives.

15. Finally, there is a tendency for humans to change toward adaptation and health, such that maladjusted adolescents are especially apt to mature as they move into adulthood.

REVIEW QUESTIONS

1. What does it mean for young people to "leave home"?

2. How may parents promote a successful disengagement of their young from the family?

3. To what extent does going away to college facilitate emotional autonomy?

4. What are some of the pros and cons of moving back home, even temporarily?

5. What is meant by "autonomous decision making"?

6. What are some ways in which young people may prepare for economic independence?

7. Why are people in the 18- to 25-year-old group so prone to loneliness?

8. In what ways are youths likely to remain the same as they move into adulthood?

9. In what ways are people likely to change as they make the transition into adulthood?

10. To what extent are you different from the person you were in high school?

Appendix: Methods of Studying Adolescents

Throughout this book you will find numerous references to research studies. In most cases the description of empirical findings is followed by a name and date in parentheses. This indicates the author or authors of the study cited, together with the publication date. The complete publication data is listed in the references at the end of the book.

For instance, in Chapter 14 on youth and drugs there are frequent references to the study by Johnston, O'Malley, and Bachman (1986). As explained in the chapter, this refers to the annual high school survey and followup studies which are funded by the National Institute of Drug Abuse. The data in this particular study are based on the survey of the seniors in the class of 1985 together with followup studies of some of the same youth several years later. The findings give us a well-rounded picture of the views and practices regarding drug abuse among high school students, college students, and other post-high school youth. Although these investigators are well aware that the information in their survey is limited to the extent that it is obtained from self-reports, they also have good reason to believe it is reliable because of the careful construction of the questionnaire, the representative sampling of students, and the promise of confidentiality.

Awareness of the different methods used in studying adolescents may help you to put the particular information in proper perspective—that is, under what conditions is it true, how representative is it, and so forth. The purpose of this section is to acquaint you with the various methods of study so that you may become more proficient in understanding the research knowledge in this field.

SCIENTIFIC INQUIRY

Essentially, scientific inquiry has to do with a way of knowing, one that is based on the systematic collection and analysis of data. Yet researchers do not go out simply looking for facts. They usually have a particular purpose in mind, expressed succinctly in a *hypothesis*. The latter may be defined as a testable idea or a statement of expected results. For example, when Margaret

Mead went to the Samoan Islands, she was intent on investigating a particular idea—namely, "Adolescent girls in Samoa will behave differently from American girls because of the cultural differences"—a hypothesis later confirmed by her findings.

Another important aspect of research concerns who is studied, which has to do with *sampling*. Because it is not feasible to study all members of the population of interest, researchers usually select a sample of subjects as a basis for understanding the population as a whole. Both the size and selection of the sample are important, with the larger and more representative the sample the better. A *random sample* is one in which every member of the population has an equal chance of being included; this is usually preferable because it provides a more objective basis of knowledge. However, because of practical considerations researchers often use other sampling techniques, such as the *representative sample,* which is based on selected variables such as age, sex, and social class. For instance, the Eighteenth Annual Gallup Poll of the Public's Attitudes toward the Public Schools conducted by Alec M. Gallup (1986)—which was cited in the chapter on the school—is based on a modified probability sample of the U.S. population. Although this sample includes only 1552 adults 18 years and older, the proportion of people in each group—such as age, sex, type of school, and type of community—roughly approximates the proportion of these people in the population as a whole. Thus, their views are considered to be representative of the public's attitudes toward public schools.

The particular characteristics, events, or behaviors of the sample being studied are called *variables*. A variable may refer to any influence that may change or vary to a greater or lesser degree, such as age or social class. The relationships between variables being investigated depend on the type of study. As we'll explain shortly, in correlational studies the aim is simply to determine whether there is any relationship betweeen two variables, and, if so, the strength of this relationship. By contrast, in experimental studies the aim is to discover the *causal* relationships between two or more variables, such as family size and academic achievement. At the same time, one must be careful in interpreting the findings from both types of studies because the interaction between any two variables may be affected by still other variables, whether included in the study or not.

METHODS OF STUDY

Sometimes people speak of the scientific method as if there were only one such method, which is misleading. Actually, there are a variety of procedures and research designs available, each with its own particular strengths and weaknesses. We will explain briefly some of the most commonly used methods: observation, the case study, survey research, correlational studies, experimental studies, and *ex post facto* studies. These methods have been arranged along a naturalistic-control dimension, according to the degree to which observations are obtained in a naturalistic as opposed to a highly controlled, experimental approach.

Observation

Observation is a primary source of psychological knowledge that can be used in several ways. In the first place, it may be a fruitful source of hypotheses and clues for further research. Charles Darwin, Sigmund Freud, and Jean Piaget all gained valuable clues for their work through intensive observations of others, including their own children. Observation is also a valuable means of collecting data on adolescent behavior and can be done in a relatively open-ended or highly structured way.

In *naturalistic observation,* the aim is to observe adolescents in those settings in which a given behavior normally occurs. For example, peer interaction could be observed on the playground, in the cafeteria, in the school hallway, or in the classroom before class begins. The cardinal rule in naturalistic observation is to stay out of the way of the behavior being observed. *Systematic observation* involves more carefully defined categories of behavior, coding schemes, and more than one observer to offset observer bias. One of the oldest forms of observation is self-observation in the form of diaries and logs. A novel approach, cited in the chapter on adolescents and their peers, involved adolescents carrying around electronic pagers for a week. For an entire week, high school students were asked to record what they were doing at various times of the day when they were interrupted by the beeper. One of the main results was that adolescents spent a much greater amount of time with their peers than with adults (Czikszentmihalyi and Larson, 1984).

Investigators who find it necessary to take a more active role in gathering data may adopt the *participant observer* approach. For example, Margaret Mead said that in her Samoan study she found it best to spend her days as an active participant in the community and then to use her evenings writing up the interviews and observations of the Samoan girls. In Piaget's case, an interest in the qualitative aspects of cognitive development led him to use a clinical method which incorporated many features of the participant-observer approach. Piaget characteristically presented the child or adolescent with a mental task, observed the individual's performance, and then questioned the subject in order to discover the underlying thought processes. Although Piaget's approach has been criticized for not being sufficiently scientific, it has been extraordinarily productive in gaining new understanding of cognitive processes.

The Case-Study Method

This is an approach to understanding adolescent development through an in-depth study of a single individual. Here the researcher exercises considerable control over the study through the choice of the subject, sources of information used, aspects of development included, and interpretation of the data. We have included several abbreviated case studies in this book, such as those in the chapter on psychological disorders. More extensive use of the case study method can be found in Nancy Ralston and Patience Thomas's book *The Adolescent: Case Studies for Analysis* (New York: Chandler Publishing Company, 1974).

Another version of the case-study approach is the use of first-person accounts of adolescents. The author has collected a number of such accounts and has used selective excerpts to illustrate various aspects of adolescent development throughout this book. A more extensive use of first-person accounts can be found in George Goethals and Dennis Klos's book *Experiencing Youth: First-Person Accounts,* 2nd edition (Boston: Little, Brown, and Company, 1976).

Still another approach is the use of retrospective autobiographical accounts of adolescence. Pertinent examples used in this book include Margaret Mead's reflections on her own adolescence, Erik Erikson's account of his adolescence, and various other accounts found in Normal Kiell's book *The Universal Experience of Adolescence* (Boston: Beacon Press, 1967).

Survey Research

Stanley Hall, one of the first investigators to adopt scientific methods for the study of adolescence, pioneered in the use of survey research. Essentially, this involves the use of questionnaires and interviews to learn something about a large number of people. Questions may be asked about opinions, attitudes, beliefs, values, preferences, and practices. It is customary, especially in pencil-and-paper questionnaires, to ask standardized questions, that is, all people are asked the same questions in the same form. Although interviews are more time-consuming, they also provide more flexibility of responses. The interviewer may explain a question that is unclear, include open-ended questions when desired, and elicit a more thorough explanation from the subject than can be obtained through the use of a questionnaire.

Investigators often combine these two strategies. For instance, the Annual Gallup Poll of the Public's Attitudes toward the Public Schools (Gallup, 1986) draws on personal, in-depth interviews as well as questionnaire data as a basis for its survey findings. Also, by comparing the results from one year to the next, the investigators may observe consistent patterns and changes in public opinion across time. A major advantage of the survey method is the use of standardized questions to query a large number of people on a variety of topics. A disadvantage is that it assumes a subject's responses are "true" when this may not be so in every case. Some people may simply give the responses they feel they are expected to give or, occasionally out of fear, they may deliberately give a deceptive response.

Correlational Studies

In this approach, investigators attempt to demonstrate that there is a statistical correlation or relationship between two or more variables, such as IQ and school grades. The characteristics being investigated are simply observed or measured to see if there is any relationship between the variables, and, if so, to determine the strength of that relationship. However, it is important to keep in mind that a correlation does not imply causality, merely that some sort of relationship exists between two or more groups of data.

Numerically, correlations may vary between $+1.00$ (a perfect positive correlation) through 0 (no correlation at all) to -1.00 (a perfect inverse correlation). In most cases, the actual correlation takes the form of numbers between 0 and 1, such as .22 or .57. The larger this number, the stronger is the relationship between the two sets of data. Correlations may also be either positive or negative, that is, inverse. A positive correlation exists whenever a change in one set of data is reflected in a similar change in the other. For instance, there is generally a positive relationship between students' class attendance and their course grade. By contrast, an inverse correlation is one in which a change in one variable or set of data is accompanied by an opposite change in the other variable. For example, there is an inverse relationship between students' course grades and the number of hours spent on a part-time job in excess of 20 hours a week, as explained in the chapter on adolescents and work.

Experimental Studies

In experimental studies, the aim is to show that there is a *causal* relationship between two or more variables. In order to do this, the experimenter manipulates some variables, known as the independent or experimental variables, and then observes the effect on the subjects' responses, measured in terms of the dependent variables. Experiments are often done in a highly controlled environment, such as the "lab" experiment. But in the case of the "field" experiment the study is conducted in the natural setting in which the behavior of interest normally occurs, such as the school, neighborhood, or work setting. Most experimental designs include a comparison between two groups—an experimental group and a control group—in order to insure that the effect measured is a function of the experimental variable rather than some characteristics of the subjects involved. Also, pre- and posttesting of both groups further insures that any differences between the results of these two groups are a result of the experimental treatment rather than extraneous influences. For example, Trebilco (1984) studied the effect of career education on career maturity among high school students in schools which had some type of career education (experimental group) with students in schools without such programs (control group). He found that students generally achieved greater career maturity when they had participated in some type of career education. Furthermore, the more innovative the career education program and the greater support it had in the community, the greater the gains in career maturity experienced by the students.

The extent to which experimental results are significant depends on statistical tests, often expressed in terms of the *p* value, with *p* standing for "probability." For instance, if the experimental results are significant at the .05 level of confidence, this means that the experimenters can safely say that the chances of such results occurring by chance alone would be only 5 in 100 times. Results at the .01 and .001 level are even more impressive. In all these instances, there is a high probability that the experimental effect was produced by the independent variable.

Ex Post Facto Designs

These are the easiest and most often used research designs. In this approach, the effects of certain events are studied after they have occurred (*ex post facto* literally means "after the fact"). Even though such studies may resemble other experimental research designs in many ways, they differ in one important respect, in that the treatment is based on selection rather than manipulation. An example would be Rumberger's study (1983) of high school dropouts discussed in the chapter on schools. Rumberger gathered a great deal of data from a wide variety of high school dropouts aged 14 to 21. His findings show that the dropout rate varies widely according to many factors, such as age, sex, race, and other socioeconomic variables, with many inner city schools in poor neighborhoods having dropout rates in the 50 percent bracket.

CROSS-SECTIONAL AND LONGITUDINAL STUDIES

There are several other experimental procedures used in studying adolescents, which differ mainly in the time span involved. Researchers interested in measuring age-related changes in development may prefer to use the cross-sectional method, the longitudinal method, or a sequential design which combines both methods.

Cross-Sectional Method

The cross-sectional method is frequently used because it is a relatively inexpensive way of detecting developmental change over a short period of time. In this approach individuals of different ages are studied at about the same time and age, and then age-related changes in development are inferred from the differences among groups. An example would be the study of adolescent sexual attitudes and practices by Laurie Zabin (1984) and her colleagues, as discussed in the chapter on adolescent sexuality. Data was gathered from over 3000 junior and senior high school students in four inner-city schools. Then the researchers discussed the sexual attitudes and practices among their subjects on the basis of certain population characteristics such as age, sex, and race.

A major drawback of this approach is that we can never be certain the changes associated with a given age group are primarily a function of that age or may be due partly to other characteristics of the group selected, such as their intelligence, family background, or values.

Longitudinal Method

By contrast, a longitudinal study has the advantage of following the same individuals over a given period of time, thus increasing the probability that the changes observed are caused primarily by developmental factors. In this approach, the researchers observe the same group of individuals (known as a *cohort*) over a period of time, and make repeated observations or measurements of the subjects at selected intervals of time. An example would be Jack Block's (1981) longitudinal study of personality development. Block observed several hundred individuals for more than 20 years, beginning when the subjects were in junior high school, then in their late teens, in their 30s, and again in middle age. One of his major findings was that there is a high degree of stability in personality development between adolescence and adulthood. However, as you'll discover in the discussion of this issue in the final chapter of this book, some traits tend to be more stable than others and some individuals exhibit more continuity of personality development than others over the years.

A major limitation of such studies is that they are time-consuming and difficult to complete. Another problem is the difficulty of maintaining the original sample size, and the distorting effects caused by subjects dropping out. Then too, subjects may become "test-wise" through familiarization with the testing procedure. Finally, there is always the possibility that some of the findings attributed to development may be the result of social and historical factors, such as social change.

Combined Cross-Sectional/Longitudinal Method

There are also research strategies that combine the cross-sectional and longitudinal approaches, thereby minimizing the disadvantages of each method. Here, the researchers take a cross-sectional sample of two or more different age groups (cohorts) and then test them sequentially on two or more occasions.

For instance, Dale Blyth (1983) and his colleagues were interested in finding out the effects of having to change schools at different ages, as seen in the impact on students' self-esteem. To accomplish this goal, the researchers obtained data from students at different grade levels in the Milwaukee schools during the same period (cross-sectional approach) and then tracked them over a 5-year period (longitudinal approach). Some of these students were in schools organized around a 6–3–3 pattern, which meant they changed schools twice during their education, while others were in schools using an 8–4 pattern, which meant they changed schools only once. One of their major findings, as discussed in the chapter on the school, is that changing schools generally has a negative impact on students' self-esteem, especially when school change coincides with the beginning of puberty, as is often the case with students in the 6–3–3 system. The use of this combined method enabled researchers to draw more comprehensive observations than use of the cross-sectional sample or longitudinal approach alone.

References

ACHENBACH, T. M. Developmental aspects of psycho-pathology in children and adolescents. In M. E. Lamb (Ed.), *Social and Personality Development.* New York: John Wiley & Sons, 1978.

———, and C. S. EDELBROCK. Behavioral problems and competencies reported by parents of normal and disturbed children aged four through sixteen. *Monographs of the Society for Research in Child Development,* 1981, Serial No. 188, Vol. 46, No.1.

ADAMS, G. R., and S. A. FITCH. Ego stage and identity status development: a cross-sequential analysis. *Journal of Personality and Social Psychology,* 1982, *43,* 574–583.

ADELSON, J., Adolescence and the generalization gap. *Psychology Today,* February 1979, 33–37.

———, and M. J. DOEHRMAN. The psychodynamic approach to adolescence. In J. Adelson (Ed.), *Handbook of Adolescent Psychology.* New York: John Wiley & Sons, 1980.

ALTMAN, S. L., and F. K. GROSSMAN. Women's career plans and maternal employment. *Psychology of Women Quarterly,* 1977, *1,* 365–375.

ALTROCCHI, J. *Abnormal Behavior.* New York: Harcourt Brace Jovanovich, 1980.

AMERICAN PSYCHIATRIC ASSOCIATION. *Diagnostic and Statistical Manual of Mental Disorders,* 3rd ed. Washington, D.C.: APA, 1980.

AMSTEY, M. Asymptomatic gonorrhea in pregnancy. *Medical Aspects of Human Sexuality,* 1981, *15,* 52–60.

ANDERSON, C. S. The search for a school climate: A review of the research. *Review of Educational Research,* Fall 1982, 368–420.

ANDERSON, S. A., and W. M. FLEMING. Late adolescents' home-leaving strategies: Predicting ego identity and college adjustment. *Adolescence,* Summer 1986, 453–459.

APTER, D., and R. VIHKO. Serum pregnenolone, progesterone, 17-hydroxy-progesterone, testosterone, and 5-dehydrotestosterone during female puberty. *Journal of Clinical Endocrinology and Metabolism,* 1977, *45,* 1039–1048.

ARCHER, S. L. The lower age boundaries of identity development. *Child Development,* 1982, *53,* 1551–1556.

ARIETI, S. *Understanding and Helping the Schizophrenic.* New York: Simon & Schuster, 1981.

ASTIN, A. W., and K. C. GREEN. *The American Freshman: The Twenty-year Trends, 1966–1985.* Los Angeles: The Higher Educational Research Institute, Graduate School of Education, University of California, Los Angeles, 1987.

AULT-RICHE, M., and D. ROSENTHAL. Family therapy with symptomatic adolescents: An integrated model. In G. K. Leigh and G. W. Peterson (Eds.), *Adolescents in Families.* Cincinnati: South-Western Publishing Co., 1986.

AUSUBEL, D. P., R. MONTEMAYOR, and P. SVAJIAN. *Theory and Problems of Adolescent Development,* 2nd ed. New York: Grune & Stratton, 1977.

BACHMAN, J. G. The American high school student: A profile based on national survey data. Paper presented at the University of California, Berkeley, June 28, 1982. In J. W. Santrock, *Adolescence,* 2nd ed. Dubuque, IA: Wm. C. Brown, 1984.

———, L. D. JOHNSTON, and P. M. O'MALLEY. *Monitoring the Future: Questionnaire Responses from the Nation's High School Seniors.* Ann Arbor, MI: Institute for Social Research, 1980.

———, and P. M. O'MALLEY. The youth in transition series: A study of change and stability in young men. In A. C. Kirckhoff (Ed.), *Research in Sociology of Education and Socialization,* vol. 1. Greenwich, CT: Jai Press, 1980.

———, P. M. O'MALLEY, and J. JOHNSTON. *Youth in Transition: Adolescence to Adulthood: Change and Stability in the Lives of Young Men,* vol. 6. Ann Arbor, MI: Institute for Social Research, 1978.

BAHR, H. M., and T. K. MARTIN. "And thy neighbor as thyself": Self-esteem and faith in people as correlates of religiosity and family solidarity among Middletown High School students. *Journal for the Scientific Study of Religion,* June 1983, 132–144.

BALTES, P. B., Life-span developmental psychology: Some converging observations on history and theory. In P. B. Baltes and O. G. Brim (Eds.), *Life-Span Development and Behavior,* vol. 2. New York: Academic Press, 1979.

BANDURA, A. *Aggression.* Englewood Cliffs, NJ: Prentice-Hall, 1973.

———. Self-efficacy: Toward a unifying theory of behavior change. *Psychological Review,* 1977, *84,* 191–215.

BARON, R. A., and D. BYRNE. *Social Psychology*, 4th ed. Boston: Allyn & Bacon, 1984.

BATSON, C. D., and W. L. VENTIS. *The Religious Experience.* New York: Oxford University Press, 1982.

BAUMASTER, R. F. Personal communication, July 2, 1984. In J. T. Tedeschi, S. Linkskold, and P. Rosenfeld, *Introduction to Social Psychology.* New York: West Publishing Company, 1985, p.85.

BECK, A. T., and J. E. YOUNG. College blues. *Psychology Today,* September 1978.

BELL, A. P., M. S. WEINBERG, and S. K. HAMMERSMITH. *Sexual Preference—Its Development in Men and Women.* Bloomington, IN: Indiana University Press, 1981.

BERNARD, H. S. Identity formation during late adolescence: A review of some empirical findings. *Adolescence,* Summer 1981, 349–358.

BERNDT, T. J. The features and effects of friendship in early adolescence. *Child Development,* 1982, *53,* 1447–1460.

———. Developmental changes in conformity to peers and parents. *Developmental Psychology,* 1979, *15,* 608–616.

BILLY, J.O.G., and J. R. UDRY. The influence of male and female best friends on adolescent sexual behavior. *Adolescence,* Spring 1985, 21–35.

BLAU, M. Why parents kick their children out. *Parents Magazine,* April 1979, 64–69.

BLEULER, M. E. The long term course of schizophrenic psychoses. In L. C. Wynne, R. L. Cromwell, and S. Matthyse, *The Nature of Schizophrenia: New Approaches to Research and Treatment.* New York: John Wiley & Sons, 1978.

BLOCK, J. Some enduring and consequential structures of personality. In A. I. Rabin et al. (Eds.), *Further Explorations of Personality.* New York: John Wiley & Sons, 1981.

BLOOM, B. S. The development of exceptional talent. Paper presented at the Biennial Meeting of the Society for Research in Child Development, Detroit, April 1983.

BLOS, P. *The Adolescent Passage: Developmental Issues.* New York: International Universities Press, 1979.

BLUMSTEIN, P., and P. SCHWARTZ. *American Couples.* New York: William Morrow, 1983.

BLYTH, D. A., R. BULCROFT, and R. G. SIMMONS. The impact of puberty on adolescents: A longitudinal study. Paper presented at the Annual Meeting of the American Psychological Association, Los Angeles, August 26, 1981.

———, J. HILL, and C. SMYTH. The influence of older adolescents on younger adolescents: Do grade-level arrangements make a difference in behaviors, attitudes, and experiences? *Journal of Early Adolescence,* 1981, *1,* 85–110.

———, J. HILL, and K. THIEL. Early adolescents' significant others: Grade and gender differences in perceived relationships with familial and non-familial adults and young people. *Journal of Youth and Adolescence,* 1982, *11,* 425–450.

———, R. SIMMONS, and S. CARLTON-FORD. The adjustment of early adolescents to school transitions. *Journal of Early Adolescence,* 1983, *III,* 105–120.

BOOTH, A., B. BRINKERHOFF, and L. K. WHITE. The impact of parental divorce on courtship. *Journal of Marriage and the Family,* 1984, *46*(Feb.), 85–94.

———, and J. N. EDWARDS, Age at marriage and marital instability. *Journal of Marriage and the Family,* 1985 (Feb.),67–75.

BORHNSTEDT, G. W., H. E. FREEMAN, and T. SMITH. Adult perspectives on children's autonomy. *The Public Opinion Quarterly,* 1981, *4,* 443–462.

BRAITHWAITE, J. The myth of social class and criminality reconsidered. *American Sociological Review,* 1981, *46,* 36–57.

BRAUNGART, R. G. Youth movements. In J. Adelson (Ed.), *Handbook of Adolescent Psychology.* New York: John Wiley & Sons, 1980.

BRIM, O. G., Jr., and J. KAGAN. *Constancy and Change in Human Development.* Cambridge, MA: Harvard University Press, 1980.

BRODY, C. J., and L. C. STEELMAN. Sibling structure and parental sex-typing of children's houshold tasks. *Journal of Marriage and the Family,* 1985, *47*(May), 265–273.

BROOKS-GUNN, J., and D. N. RUBLE. The development of menstrual-related beliefs and behavior during adolescence. *Child Development,* 1982, *53,* 1567–1577.

BRUCH, H. *The Golden Cage: The Enigma of Anorexia Nervosa.* Cambridge, MA: Harvard University Press, 1978.

BULLOUGH, V. L. Age at menarche: A misunderstanding. *Science,* 1981, *213,* 365–366.

BURKE, R. J., and T. WEIR. Benefits to adolescents of informal helping relationships with their parents and peers. *Psychological Reports,* 1978, *42,* 1175–1184.

BURTON, N. W., and L. V. JONES. Recent trends in achievement levels of black and white youth. *Educational Researcher,* 1982, *11,* 10–17.

Cancer Facts and Figures, 1985. New York: American Cancer Society, 1985.

CASH, T. F., B. A. WINSTEAD, and L. H. JANDA. The great American shape-up. *Psychology Today,* April 1986, 30–37.

CHERLIN, A., and J. McCARTHY. Remarried couple households: Data from the June 1980 current population survey. *Journal of Marriage and the Family,* 1985, *47*(Feb.), 23–30.

CLIFFORD, R. Development of masturbation in college

women. *Archives of Sexual Behavior,* 1978, *7,* 559–573.

COLBY, A., L. KOHLBERG, J. GIBBS, and M. LIEBERMAN. *A Longitudinal Study of Moral Judgment.* Unpublished manuscript. Harvard University, 1980. In J. W. Santrock, *Adolescence,* 2nd ed. Dubuque, IA: Wm. C. Brown, 1984.

COLE, S. *Working Kids on Working.* New York: Lothrop, Lee, and Shephard, 1981.

COLEMAN, J. C. Friendship and the peer group in adolescence. In J. Adelson (Ed.), *Handbook of Adolescent Psychology.* New York: John Wiley & Sons, 1980.

COLEMAN, J. S., T. HOFFER, and S. KILGORE. *Public and Private Schools: A Report to the National Center for Education Statistics by the National Opinion Research Center.* University of Chicago, March, 1981.

COLEMAN, M., L. H. GANONG, and P. ELLIS. Family structure and dating behavior of adolescents. *Adolescence,* Fall 1985, 537–543.

Connecticut Juvenile Justice Plan for 1978. Hartford, CT: Connecticut Justice Commission, 1978.

COOKE, C., A. GINSBURG, and M. SMITH. The sorry state of education statistics. *Basic Education,* January 1985, 308.

COONS, P. M. Multiple personality: Diagnostic considerations. *Journal of Clinical Psychology,* 1980, *41,* 330–336.

COREY, L. Genital herpes simplex virus infections: Current concepts in diagnosis, therapy, and prevention. *Annals of Internal Medicine,* 1983, *98,* 973–983.

COSTA, P. T., Jr., and R. R. McCRAE. Still stable after all these years: Personality as a key to some issues in adulthood and old age. In P. B. Baltes and G. B. Orville, Jr. (Eds.), *Life-Span Development and Behavior,* vol. 3. New York: Academic Press, 1980.

CRITTENDEN, A. One life, 10 jobs. *The New York Times,* November 23, 1980.

CROOK, R. H., C. C. HEALY, and D. W. O'SHEA. The linkage of work achievement to self-esteem, career maturity, and college achievement. *Journal of Vocational Behavior,* August 1984, 70–79.

CROOKS, R., and K. BAUER. *Our Sexuality.* Menlo Park, CA: Benjamin/Cummings Publishing Company, 1983.

CROSS, H. J., and R. R. KLEINHESSELINK. Psychological perspectives on youth. In J. F. Adams (Ed.), *Understanding Adolescence,* 4th ed. Boston: Allyn and Bacon, 1980.

CSIKSZENTMIHALYI, M., and R. LARSON. *Being Adolescent.* New York: Basic Books, 1984.

CUTRONA, C. Transition to college: Loneliness and the process of social adjustment. In L. Peplau and D. Perlman (Eds.), *Loneliness: A Sourcebook of Current Theory, Research, and Therapy.* New York: John Wiley & Sons, 1982.

DANNER, F. W., and M. C. DAY. Eliciting formal operations. *Child Development,* 1977, *48,* 1600–1606.

DAVIES, M., and D. B. KANDEL. Parental and peer influence on adolescents' educational plans: Some further evidence. *American Journal of Sociology,* 1981, *87,* 363–387.

DAVIS, K. E. Near and dear: Friendship and love compared. *Psychology Today,* February 1985, 22–30.

DAVISON, G. C., and J. M. NEALE. *Abnormal Psychology,* 3rd ed. New York: John Wiley & Sons, 1982.

de TURCK, M. A., and G. R. MILLER. Adolescent perception of parental message strategies. *Journal of Marriage and the Family,* 1983, *45*(Aug.), 543–552.

DEAN, R. A. Youth: Moonies' target population. *Adolescence,* 1982, 567–574.

DEARMAN, N. B., and V. W. PLISKO. *The Condition of Education.* Washington, D.C.: U.S. Government Printing Office, 1981.

————, and V. W. PLISKO (Eds.). *The Condition of Education: 1979 Edition.* Washington D.C.: National Center for Educational Statistics, 1979.

DeLAMATER, J., and P. MacCORQUODALE. *Premarital Sexuality.* Madison, WI: University of Wisconsin Press, 1979.

DENY, S. J., and D. A. MURPHY. Designing systems that train learning abilities: From theory to practice. *Review of Educational Research,* Spring 1986, 1–39.

DeVAUS, D. A. The relative importance of parents and peers for adolescent religious orientation: An Australian study. *Adolescence,* Spring 1983, 147–158.

DIENSTBIER, R. A., L. R. KAHLE, K. A. WILLIS, and G. B. TUNNELL. The impact of moral theories on cheating: Studies of emotion attribution and schema activation. *Motivation and Emotion,* 1980, *4,* 193–216.

DOANE, J., K. WEST, M. J. GOLDSTEIN, E. RODNICK, and J. JONES. Parental communication deviance and affective style as predictors of subsequent schizophrenia spectrum disorders in vulnerable adolescents. *Archives of General Psychiatry,* 1981, *38,* 679–685.

DORNBUSH, S. M., L. CARLSMITH, R. T. GROSS, J. A. MARTIN, D. JENNING, A. ROSENBERG, and D. DUKE. Sexual development, age, and dating: A comparison of biological and sociological influence upon the set of behaviors. *Child Development,* 1981, *52,* 179–185.

DOUGLAS, J. D., and A. C. WONG. Formal operations: Age and sex differences in Chinese and American children. *Child Development,* 1977, *48,* 689–692.

DOUGLAS, K., and D. ARENBERG. Age changes, cohort differences, and cultural changes on the Guilford-Zimmerman Temperament Survey. *Journal of Gerontology,* 1978, *33*(5), 737–747.

DOYLE, L. Drug hotline aide: "Crack" spread fast. *The Philadelphia Inquirer,* August 9, 1986, p. 1–D.

DREYER, P. Sexuality during adolescence. In B. Wolman (Ed.), *Handbook of Developmental Psychology.* Englewood Cliffs, NJ: Prentice-Hall, 1982.

DUDLEY, R. L., and M. G. DUDLEY. Transmission of religious values from parents to adolescents. *Book of Abstracts: Annual Meeting 1984.* Chicago, IL: Society for the Scientific Study of Religion, 1984.

DUNNE, F., R. ELLIOT, and W. S. CARLSEN. Sex differences in the educational and occupational aspirations of rural youth. *Journal of Vocational Behavior,* 1981, *18,* 56–66.

DUNPHY, D. C. Peer group socialization. In R. Muuss (Ed.), *Adolescent Behavior and Society,* 3rd ed. New York: Random House, 1980.

———, The social structure of urban adolescent peer groups. *Sociometry,* 1963, *26,* 230–246.

EAGLY, A. H. Gender and social influence: A social psychological analysis. *American Psychologist,* 1983, *38,* 971–981.

———, and L. L. CARLI. Sex of researchers and sex-typed communications as determinants of sex differences in influenceability: A meta-analysis of social influence studies. *Psychological Bulletin,* 1981, *90,* 1–20.

EGELAND, J. A., and A. M. HOSTETTER. Amish study, I: Affective disorders among the Amish. *American Journal of Psychiatry,* 1983, *140,* 56–61.

EISENBERGER, R., and F. A. MASTERSON. Required high effort increases subsequent persistence and reduces cheating. *Journal of Personality and Social Psychology,* 1983, *44,* 593–599.

ELDER, G. H. Adolescence in historical perspective. In J. Adelson (Ed.), *Handbook of Adolescent Psychology.* New York: John Wiley & Sons, 1980.

ELIFSON, K. W., D. M. PETERSEN, and C. K. HADAWAY. Religion and delinquency: A contextual analysis. *Criminology,* 1983, *21,* 505–527.

ELKIND, D. *The Hurried Child.* Reading, MA: Addison-Wesley, 1981.

———. Understanding the young adolescent. *Adolescence,* 1978, *13,* 127–134.

ELLIOTT, D. S., and S. S. AGETON. Reconciling race and class differences in self-reported and official estimates of delinquency. *American Sociological Review,* 1980, *45,* 95–110.

ENRIGHT, R. D., D. K. LAPSLEY, and D. G. SLUKLA. Adolescent egocentrism in early and late adolescence. *Adolescence,* 1979 (Winter), *14*(56), 687–695.

ERIKSON, E. H. Youth and the life cycle. In R. E. Muuss (Ed.), *Adolescent Behavior and Society,* 3rd ed. New York: Random House, 1980.

———. *Identity and the Life Cycle.* New York: W. W. Norton, 1980.

———. *Identity: Youth and Crisis.* New York: W. W. Norton, 1968.

———. *Life History and the Historical Moment.* New York: W. W. Norton, 1975.

EVERSOLL, D. A. A two-generational view of fathering. *Family Coordinator,* 1979, *28,* 503–508.

FALKOWSKI, C., and W. W. FALK. Homemaking as an occupational plan: Evidence from a national longitudinal study. *Journal of Vocational Behavior,* April 1983, 227–242.

FAUST, M. S. Alternative constructions of adolescent growth. In J. Brooks-Gunn and A. C. Peterson (Eds.), *Girls at Puberty: Biological, Psychological, and Social Perspectives.* New York: Plenum, 1983.

FEATHER, N. T. Values in adolescence. In J. Adelson (Ed.), *Handbook of Adolescent Psychology.* New York: John Wiley & Sons, 1980.

FEENEY, S. *Schools for Young Adolescents.* Carrboro, NC: Center for Early Adolescence, 1980.

FELDMAN, R. S. *Social Psychology.* New York: McGraw-Hill, 1985.

FETTERS, W. B., G. H. BROWN, and J. A. OWINGS. *High School Seniors: A Comparative Study of the Classes of 1972 and 1980.* National Center for Educational Statistics. Washington, D.C.: U.S. Government Printing Office, 1984.

FISCHER, J. L. Reciprocity, agreement, and family style in family systems with a disturbed and nondisturbed adolescent. *Journal of Youth and Adolescence,* 1980, *9,* 391–406.

FORISHA, B. Creativity and imagery in men and women. *Perceptual and Motor Skills,* 1978, *47,* 1255–1264.

FOX, M. F., and S. HESSE-BIBER. *Women at Work.* Palo Alto, CA: Mayfield, 1984.

FREEDMAN, S. G. Survey finds students more worried on jobs. *The New York Times,* May 9, 1982, p.12.

FREEMAN, D. *Margaret Mead and Samoa.* Cambridge, MA: Harvard University Press, 1983.

FREEMAN, R., and D. WISE (Eds.). *The Youth Labor Market Problem: Its Nature, Causes, and Consequences.* Chicago: University of Chicago Press, 1982.

FREUD, A. Adolescence. In R. S. Eissler et al. (Eds.), *The Psychoanalytic Study of the Child.* New York: International Universities Press, 1958, Vol. XIII.

———. Adolescence as a developmental disturbance. In G. Kaplan and S. Lobovici (Eds.), *Adolescence: Psychosocial Perspectives.* New York: Basic Books, 1969.

FREUD, S. *New Introductory Lectures on Psychoanalysis.* Trans. by James Strachey. New York: W. W. Norton, 1964.

FRISCH, R. E. Fatness, puberty, and fertility. In J. Brooks-Gunn and A. C. Petersen (Eds.), *Girls at Puberty: Biological, Psychological, and Social Perspectives.* New York: Plenum, 1983.

FURSTENBERG, F. F., R. HERVEG-BROWN, J. SHEA, and D. WEBB. Family communication and teenagers' contraceptive use. *Family Planning Perspectives,* July/August 1984, 163–170.

FURSTENBERG, F., Jr. *Unplanned Parenthood*. New York: Free Press, 1976.

GALLUP. A. M. The 18th Annual Gallup Poll of the Public's Attitudes toward the Public Schools. *Phi Delta Kappan,* September 1986, 43–59.

———. The 17th Annual Gallup Poll of the Public's Attitudes toward the Public Schools. *Phi Delta Kappan,* September 1985, 35–47.

GALLUP. G. H. Drug abuse called biggest problem facing teenagers. *Daily Intelligencer/Montgomery County Record,* September 22, 1983, p.15.

———. Gallup youth survey. *Denver Post*, November 20, 1979, p. 36.

———. *The Gallup Poll: Public Opinion 1978*. Wilmington, DE: Scholarly Resources, Inc., 1979.

———. 1978 Gallup Poll cited in C. Norback (Ed.), *The Complete Book of American Surveys*. New York: New American Library, 1980.

———, and D. POLING. *The Search for America's Faith*. New York: Abington, 1980.

GARBARINO, J., J. SEBES, and C. SCHELLENBACH. Families at risk for destructive parent-child relations in adolescence. Unpublished manuscript. State College: Pennsylvania State University. In Laurence Steinberg, *Adolescence*. New York: Knopf, 1985.

———, and C. E. ASP. *Successful Schools and Competent Students*. Lexington, MA: Lexington Books, 1981.

GARFINKEL, B. C., A. FROESE, and J. HOOD. Suicide attempts in children and adolescents. *American Journal of Psychiatry*, 1982, *139*, 1257–1261.

GAYLIN. J. What boys look for in girls. *Seventeen*, March 1978, 107–113.

GESELL, A. L., F. L. ILG. and L. B. AMES. *Youth: The Years from Ten to Sixteen*. New York: Harper, 1956.

GETZELS, J. W., and M. CSIKSZENTIMIHALYI. *The Creative Vision*. New York: John Wiley & Sons, 1976.

GILLIAM, D. Why do we insist on rushing kids into growing up? *The Philadelphia Inquirer*, September 27, 1983.

GILLIGAN, C. Remapping Development. Paper presented at the biennial meeting of the Society for Research in Child Development, Toronto, 1985.

GINZBERG, E. The job problem. *Scientific American*, 1977, *237*, 43–51.

———. Toward a theory of occupational choice: A restatement. *Vocational Guidance Quarterly*, 1972, *20*, 169–176.

GISPERT, M., K. WHEELER, L. MARSH. and M. S. DAVID. Suicidal adolescents: Factors in evaluation. *Adolescence*, Winter 1985, 753–762.

GLENN, N. D., and K. K. KRAMER. The psychological well-being of adult children of divorce. *Journal of Marriage and the Family*, 1985, 47(Nov.), 905–912.

GOLD, M., and R. J. PETRONIO. Delinquent behavior in adolescence. In J. Adelson (Ed.), *Handbook of Adolescent Psychology*. New York: John Wiley & Sons, 1980.

GOLDSTEIN, B. *Human Sexuality*. New York: McGraw-Hill, 1976.

GOLEMAN, D. Leaving home: Is there a right time to go? *Psychology Today*, August 1980, 52–61.

———. What colleges have learned about suicide. *The New York Times*, February 23, 1986, p. 22E.

GORDON, T. *P.E.T.: Parent Effectiveness Training*. New York: New American Library, 1975.

GORMAN, M. E. Using the *Eden Express* to teach introductory psychology. *Teaching of Psychology*, 1984, *11*, 39–40.

GOULD, R. *Transformations*. New York: Simon & Schuster, 1978.

GRABE, M. School size and the importance of school activities. *Adolescence*, 1981, *16*(61), 21–31.

GRANT, W. V., and T. D. SNYDER. *Digest of Education Statistics 1983–84*. National Center for Education Statistics. Washington, D.C.: U.S. Government Printing Office, 1984.

GRATTAN, T. C. *Civilized America*, vol. II. London: Bradbury & Evans, 1859. In F. P. Rice, *The Adolescent*, 5th ed. Boston: Allyn & Bacon, 1987.

GREENBERG, J. Suicide linked to brain chemical deficit. *Science News*, 1982, *121*, 355.

GREENBERGER, E., and L. STEINBERG. Sex differences in early work experience: Harbinger of things to come? *Social Forces*, 1983, *62*, 467–486.

GREGG, C. H. Sexuality education: Who should be teaching the children? *SIECUS Report*, September 1982, *11*(5), 1–4.

GREIF, G. L. Single fathers rearing children. *Journal of Marriage and the Family*, 1985, 47(Feb.), 185–191.

GRESSARD. C. F. Treatment outcome of criminally involved drug abusers: An analysis. Ph.D. dissertation, University of Iowa, 1979. In G. K. Leight and G. W. Peterson (Eds.), *Adolescents in Families*. Cincinnati: South-Western Publishing Co., 1986.

GRIFFIN. N., L. CHASSIN. and R. D. YOUNG. Measurement of global self-concept versus multiple role-specific self-concepts in adolescents. *Adolescents*, Spring 1981, 49–56.

GROTEVANT, H. D., and M. DURRETT. Occupational knowledge and career development in adolescence. *Journal of Vocational Behavior*, 1980, *17*, 171–182.

———, and W. L. THORBECKE. Sex differences in styles of occupational identity formation in late adolescence. *Developmental Psychology*, 1982, *18*, 396–405.

GURMAN, A. S. Family therapy research and the "new epistemology." *Journal of Marital and Family Therapy*, 1983, *9*(3), 227–234.

ALAN GUTTMACHER INSTITUTE. *Teenage Pregnancy: The Prob-*

lem That Hasn't Gone Away. New York: Alan Gutt-macher Institute, 1981.

HAAS, L. Domestic role sharing in Sweden. *Journal of Marriage and the Family,* 1979, *43*(Nov.), 957–967.

HADAWAY, C. K., K. W. ELIFSON, and D. M. PETERSON. Religious involvement and drug use among urban adolescents. *Journal for the Scientific Study of Religion,* June 1984, 109–128.

HALEY, J. *Leaving Home.* New York: McGraw-Hill, 1980.

HALL, G. S. *Adolescence.* 2 vols. New York: D. Appleton & Company, 1904.

———. *Life and Confessions of a Psychologist.* New York: D. Appleton & Company, 1923. In N. Kiell, *The Universal Experience of Adolescence.* Boston: Beacon Press, 1967, p. 184.

HANEY, B., and M. GOLD. The juvenile delinquent nobody knows. *Psychology Today,* September 1973.

HANKS, M., and B. K. ECKLAND. Adult voluntary association and adolescent socialization. *Sociological Quarterly,* 1978, *19*(3), 481–490.

HANSEN, S. L. Dating choices of high school students. *Family Coordinator,* 1977, *26*, 133–138.

HARRIS, L. AND ASSOCIATES, INC. *Public Awareness of the National Institute on Alcohol Abuse and Alcoholism Advertising Campaign and Public Attitudes toward Drinking and Alcohol Abuse.* Reports prepared for the National Institute on Alcohol Abuse and Alcoholism. Phase Four Report and Overall Summary, 1974. In F. P. Rice, *The Adolescent.* Boston: Allyn and Bacon, 1981.

HARTER, S. Developmental perspectives on the self-system. In P. H. Mussen (Ed.), *Handbook of Child Psychology,* vol. 4. New York: John Wiley & Sons, 1983.

HARTUP, W. W. *The peer system.* In P. H. Mussen (Ed.), *Carmichael's Manual of Child Psychology,* 4th ed., vol. 4. New York: John Wiley & Sons, 1983

The Harvard Medical School Health Letter, AIDS: Update, Part I, November 1985, pp. 1–4; Part II, December 1985, pp. 2–5.

The Harvard Medical School Health Letter, April 1981, p. 2.

HASKELL, M. R., and L. YABLONSKY. *Juvenile Delinquency,* 3rd ed. Boston: Houghton Mifflin, 1982.

HASS, A. *Teenage Sexuality.* New York: Macmillan, 1979.

HASSETT, J. But that would be wrong . . . *Psychology Today,* November 1981, 34–50.

HAVIGHURST, R. J. *Developmental Tasks and Education,* 3rd ed. New York: David McKay, 1972.

HENSHAW, S. K., N. J. BINKIN, E. BLAINE, and J. C. SMITH. A portrait of American women who obtain abortions. *Family Planning Perspectives,* March/April 1985, 90.

HETHERINGTON, E. Children and divorce. In R. Henderson (Ed.), *Parent-child Interaction: Theory, Research, and Prospects.* New York: Academic Press, 1981.

HIRSCHFELD, R. M. A., and C. K. CROSS. Epidemiology of affective disorders: psychosocial risk factors. *Archives of General Psychiatry,* 1982, *39*, 35–46.

HODGSON, J. W., and J. L. FISCHER. Sex differences in identity and intimacy development in college youth. *Journal of Youth and Adolescence,* 1979, *8*, 37–50.

HOFFMAN, M. L. Moral development in adolescence. In J. Adelson (Ed.), *Handbook of Adolescent Psychology.* New York: John Wiley and Sons, 1980.

HOLMES, K. Natural history of herpes: Current trends in treatment. Paper presented at the Herpes Symposium, Oregon Health Sciences University, Portland, April 2, 1982. In R. Crooks and K. Bauer, *Our Sexuality,* 2nd ed. Menlo Park, CA: The Benjamin/Cummings Publishing Co., 1983.

HOLMES, L. D. *Quest for the Real Samoa.* South Hadley, MA: Bergin & Garvey, 1986.

HOWARD, K. I., S. M. KOPTA, M. S. KRAUSE, and D. E. ORLINSKY. The dose-effect relationship in psychotherapy. *American Psychologist,* February 1986, 159–164.

HUNTER, F., and J. YOUNISS. Changes in functions of three relations during adolescence. *Developmental Psychology,* 1982, *18*, 806–811.

HUSTON, T. L., M. RUGGIERO, R. CONNER, and G. GEIS. Bystander intervention into crime: A study based on naturally occurring episodes. *Social Psychology Quarterly,* 1981, *44*, 14–23.

INHELDER, B., and J. PIAGET. *The Growth of Logical Thinking.* New York: Basic Books, 1958.

JERSILD, A. T., J. S. BROOK, and D. W. BROOK. *The Psychology of Adolescence,* 3rd ed. New York: Macmillan, 1978,.

JESSOP, D. J. Family relationships as viewed by parents and adolescents: A specification. *Journal of Marriage and the Family,* 1981, *43*(Feb.), 95–107.

JESSOR, R. and S. L. JESSOR. Adolescent development and the onset of drinking. In R. E. Muuss (Ed.), *Adolescent Behavior and Society,* 3rd ed. New York: Random House, 1980.

JOHNSON, C. A., J. W. GRAHAM, and W. B. HANSEN. Drug use by peer leaders. Paper presented at the annual meeting of the American Psychological Association, Los Angeles, August 25, 1981. In J. J. Conger and Anne C. Petersen, *Adolescence and Youth,* 3rd ed. New York: Harper, 1984.

JOHNSON, R., and M. M. CARTER. Flight of the young: Why children run away from their homes. *Adolescence,* 1980, *15*(58), 483–489.

JOHNSTON, L. D., and P. M. O'MALLEY. Why do the nation's students use drugs and alcohol?: Self-reported reasons from nine national surveys. *Journal of Drug Issues,* 1986, *16*, 29–66.

———, P. M. O'MALLEY, and L. K. EVELAND. Drugs and delinquency: A search for causal connections. In D. G. Kendel

(Ed.), *Longitudinal Research on Drug Use*. Washington, D.C.: Hemisphere, 1978.

———, P. M. O'MALLEY, and J. D. BACHMAN. *Drug Use Among American High School Students, College Students, and Other Young Adults*. National Institute on Drug Abuse. Washington, D.C.: U.S. Government Printing Office, 1986.

JONES, E., M.D. *The Life and Work of Sigmund Freud*, vol. 1. New York: Basic Books, 1953.

JONES-WITTERS, P., and W. L. WITTERS. *Drugs and Society*. Monterey, CA: Wadsworth Health Science, 1983.

JURICH, A. P., C. J. POLSON, J. A. JURICH, and R. A. BATES. Family factors in the lives of drug users and abusers. *Adolescence*, Spring 1985, 143–159.

KACERGUIS, M. A., and G. R. ADAMS. Erikson stage and resolution: The relationships between identity and intimacy. *Journal of Youth and Adolescence*, 1980, *9*, 117–126.

KAHOE, R. D., and M. J. MEADOW. A developmental perspective on religious orientation dimensions. *Journal of Religion and Health*, 1981, *20*, 8–17.

KALTER, N. Children of divorce in an outpatient psychiatric population. *American Journal of Orthopsychiatry*, 1977, *47*, 40–51.

KANNER, ALAN. Curse of unknown origin. *Family Therapy*, 1984, *11*(1), 49–53.

KAUFMAN, I. P. Juvenile justice: A plea for reform. *New York Times Magazine*, October 14, 1979, 42–60.

KELLY, C., and G. C. GOODWIN. Adolescents' perception of three styles of parental control. *Adolescence*, Fall 1983, 567–571.

KENNEDY, S. Youth programs stress non-alcoholic activities. *The Philadelphia Inquirer*, May 28, 1984, p. 8A.

KEPHART, W. M. *The Family, Society and the Individual*, 4th ed. Boston: Houghton-Mifflin, 1977.

KESSLER, G. R., F. A. IBRAHIM, and H. KAHN. Character development in adolescents. *Adolescence*, Spring 1986, 1–9.

KETT, J. F. *Rites of Passage: Adolescence in America 1790 to the Present*. New York: Basic Books, 1977.

KIELL, N. (Ed.). *The Universal Experience of Adolescence*. Boston: Beacon Press, 1967.

KINGSTON, P. W., and S. L. NOCK. Consequences of the family work day. *Journal of Marriage and the Family*, 1985, *47*(August), 619–629.

KIRBY, D., J. ALTER, and P. SCALES. *An Analysis of U.S. Sex Education Programs and Evaluation Methods*. Atlanta, GA: U.S. Dept. of Health, Education and Welfare. Report #CDC 2021-79-DK-FR, 1979. In W. H. Masters, V. E. Johnson, and R. C. Kolodny, *Human Sexuality*, 2nd ed. Boston: Little, Brown, 1985.

KISKER, E. E. Teenagers talk about sex, pregnancy and contraception. *Family Planning Perspectives*, March/April 1985, 83–90.

KLEIN, J. R. and I. F. LITT. Menarche and Dysmenorrhea. In J. Brooks-Gunn and A. C. Petersen (Eds.), *Girls at Puberty: Biological, Psychological, and Social Perspectives*. New York: Plenum, 1983.

KOHLBERG, L. *The Psychology of Moral Development*, vol. II. San Francisco: Harper & Row, 1984.

KOHN, M. L., and C. SCHOOLER. *Work and Personality*. Norwood, NJ: Ablex, 1983.

KOHUT, H. *How Does Psychoanalysis Cure?* Chicago: University of Chicago Press, 1984.

KRONICK, D. An examination of psychosocial aspects of learning disabled adolescents. *Learning Disability Quarterly*, 1978, *1*(4), 86–93.

KUHN, D., and J. ANGELEV. An experimental study of the development of formal operational thought. *Child Development*, 1976, *47*, 697–706.

KULKA, R. A., and H. WEINGARTEN. The long-term effects of parental divorce in childhood on adult adjustment. *Journal of Social Issues*, 1979, *33*(4), 50–78.

LAMB, D. H., and G. D. REEDER. Reliving golden days. *Psychology Today*, June 1986, 22–30.

LARSON, L. E. The influence of parents and peers during adolescence: The situation hypothesis revisited. In R. E. Muuss (Ed.), *Adolescent Behavior and Society*, 3rd ed. New York: Random House, 1980.

LEO, J. The revolution is over. *Time*, April 9, 1984, 74–83.

LERNER, B. Self-esteem and excellence: The choice and the paradox. *American Educator*, 1985, *9*(4), 10–16.

LERNER, R. M., J. B. ORLOS, and J. R. KNAPP. Physical attractiveness, physical effectiveness, and self-concept in late adolescence. *Adolescence*, 1976, *11*, 313–326.

LEVINE, E., and C. KOZAK. Drug and alcohol use, delinquency, and vandalism among upper middle class pre- and post-adolescents. *Journal of Youth and Adolescence*, 1979, *8*(1), 91–101.

LEVINSON, D. *The Seasons of a Man's Life*. New York: Knopf, 1978.

LEWIN, T. A new push to raise women's pay. *New York Times*, January 1, 1984, p. 15.

LEWIN-EPSTEIN, N. *Youth Employment During High School*. National Center for Education Statistics. Washington, D.C.: U.S. Government Printing Office, 1981.

LEWINE, R.R.J. Sex differences in schizophrenia: Timing or subtypes? *Psychological Bulletin*, 1981, *90*, 432–444.

LEWIS, C. How adolescents approach decisions: Changes over grades seven to twelve and policy implications. *Child Development*, 1981, *52*, 538–544.

LIPSITZ, J. S. *Successful Schools of Young Adolescents*. Chapel Hill, NC: The Center for Early Adolescence, 1983.

LIVSON, N., and H. PESKIN. Perspectives on adolescence from longitudinal research. In J. Adelson (Ed.), *Handbook of Adolescent Psychology*. New York: John Wiley, 1980.

LLOYD, D. Prediction of school failure from third-grade data. *Educational and Psychological Measurement,* 1978, *38,* 1193–1200.

LOCKARD, J. S., B. C. KIRKEVOLD, and D. F. KALK. Cost-benefit indexes of deception in nonviolent crime. *Bulletin of the Psychonomic Society,* 1980, *16,* 303–306.

LONG, T. E., and J. K. HADDEN. Religious conversion and the concept of socialization: Integrating the brainwashing and drift models. *Journal for the Scientific Study of Religion,* March 1983, 1–14.

LOUDEN, D. M. A comparative study of self-esteem among minority group adolescents in Britain. *Journal of Adolescence,* 1980, *3,* 17–33.

LUEPNITZ, D. A. Which aspects of divorce affect children? *Family Coordinator,* 1979, *28,* 79–85.

LUNZER, E. A. Formal reasoning: A reappraisal. In B. Z. Presseisen, D. Goldstein, and M. Appel (Eds.), *Topics in Cognitive Development.* New York: Plenum, 1978.

McALLISTER, E.W.C. Religious attitudes of women college students: A follow-up study. *Adolescence,* Winter 1985, 797–804.

McCABE, M.P., and J. K. COLLINS. Sex role and dating orientation. *Journal of Youth and Adolescence,* 1979, *8,* 407–425.

McCORMICK, M. C., S. SHAPIRO, and B. STARFIELD. High risk young mothers: Change in infant mortality and morbidity in four areas in the United States, 1973–1978. Presented at the Annual Meeting of the American Pediatric Association, San Francisco, CA, April 30, 1981. In W. H. Masters, V. E. Johnson, and R. C. Kolodny, *Human Sexuality,* 2nd ed. Boston: Little, Brown, 1985.

McCRARY, E. Loneliness, danger and anxiety: The lot of many a latchkey child. *The Philadelphia Inquirer,* June 4, 1984, p. 20–BN.

McGEE, E. A. *Too Little, Too Late.* New York: Ford Foundation, 1982.

McKINNON, J. W. The college student and formal operations. In Renner, Stafford, et al., *Research, Teaching, and Learning with the Piaget Model.* Norman, OK: University of Oklahoma Press, 1976.

McCLAUGHLIN, S. D., W. R. GRADY, J. O. G. BILLY, N. S. LANDALE, and L. D. WINGES. The effects of the sequencing of marriage and first birth during adolescence. *Family Planning Perspectives,* January/February 1986, 12–18.

MACCOBY, E. E., and C. N. JACKLIN. *The Psychology of Sex Differences.* Stanford, CA: Stanford University Press, 1974.

MANN, D. W. *When Delinquency is Defensive: Self-esteem and Deviant Behavior.* Unpublished doctoral dissertation, University of Michigan, Ann Arbor, 1976. In Delinquent behavior in adolescence, by M. Gold and R. J. Petronio. In J. Adelson (Ed.), *Handbook of Adolescent Psychology.* New York: John Wiley & Sons, 1980.

MARCIA, J. Identity in adolescence. In J. Adelson (Ed.), *Handbook of Adolescent Psychology.* New York: John Wiley & Sons, 1980.

MARTORANO, S. C. A developmental analysis of performance on Piaget's formal operations tasks. *Developmental Psychology,* 1977, *13,* 666–672.

MASLOW, A. H. *The Farther Reaches of Human Nature.* New York: Viking Press, 1971.

MASTERS, W. H., V. E. JOHNSON, and R. C. KOLODNY. *Human Sexuality,* 2nd ed. Boston: Little, Brown, 1985.

MEAD, M. The American family: An endangered species. *TV Guide,* December 30–January 5, 1978–1979.

———. *Blackberry Winter.* New York: William Morrow, 1972.

———. *Coming of Age in Samoa.* New York: New American Library, 1950.

———. *Growing up in New Guinea.* New York: New American Library, 1953.

MEADOW, M. J., and R. D. KAHOE. *Psychology of Religion.* New York: Harper & Row, 1984.

MENAGHAN, E. G. Seeking help for parental concerns in the middle years. *American Journal of Community Psychology,* 1978, *6*(5), 477–488.

MILLER, P. Y., and W. SIMON, The development of sexuality in adolescence. In J. Adelson (Ed.), *Handbook of Adolescent Psychology.* New York: John Wiley & Sons, 1980.

MILLER, W. B. Statement in *Serious Youth Crime.* Hearings before the subcommittee to investigate juvenile delinquency of the Committee on the Judiciary, United States Senate Ninety-fifth Congress, Second Session. Washington, D.C.: U.S. Government Printing Office, 1978.

MINUCHIN, P. P., and E. K. SHAPIRO. The school as a context for social development. In P. H. Mussen (Ed.), *Carmichael's Manual of Child Psychology,* 4th ed.. New York: John Wiley & Sons, 1983.

MINUCHIN, S. *Psychosomatic Family: Anorexia in Context.* Cambridge, MA: Harvard University Press, 1978.

———, and H. C. FISHMAN. *Family Therapy Techniques.* Cambridge, MA: Harvard University Press, 1981.

MONEY, J. *Love and Love Sickness: The Science of Sex, Gender Difference and Pair-Bonding.* Baltimore: Johns Hopkins University Press, 1980.

MONTEMAYOR, R. The relationship between parent-adolescent conflicts and the amount of time adolescents spend with their parents, peers, and alone. *Child Development,* 1982, *53,* 1512–1519.

———, and M. EISEN. The development of self-conceptions from childhood to adolescence. *Developmental Psychology,* 1977, *13,* 314–319.

MOORE, A. Programs on gangs may die. *The Philadelphia Inquirer,* March 31, 1979, p. 1–C.

MORASH, M. A. Working class membership and the adolescent identity crisis. *Adolescence,* 1980, *15*(58), 313–320.

MORGAN, E., and B. A. FARBER. Toward a reformulation of the Eriksonian model of female identity development. *Adolescence,* 1982, *17*(65), 199–211.

MORRISON, A., P. GJERDE, and J. H. BLOCK. A prospective study of divorce and its relation to family functioning. Paper presented at the biennial meeting of the Society for Research in Child Development, Detroit, April 1983. In John W. Santrock, *Adolescence.* Dubuque, IA: Wm. C. Brown, 1984.

MORTIMER, J. T., and D. KUMKA. A further examination of the "Occupational Linkage Hypothesis." *The Sociological Quarterly,* 1982, *23,* 3–16.

MUNRO, G., and G. ADAMS. Egoidentity formation in college students and working youth. *Developmental Psychology,* 1977, *13,* 523–524.

MUUSS, R. E. Adolescent eating disorder: bulimia. *Adolescence,* Summer 1986, 257–267.

———. Social cognition: Robert Selman's theory of roletaking. *Adolescence,* Fall 1982, 499–525.

NAEDELE, W. F. Researcher sees private schools eroding into ethical wastelands. *The Philadelphia Inquirer,* May 5, 1982, p. 5–B.

NATIONAL COMMISSION ON EXCELLENCE IN EDUCATION. *A Nation at Risk: The Imperative for Educational Reform.* Washington, D.C.: U.S. Department of Education, 1983.

NATIONAL COMMISSION ON YOUTH. *The Transition of Youth to Adulthood: A Bridge Too Long.* A report to educators, sociologists, legislators, and youth policymaking bodies. Boulder, CO: Westview Press, 1980.

NATRIELLO, G., and S. M. DORNBUSH. Bringing behavior back in: The effects of student characteristics and behaviors on the classroom behavior of teachers. *American Educational Research Journal,* Spring 1983, 28–43.

NEIMARK, E. D. Adolescent thought: Transition to formal operations. In B. B. Wolman (Ed.), *Handbook of Developmental Psychology.* Englewood Cliffs, NJ: Prentice-Hall, 1982.

———. Intellectual development during adolescence. In F. D. Horowitz (Ed.), *Review of Child Development Research,* vol. 4. Chicago: University of Chicago Press, 1975.

NESBIT, J. Number of cocaine users has increased, survey says. *The Philadelphia Inquirer,* October 10, 1986, p. 12–C.

NEWMAN, J. Adolescents: Why they can be so obnoxious. *Adolescence,* Fall 1985, 635–646.

NEWTON, D. E. The status of programs in human sexuality: A preliminary study. *The High School Journal,* 1982, *6,* 232–239.

NIAID STUDY GROUP. *Sexually Transmitted Diseases: 1980 Status Report.* NIH Publication No. 81–2213. Washington, D.C.: U.S. Government Printing Office, 1981.

NICHOLI, A. Will the losses shock us into strong action? *The New York Times,* July 6, 1986, p. 2S.

NICOL, S. E., and I. I. GOTTESMAN. Clues to the genetics and neurobiology of schizophrenia. *American Scientist,* 1983, *71,* 398–404.

NISSIM-SABAT, D. Relationships between Piaget's cognitive stages and social orientation. *Psychological Reports,* 1978, *43,* 1315–1318.

NORMAN, J., and M. W. HARRIS. *The Private Life of the American Teenager.* New York: Rawson, Wade, 1981.

O'BRIEN, R., and M. CHAFETZ, M.D. *The Encyclopedia of Alcoholism.* New York: Facts on File Publications, 1982.

———, and S. COHEN, M.D. *The Encyclopedia of Drug Abuse.* New York: Facts on File Publications, 1984.

OFFER, D., and J. OFFER. *From Teenage to Young Manhood.* New York: Basic Books, 1975.

———, E. OSTROV, and K. I. HOWARD. *The Adolescent.* New York: Basic Books, 1981.

OLMEDO, E. L. Testing linguistic minorities. *American Psychologist,* 1981, *36,* 1078–1085.

OLSON, D. H., C. S. RUSSELL, and D. H. SPRENKLE. Circumplex model of marital and family systems: IV. Theoretical update. *Family Process,* 1983, *22,* 69–83.

O'MALLEY, P. M., and J. G. BACHMAN. Self-esteem: Change and stability between ages 13 and 23. *Developmental Psychology,* 1983, *19*(2), 257–268.

On Campus. Teen suicide: The alarming statistics. Published by the American Federation of Teachers, February 1985, p. 7.

OPENSHAW, D. K., B. C. ROLLINS, and D. L. THOMAS. Parental influences on adolescent self-esteem. *Journal of Early Adolescence,* 1984, *4*(2), 259–274.

———, and D. L. THOMAS. The adolescent self and the family. In G. K. Leigh and G. W. Peterson, *Adolescents in Families.* Cincinnati: South-Western Publishing Co., 1986.

———, D. L. THOMAS, and B. C. ROLLINS. Socialization and adolescent self-esteem: Symbolic interaction and social learning explanations. *Adolescence,* Summer 1983, 317–329.

OPPENHEIMER, M. What you should know about herpes. *Seventeen,* October 1982, 154–155, 170.

ORR, M. T. Sex education and contraceptive education in U.S. public high schools. *Family Planning Perspectives,* 1982, *14,* 304–313.

OSBORN, S. G., and D. J. WEST. Do delinquents really reform? *Journal of Adolescence,* 1980, *3,* 99–114.

OSTROW, R. J. Report says legal drugs are the worst killers. *The Philadelphia Inquirer,* November 10, 1982, p. 13–A.

PARDECK, J. T., V. WOLF, S. KILLION, and G. SILVERSTEIN. Individual therapy vs. family therapy: Which is more effective? *Family Therapy*, 1983, *10*(2), 173–181.

PARISH, J., and T. S. PARISH. Children's self-concept as related to family structure and family concept. *Adolescence*, Fall 1983, 649–658.

PARISH, T., and J. W. DOSTAL. Evaluations of self and parent figures by children from intact, divorced, and reconstituted families. *Journal of Youth and Adolescence*, 1980, *9*, 347–351.

PARLEE, M. B. et al. The friendship bond. *Psychology Today*, October 1979, 43–54, 113.

PATTERSON, G. *Coercive Family Processes*. Eugene, OR: Castala Publishing, 1981.

PEPLAU, L. A., Z. RUBIN, and C. T. HILL. Sexual intimacy in dating relationships. *Journal of Social Issues*, 1977, *33*, 2.

PETER, J. B., Y. BRYSON, and M. A. LOVETT. Genital herpes: Urgent questions, elusive answers. *Diagnostic Medicine*, March/April, 1982, 71–74, 76–88.

PETERSEN, A. C., and A. M. BOXER. Adolescent sexuality. In T. Coates, A. C. Petersen, and C. Perry (Eds.), *Promoting Adolescent Health: A Dialog on Research and Practice*. New York: Academic Press, 1982.

———, and B. TAYLOR. The biological approach to adolescence. In J. Adelson (Ed.), *Handbook of Adolescent Psychology*. New York: John Wiley & Sons, 1980.

The Philadelphia Inquirer. Study of crime suspects finds 56% use illegal drugs. June 4, 1986, p. 14–A.

PHILLIPS, D. P. Newsline. *Psychology Today*, January 1981.

PIAGET, J. Intellectual evolution from adolescence to adulthood. *Human Development*, 1972, *15*, 1–12.

PLANNED PARENTHOOD FEDERATION OF AMERICA. 11 million teenagers: What can be done about the epidemic of adolescent pregnancies in the United States. New York: The Alan Guttmacher Institute, 1977.

PLECK, JOSEPH. *The Myth of Masculinity*. Cambridge, MA: MIT Press, 1981.

PLISKO, V. W., and J. D. STERN (Eds.). *The Condition of Education*, 1985 edition. National Order for Educational Statistics. Washington, D.C.: U.S. Government Printing Office.

POLIT-O'HARA, D., and J. R. KAHN. Communication and contraceptive practices in adolescent couples. *Adolescence*, Spring 1985, 33–43.

POLOVY, P. A study of moral development and personality relationships in adolescents and young adult Catholic students. *Journal of Clinical Psychology*, 1980, *36*(3), 752–757.

POOLE, M. E. The schools adolescents would like. *Adolescence*, Summer 1984, 447–458.

POSTMAN, N. *The Disappearance of Childhood*. New York: Delacorte, 1982.

PRESCOTT, P. *The Child Savers*. New York: Knopf, 1981.

PUTNAM, F. Traces of Eve's faces. *Psychology Today*, October 1982.

RENNER, J. W., and D. G. STAFFORD. The operational levels of secondary school students. In J. W. Renner, D. G. Stafford, et al., *Research, Teaching, and Learning with the Piaget Model*. Norman, OK: University of Oklahoma Press, 1976.

RICE, F. P. *Contemporary Marriage*. Boston: Allyn & Bacon, 1983.

RICHMOND, L. H. Some further observations on private practice and community clinic adolescent psychotherapy groups. *Corrective and Social Psychiatry and Journal of Behavior Technology, Methods and Therapy*, 1978, *24*, 57–61.

ROAZEN, P. *Erik H. Erikson*. New York: The Free Press, 1976.

ROBBINS, C., H. B. KAPLAN, and S. S. MARTIN. Antecedents of pregnancy among unmarried adolescents. *Journal of Marriage and the Family*, April 1985, 567–583.

ROBBINS, S. L., M.D., M. ANGELL, M.D., and V. KUMAR, M.D. (Eds.). *Basic Pathology*, 3rd ed. Philadelphia: Saunders, 1981.

ROBINS, L. N., J. E. HELZER, M. M. WEISSMAN, H. ORVASCHEL, E. GRUENBERG, J. D. BURKE, Jr., and D. A. REGIER. Lifetime prevalence of specific psychiatric disorders in three sites. *Archives of General Psychiatry*, 1984, *41*(10), 949–958.

ROBINSON, I. E., and F. D. JEDLICKA. Changes in sexual attitudes and behavior in college students from 1965 to 1980: A research note. *Journal of Marriage and the Family*, February 1982, 237–240.

ROLL, S., and L. MILLEN. The friend as represented in the dreams of late adolescents: Friendship without rose-colored glasses. *Adolescence*, 1979, *14*(54), 255–275.

ROSCOE, B., and K. L. PETERSON. Older adolescents: A self-report of engagement in developmental tasks. *Adolescence*, Summer 1984, 391–396.

ROSENBERG, F. R., and M. ROSENBERG. Self-esteem and delinquency. *Journal of Youth and Adolescence*, 1978, *7*, 279–291.

ROTHMAN, K. M. Multivariate analysis of the relationship of person concerns to adolescent ego identity status. *Adolescence*, Fall 1984, 713–727.

RUBENSTEIN, C. Survey report: Money and self-esteem, relationships, secrecy, envy, satisfaction. *Psychology Today*, May 1981, 29–44.

———. The modern art of courtly love. *Psychology Today*, July 1983, 40–49.

———, P. SHAVER, and L. A. PEPLAU. Loneliness. In N. Jackson (Ed.), *Personal Growth and Behavior*, 82/83. Guilford, CT: Dushkin Publishing Group, 1982.

RUBLE, D. N., and J. BROOKS-GUNN. The experience of menarche. *Child Development*, 1982, *53*, 1557–1566.

RUBY, T., and R. LAW. School dropouts—They are not what

they seem to be. *Children and Youth Services Review,* 1982, *3,* 279–291.

RUMBERGER, RUSSELL W. Dropping out of high school: The influence of race, sex, and family background. *American Education Research Journal,* Summer 1983, 199–220.

RUST, J. O., and A. McCRAW. Influence of masculinity-femininity on adolescent self-esteem and peer acceptance. *Adolescence,* Summer 1984, 359–366.

RUTTER, M. School effects on pupil progress: Research findings and policy implications. *Child Development,* 1983, *54,* 1–29.

———, and H. GILLER. *Juvenile Delinquency.* Baltimore: Penguin Books, 1983.

ST. CLAIR, S., and H. D. DAY. Ego identity status and values among high school females. *Journal of Youth and Adolescence,* 1979, *8*(3), 317–326.

SANIK, M. M., and K. STAFFORD. Adolescents' contribution to household production: Male and female differences. *Adolescence,* Spring 1985, 207–215.

SARASON, I. G., and B. R. SARASON. *Abnormal Psychology,* 4th ed. Englewood Cliffs, NJ: Prentice-Hall, 1984.

SAXE, J. G. *The Blind Man and the Elephant.* New York: McGraw-Hill, 1963.

SCALES, P., and K. EVERLY. A community sex education program for parents. *The Family Coordinator,* 1977, *26,* 37–45.

SCHMID, R. E. Survey finds 4 women in 5 try sex before marriage. *The Philadelphia Inquirer,* April 13, 1985, 2–A.

———. 2 million latch-key children. *The Philadelphia Inquirer,* February 6, 1987, 7–A.

SCHILL, W. J., R. McCARTIN, and K. MEYER. Youth employment: Its relationship to academic and family variables. *Journal of Vocational Behavior,* April 1985, 155–163.

SCHOFIELD, J. Complementary and conflicting identities: Images and interaction in an interracial school. In S. Asher and J. Gottman (Eds.), *The Development of Children's Friendships.* Cambridge: Cambridge University Press, 1981.

SCHREIBER, F. *Sybil.* New York: Warner, 1974.

SCHULENBERG, J. E., F. W. VONDRACEK, and A. C. CROUTER. The influence of the family on vocational development. *Journal of Marriage and the Family,* February 1984, 129–143.

SCOTT, E. M. Young drug abusers and non-abusers: A comparison. *International Journal of Offender Therapy and Comparative Criminology,* 1978, *22*(2), 105–114.

SCOTT, V. Actress: Films distort teen life. *The Philadelphia Inquirer,* September 2, 1985, 6–E.

SEBALD, H. Adolescents' shifting orientation toward parents and peers: A curvilinear trend over recent decades. *Journal of Marriage and the Family,* February 1986, 5–13.

———. Adolescents' concepts of popularity and unpopularity, comparing 1960 with 1976. *Adolescence,* 1981, *16*(61), 187–193.

SELMAN, R. L. *The Growth of Interpersonal Understanding.* New York: Academic Press, 1980.

SHAPIRO, H. S., and E. LOUNSBERRY. Nesting. *The Philadelphia Inquirer,* February 22, 1983, 1–D.

SHARABANY, R., R. GERSHONI, and J. HOFMAN. Girlfriend, boyfriend: Age and sex differences in intimate friendship. *Developmental Psychology,* 1981, *17,* 800–808.

SHIGETOMI, D. C., D. P. HARTMANN, and D. M. GELFAND. Sex differences in children's altruistic behavior and reputation for helpfulness. *Developmental Psychology,* 1981, *17,* 377–386.

SIEG, A. Why adolescence occurs. In H. D. Thornburg (Ed.), *Contemporary Adolescence: Readings,* 2nd ed. Monterey, CA: Brooks/Cole Publishing Company, 1975.

SIFFORD, D. Cults: Educating children is the best prevention. *The Philadelphia Inquirer,* October 10, 1983, 4–C.

SILVESTRI, G. T., and J. M. LUKASIEWICZ. Occupational employment projections: The 1984–95 outlook. In *Monthly Labor Review,* U. S. Department of Labor (Washington, DC: U.S. Government Printing Office), November 1985.

SINGER, M., and R. ISRALOWITZ. Probation: A model for coordinating youth services. *Juvenile and Family Court Journal,* 1983, *1,* 35–41.

SLOCUM, J. W., and W. L. CRON. Job attitudes and performance during three career stages. *Journal of Vocational Behavior,* April 1985, 126–145.

SMILGIS, MARTHA. The big chill: Fear of AIDS. *Time,* February 16, 1987.

SMITH, E. T. Adolescent reactions to attempted parental control and influence techniques. *Journal of Marriage and the Family,* 1983, *45*(Aug.), 533–542.

SMITH, G. M. Relations between personality and smoking behavior in preadult subjects. In R. E. Muuss (Ed.), *Adolescent Behavior and Society,* 3rd ed. New York: Random House, 1980.

SORENSON, R. C. *Adolescent Sexuality in Contemporary America.* New York: World Publishing Company, 1973.

SPENCE, J., and R. HELMREICH. *Masculinity and Femininity.* Austin, TX: University of Texas Press, 1979.

STAPLES, R. Changes in black family structure: The conflict between family ideology and structural conditions. *Journal of Marriage and the Family,* 1985, *47*(Nov.), 1005–1013.

STEINBERG, L., E. GREENBERGER, L. GARDUQUE, M. RUGGIERO, and A. VAUX. Effects of working on adolescent development. *Developmental Psychology,* 1982, *4,* 385–395.

STEPHAN, C. S., and J. CORDER. The effects of dual-career families on adolescents' sex-role attitudes, work and family plans, and choice of important others. *Journal of Marriage and the Family,* 1985, *47*(Nov.), 921–929.

STERN, M., J. E. NORTHMAN, and M. R. VAN SLYCK. Father absence and adolescent "problem behaviors": Alcohol consumption, drug use, and sexual activity. *Adolescence,* Summer 1984, 301–312.

STERNBERG, R. J. *Beyond IQ.* Cambridge: Cambridge University Press, 1985.

———. How to teach intelligence. *Educational Leadership,* September 1984, 38–48.

STORY, M. D. A comparison of university student experiences with various sexual outlets in 1974 and 1980. *Adolescence,* Winter 1982, 737–747.

STOTT, D. *Delinquency.* New York: SP Medical and Scientific Books, 1982.

SULLIVAN, K., and A. SULLIVAN. Adolescent-parent separation. *Developmental Psychology,* 1980, *16*(2), 93–99.

SUPER, D. E. A life-span life-space approach to career development. *Journal of Vocational Behavior,* 1980, *16* (3), 282–298.

———, and D. T. HALL. Career development: Exploration and planning. *Annual Review of Psychology,* 1978, *29,* 333–372.

SWOPE, G. W. Kids and cults: Who joins and why? *Media and Methods,* 1980, *16,* 18–21.

TANNER, J. M. *Fetus into Man: Physical Growth from Conception to Maturity.* Cambridge, MA: Harvard University Press, 1978.

TAUBE, C. A., and S. A. BARRETT (Eds.). *Mental Health, United States, 1985.* Department of Health and Human Services, Publication No. (ADM) 85-1378. Washington, D.C.: National Institutes of Mental Health, 1985.

TENZER, A. Parental influences on the occupational choices of career women in male-dominated and traditional occupations. *Dissertation Abstracts International,* 1977, *38,* 2014.

THIRER, JOEL, and STEPHEN D. WRIGHT. Sport and social status for adolescent males and females. *Sociology of Sport Journal,* 2, 1985, 164–171.

THORNBURG, H. D. Sources of sex education among early adolescents. *Journal of Early Adolescence,* 1981, *1,* 171–184.

———. Adolescent delinquency and families. In G. K. Leigh and G. W. Peterson (Eds.), *Adolescents in Families.* Cincinnati: South-Western Publishing Co., 1986.

THORNE, C. R., and R. R. DEBLASSIE. Adolescent substance abuse. *Adolescence,* Summer 1985, 335–347.

TREBLICO, G. R. Career education and career maturity. *Journal of Vocational Behavior,* October 1984, 191–202.

TRICKETT, E. Toward a social ecological conception of adolescent socialization: Normative data on contrasting types of public school classrooms. *Child Development,* 1978, *49,* 408–414.

TURANSKI, J. J. Reaching and treating youth with alcohol-related problems: A comprehensive approach. *Alcohol Health and Research World,* 1982, *7, 3–9.*

United Crime Reports for the United States, 1985. Washington, D.C.: Federal Bureau of Investigation, Government Printing Office, 1985.

U.S. BUREAU OF THE CENSUS. *Statistical Abstract of the United States, 1987,* 107th ed. Washington, D.C.: U.S. Government Printing Office, 1986.

U.S. BUREAU OF THE CENSUS. *Statistical Abstract of the United States, 1986,* 106th ed. Washington, D.C.: U.S. Government Printing Office, 1985.

U.S. DEPARTMENT OF HEALTH AND HUMAN SERVICES. *Schizophrenia, is there an answer?* Rockville, MD: U.S. Department of Health and Human Services, 1981.

U.S. DEPARTMENT OF LABOR, BUREAU OF LABOR STATISTICS. *Occupational Outlook Handbook,* 1986–1987 edition. Washington, D.C.: U.S. Government Printing Office, 1986.

U.S. DEPARTMENT OF LABOR. *Dramatic new occupational inroads scored by women.* Washington, D.C.: U.S. Government Printing Office, November 1982.

U.S. News and World Report. America's cults gaining ground again. July 5, 1982, 37–41.

U.S. News and World Report. What world's teenagers are saying. June 30, 1986, 68.

VELDMAN, D. J., and JULIE P. SANFORD. The influence of class ability level on student achievement and classroom behavior. *American Educational Research Journal,* Fall 1984, 629–644.

VONNEGUT, MARK. *The Eden Express: A Personal Account of Schizophrenia.* New York: Praeger, 1975.

WALKER, J. I., M.D. *Everybody's Guide to Emotional Well-being.* San Francisco: Harbor Publishing, 1982.

WALLERSTEIN, J. S. Children of divorce: Preliminary report of a ten-year followup of young children. *American Journal of Orthopsychiatry,* 1985, *54*(July), 444–458.

———, and J. B. KELLY. *Surviving the Breakup: How Children and Parents Cope with Divorce.* New York: Basic Books, 1980.

WARSHAK, R. A., and J. W. SANTROCK. Children of divorce: Impact of custody disposition on social development. In E. J. Callahan and K. A. McCluskey (Eds.), *Life-Span Developmental Psychology: Normative Life Events.* New York: Academic Press, 1983.

WATERMAN, A. Identity development from adolescence to adulthood: An extension of theory and a review of research. *Developmental Psychology,* 1982, *18,* 341–358.

WATT, N., J. ANTHONY, L. WYNNE, and J. ROLF. *Children at Risk for Schizophrenia: A Longitudinal Perspective.* New York: Cambridge University Press, 1984.

WEIL, A., M.D., and W. ROSEN. *Chocolate to Morphine.* Boston: Houghton Mifflin, 1983.

WEINER, I. B. Psychopathology in adolescence. In J. Adelson (Ed.), *Handbook of Adolescent Psychology.* New York: John Wiley & Sons, 1980.

WEISHAAR, M. E., B. J. GREEN, and L. W. CRAIGHEAD. Primary influences of initial vocational choices for college women. *Journal of Vocational Behavior,* 1981, *18,* 67–78.

WEISS, R. S., The impact of marital dissolution on income and consumption in single-parent households. *Journal of Marriage and the Family.* 1984, *46*(Feb.), 115–127.

WELSH, G. S. Personality correlates of intelligence and creativity in gifted adolescents. In J. C. Stanley, W. C. George, and C. H. Solano (Eds.), *The Gifted and the Creative.* Baltimore, MD: Johns Hopkins University Press, 1977.

WESTERNDORP, F., K. L. BRINK, M. K. ROBERTSON, and I. E. ORTIZ. Variables which differentiate placement of adolescents into juvenile justice or mental health systems. *Adolescence,* Spring 1986, 23–37.

WESTOFF, C. F., G. CALOT, and A. D. FOSTER. Teenage fertility in developed nations: 1971–1980. *Family Planning Perspectives,* May/June 1983, 105–110.

WINER, J. A. et al. Sexual problems in users of a student mental health clinic. *Journal of Youth and Adolescence,* 1977, *6,* 117–126.

WOLF, F. M., and G. L. LARSON. On why adolescent formal operators may not be creative thinkers. *Adolescence,* 1981, *16,* 345–348.

WOODROOF, J. T. Premarital sexual behavior and religious adolescents. *Journal for the Scientific Study of Religion,* December 1985, 343–366.

WRIGHT, L. S. Parental permission to date and its relationship to drug use and suicidal thoughts among adolescents. *Adolescence,* Summer 1982, 409–418.

——. Suicidal thoughts and their relationships to family stress and personal problems among high school seniors and college undergraduates. *Adolescence,* Fall 1985, 575–580.

WRIGHT, S. A. Post-involvement attitudes of voluntary defectors from controversial new religious movements. *Journal for the Scientific Study of Religion,* June 1984, 172–182.

YANKELOVICH, D. *New Rules.* New York: Bantam Books, 1981.

——. Telephone survey results. In E. Thomas, Growing pains at 40. *Time,* May 19, 1986, 22–41.

YOUNISS, J., and A. DEAN. Judgment and imagery aspects of operations: A Piagetian study with Korean and Costa Rican children. *Child Development,* 1974, *45,* 1020–1031.

YUSSEN, S. R. Characteristics of moral dilemmas written by adolescents. *Developmental Psychology,* 1977, *13,* 162–163.

ZABIN, L. S., M. B. HIRSH, E. A. SMITH, and J. B. HARDY. Adolescent sexual attitudes and behavior: Are they consistent? *Family Planning Perspectives,* July/August 1984, 181–186.

ZELNIK, M., J. F. KANTNER, and K. FORD. *Sex and Pregnancy in Adolescence.* Beverly Hills, CA: Sage Publications, 1981.

——, and Y. J. KIM. Sex education and its association with teenage sexual activity, pregnancy, and contraceptive use. *Family Planning Perspectives,* 1982, *14,* 117–126.

——, and F. K. SHAH. First intercourse among young Americans. *Family Planning Perspectives,* March/April 1983, 64–70.

ZERN, D. S. The expressed preference of different ages of adolescents for assistance in the development of moral values. *Adolescence,* Summer 1985, 405–423.

ZIMBARDO, P., P. PILKONIS, and R. NORWOOD. *The Silent Prison of Shyness.* Glenview, IL: Scott, Foresman, 1974.

ZUCKERMAN, M. A biological theory of sensation seeking. In M. Zuckerman (Ed.), *Biological Bases of Sensation Seeking, Impulsivity, and Anxiety.* Hillsdale, NJ: Lawrence Erlbaum, 1983.

Subject Index

Adolescence
  abbreviated, 5–6
  boundaries, 11–12
  defined, 2, 11–12
  developmental patterns, 39
  early adolescence, 12
  historical perspective, 2
  hurried child syndrome, 13–14
  late adolescence, 12
  midadolescence, 12
  prolonged, 9
  psychosocial moratorium, 9
  rites of passage, 4–5
  social change, 7–8
  as social invention, 6
  theories, 20–27
  universal experience, 2–4
  and youth, 12
Adolescents
  employment, 234–39
  growth patterns, 39
  identity crisis, 138–41
  peer groups, 161–65
  population trends, 14–15
  storm and stress, 22, 27, 34, 38–40
AIDS
  sexual attitudes, 209, 226
  syndrome, 225–26
  transmission, 225–26
Alcohol, 320–24
  abuse, 322–24
  automobile accidents, 324
  effects, 320–21
  psychological, social factors, 323–24
  use, 321–24
Amenorrhea, 57
Anorexia, 354–56
Antisocial personality, 300–302, 368
Anxiety disorders, 351–53
  anxiety reactions, 352
  defined, 351
  obsessive-compulsive disorders, 353–54
  phobias, 352–53
Attitudes
  college students, 16–17
  parental attitudes, 117–18
  sexual, 208–14
  toward parents, 111–12, 114–15, 117–18
  toward school, 187, 200–202
Autonomous decision making, 125–27, 381–83
Autonomy
  behavioral, 124
  college experience, 127
  decision making, 125–27, 381–83
  emotional, 125
  satellization theory, 122–24

Barbiturates, 334–35
Behavioral autonomy, 124
Biological processes. See Physical development
Birth control, 219–22. See also Contraceptive use
Blood-alcohol levels, 321
Body image, 51–53
  body characteristics, 52
  sex differences, 51
Bulimia, 356–57

Career choice
  aspirations, 246–47
  career education, 250–53
  career interest inventories, 242–44
  career maturity, 241
  college 253–54
  distribution of sexes, 249
  Holland's personality-occupational types, 242–43
  influences, 244–50
  stages, 239–41
  vocational identity, 240–41
Career choice, influences on
  career education, 250–53
  college, 253–54
  family, 244–46
  sex differences, 247–50
  socioeconomic, 246–47
Career education, 250–53
Career outlook, 255–57

Cocaine, 331–34
  age trends, 332
  crack, 333–34
  extent of use, 331–32
Cognitive development, 67–81
  accommodation, 68
  assimilation, 68
  cognitive maturity, 80–81
  concrete operational stage, 70
  creativity, 81–85
  egocentrism, 78–80
  equilibration, 68–69
  formal operational stage, 70–71
  and learning, 71–76
  preoperational stage, 70
  sensorimotor stage, 70
  social cognition, 79
  and socialization, 77–81
  stages, 69–71
  variables, 71
College
  adjustment, 378–79
  advantages, 253–54
  career choice, 253–54
  career outlook, 255–57
  career-minded students, 15, 17, 253–54
  leaving home, 377–79
Compulsory education, 179–81
Conformity
  age, 164
  sex differences, 165
  types, 164
  variables, 164–65
Consistency of behavior
  moral, 274–76, 282–84
  sexual, 210–12
Contraceptive use, 219–22, 229–30
  couple communication, 220
  family communication, 221
  methods, 220
  premarital pregnancy, 222
  reasons not used, 221–22
  sex education, 229–30
Creativity
  achievement, 84
  creative adolescents, 82–83
  defined, 81
  intelligence, 84
  schools, 81, 84–85
  self-actualizing creativeness, 81–82
  test items, 82
  types, 81
Critical stage theory, 21
Cults, religious, 279–81
  deprogramming, 281
  membership, 279–80
  reasons for joining, 280–81

types, 280
Cultural theories of adolescence, 33–37

Dating
  ages, 171
  going steady, 123
  patterns, 171–73
  purposes, 171
  sex differences, 173
Decision making, 125–27, 381–83
Defense mechanisms, 28–29
Delinquency
  age trends, 288–90
  conduct disorders, 300–302
  defined, 288
  and drugs, 302–304
  family influence, 295–96
  family therapy, 308
  gangs, 299
  juvenile court, 305–306
  juvenile facilities, 306–308
  personality factors, 299–300
  prevention, 308–309
  psychopathology, 300–302
  sex differences, 292–94
  socioeconomic factors, 296–99
  types of offenses, 290–92
Depression
  incidence, 357–58
  manifestations, 358–59
  masked, 358–59
  and suicide, 360, 362
  types, 359–60
Development. *See specific types of development*
Developmental stages
  Erikson's, 137
  Freud's, 26–27
  Gesell's, 22–23
  Kohlberg's, 262–64
  Piaget's, 69–71
Developmental theories
  cognitive, 68–71
  maturational, 22–24
  moral, 261–66
  psychoanalytic, 25–30
  psychosocial, 31–32, 138–42
  religious, 276–79
Divorce
  adolescents' coping, 99–100
  and early marriage, 175–756
  effects, 99–101
  emotional conflicts, 98–99
  long-term effects, 100–101
Dropouts, 197–99
  rate, 197
  reasons, 197–99
Drug abuse, 313, 319, 339–41

alternatives to drugs, 341–45
drug dependence, 318–19
treatment, 340–41
Drug dependence
    defined, 318
    physical dependence, 319
    psychological dependence, 319
Drugs, illegal
    cocaine, 331–34
    defined, 313
    hallucinogens, 335–36
    heroin, 336
    LSD, 336
    marijuana, 327–31
    PCP, 336
Drug use
    abuse of drugs, 313, 319, 339–41
    age, 316–18
    alcohol, 320–24
    alternatives to drugs, 341–45
    among high school seniors, 314–15
    amphetamines, 331
    barbiturates, 334–35
    cocaine, 331–34
    comparative effects, 338
    and delinquency, 302–304
    development of drug use, 316–18
    drug dependence, 318–19
    extent of drug use, 313–16
    hallucinogens, 335–36
    heroin, 336
    inhalants, 335
    marijuana, 317–31
    relationships with drugs, 337–40
    suicide attempts, 363
    tobacco, 324–26
    tranquilizers, 335
    treatment, 340–41
    why youth use drugs, 318–20
Dual-career families
    household chores, 95–97
    latchkey teens, 97–98
    working mothers, 93–95
Dyslexia, 194
Dysmenorrhea, 57

Early adult transition, 375
Early maturers
    girls, 61–62
    boys, 63
Eating disorders, 354–57
    anorexia, 354–56
    bulimia, 356–57
Education
    attitudes toward, 200–201
    college, 17, 253–57, 377–79
    compulsory, 8, 179–81
    high school curricula, 181–82

mainstreaming, 194–95
middle schools, 182–84
moral education, 266, 275–76
new directions, 202–203
and women's movement, 17
Egocentrism, 78–80
    and cognitive maturity, 80–81
    imaginary audience, 78–79
    personal fable, 79–80
Emotional autonomy, 125
Empathy, 271–72
Erikson's developmental stages, 137

Family
    and career choice, 244–46
    changes, 89–93
    changing functions, 89
    defined, 88
    and delinquency, 295–96
    divorce, 98–101
    dual-career, 93–98
    household chores, 95–97
    parent characteristics, 92
    remarried families, 106–107
    roles, 91–93
    single-parent families, 101–106
    size, 89
    stepparents, 106–107
    therapy, 308, 369–70
    working mothers, 93–95
Family therapy
    and delinquency, 308
    psychological disorders, 369–70
Formal operational thinking
    defined, 70–71
    facilitating, 76
    and learning, 71–76
    and socialization, 77–81
Friendship
    age differences, 167–69
    love, 385–87
    opposite-sex, 169–70
    parent/peer intimacy, 166–67
    same-sex, 169–70
    sex differences, 169–70

Gangs, 299
Going steady, 173, 217
Groups homes, 307, 317
Growth spurt
    boys, 48
    girls, 48
    pubescence, 44

Hallucinogens, 335–36
Helping behavior
    bystander effect, 273–74
    empathy, 271–72

Helping behavior (*cont.*)
   moral values, 274–76
   variables, 272–74
Heroin, 336
High school
   changes desired by students, 203–204
   class size, 186
   compulsory education, 179–81
   curricula, 181–82
   dropouts, 197–99
   exceptional students, 193–95
   extracurricular activities, 15
   family background, 196
   graduates, 180
   homework, 190
   learning ability, 191–93
   learning climate, 189–90
   and outside jobs, 235–37
   problems, 202
   rating of public schools, 200–201
   reunions, 391
   school size, 185–86
   socioeconomic factors, 196–97
   teachers, 186–89
Homosexuality, 217–19
   adolescent experimentation, 218–19
   causes, 219
   incidence, 218
Household chores, 95–97
   sex differences, 96–97
Hormonal changes, 45–46
Hotlines, 370–71
Hurried child syndrome, 13–14

Identity
   adolescent identity crisis, 138–41
   changes during adolescence, 145–46
   defined, 139
   Erikson's theory, 137–38
   Erikson's youth, 140
   identity statuses, 142–45
   and intimacy, 149–50
   and marriage, 175
   negative, 141–42
   sex differences, 148–50
   and social change, 147
Identity status
   achievement, 143–44
   changes during adolescence, 145–46
   confusion, 142–43, 145
   foreclosure, 145
   moratorium, 144–45
   variables, 147–50
Imaginary audience, 78–79
Intelligence
   and achievement, 84
   and creativity, 84, 191–92

defined, 68, 191
   and learning ability, 191–93
   Sternberg's theory, 191–93
   tests, 191, 193
Intimacy
   adolescent-parent, 166–67
   developmental task, 385–88
   friendship, 167–69
   and identity, 148–50

Juvenile justice system
   juvenile court, 305–306
   juvenile court act, 8
   juvenile facilities, 306–308
   police, 305
   probation, 307–308
Juvenile delinquency, 287–309

Kohlberg's theory of moral development,
   261–67
   attainment of moral reasoning, 264–66
   critique, 266–67
   stages, 262–64

Latchkey teens, 97–98
Late maturers
   boys, 63
   girls, 61–62
Leaving home, 375–79
   college, 377–79
   family, 376–77
   psychological factors, 375–77
Living arrangements of youth, 379
Loneliness, 168, 385
Love
   friendship, 385–87
   love withdrawal, 112–13
   marriage, 386–88
   relationships, 12, 212–14, 217, 385–87
   sex, 212–14
Marijuana, 327–31
   effects, 327–28
   health hazard, 328–29
   patterns of use, 329–31
Marriage
   divorce rate, 175
   early, 174–76
   incidence, 174
   love, 386–88
   marital instability, 175
   median age, 173–74
   pregnancy, 174–76
   relationships, 387–88
Marriage, early
   divorce rate, 175
   reasons, 174–75

Masturbation, 215–16
Menarche
  adjustment, 54–57
  age, 47, 57
  amenorrhea, 57
  attitudes, 55–57
  cycle, 57
  dysmenorrhea, 57
Methods of study
  case-study, 397–98
  correlational studies, 399
  cross-sectional studies, 400–401
  ex post facto studies, 400
  experimental studies, 399–400
  hypotheses, 395–96
  longitudinal studies, 401
  observation, 397
  sampling, 396
  survey research, 398
  variables, 396
Middle schools, 182–84
Modeling, 32–33
Moral behavior
  cheating on tests, 269–71
  defined, 267
  guilt, 270–71
  moral values, 274–76
  religion, 282–84
  resistance to wrongdoing, 268–71
  unethical behavior, 268–69
Moral development
  empathy, 271–72
  moral behavior, 267–73
  moral education, 266, 275–76
  moral reasoning, 261–66
  sex differences, 267
  stages, 262–64
Moral reasoning, 261–66
  age differences, 265
  attainment, 264–66
  moral dilemmas, 261–63

Negative identity, 141–42
Nicotine, 326
Nocturnal emission, 60

Observation, 397
Occupation. *See* Career
Oral stage, 26
Ovaries, 44–45

Parent-child communication
  agreement/disagreements, 117–18
  attitudes, 116–17
  empathetic listening, 121
  improving, 120–22
  values, 117–18

Parent-adolescent conflict
  developmental differences, 119
  sex differences, 119
  variables, 119–20
Parent-adolescent relationships
  adolescents' perceptions, 111–12,
    114–16
  communication, 116–18, 120–22
  conflicts, 118–20
  parental control, 112–14
  parental influence, 154–57
Parent/peer influence
  adolescents' perceptions, 114–16
  age differences, 154
  authoritarian, 112–15
  changing patterns, 155–57
  democratic, 112–15
  love withdrawal, 112–15
  sex differences, 154–57
  variables, 115–16
Peer groups
  changes during adolescence, 162–64
  cliques, 161, 162–63
  conformity, 164–65
  crowds, 162–63
Peer relations
  age differences, 154
  conformity, 164–65
  dating, 163, 170–73
  friendship, 165–70
  function of peers, 157–58
  loneliness, 168
  peer groups, 161–64
  peers versus parents, 154–57
  popularity, 159–61
  shyness, 172
  social acceptance, 158–61
Personal fable, 79–80
Physical development, 43–64
  body image, 51–53
  critical body weight, 46
  earlier maturation, 46–47
  early versus late maturers, 61–62
  growth spurt, 44, 48
  height, 46–48
  hormonal changes, 45–46
  individual differences, 63–64
  muscular changes, 49–51
  puberty, 24–25, 44–46
  pubescence, 44
  sequence, 55, 57, 59
  sexual maturation, 53–60
  skeletal changes, 49–51
  weight, 46–49
Piaget's theory of cognitive development,
  68–70
Popularity
  activities, 160

Popularity (*cont.*)
    personality, 159–60
    sex differences, 160–61
Pregnancy, premarital, 222–24
    early marriage, 175
    incidence, 222–23
    marriage, 223–24
    options, 223–24
    psychological consequences, 223
    sex differences, 230
Privacy, demand for, 123
Probation, 307–308
Psychoanalytical theory, 25–30
Psychological disorders
    antisocial personality, 268, 300–302
    anxiety disorders, 351–53
    delinquency, 300–302
    depression, 357–60
    eating disorders, 354–57
    incidence, 349–51, 368
    multiple personality, 354
    schizophrenia, 364–68
    treatment, 368–71
Psychotherapy, 368–71
    family, 369–70
    individual, 369
    group, 369
Puberty
    defined, 2, 44
    direct-effects model, 24
    mediated-effects model, 24
    physical maturation, 48–51
    rites of passage, 4–5
    sexual maturation, 53–60
    uneven development, 2–3
Pubescence, 44

Religion
    autonomous religiousness, 278
    Bar mitzvah, 5
    Bat mitzvah, 5
    church attendance, 276, 283
    confirmation, 5
    cults, 279–81
    defined, 276
    developmental sequence, 276–79
    extrinsic religious orientation, 277
    helping behavior, 283–84
    intrinsic religious orientation, 278
    moral behavior, 282–84
    religious observance, 277–78
Remarried families, 106–107
    stepparents, 106–107
Resatellization, 123
Research design, 399–402
Rites of passage
    in American society, 5
    in preindustrial societies, 4–5

Role-taking, 79
Runaways, 290–92

Schizophrenia
    adolescent, 365
    contributing factors, 365–67
    incidence, 364–65
    outlook, 367–68
    treatment, 367–68
School
    compulsory education, 179–81
    curricula, 181–82
    dropouts, 197–99
    graduates, 180
    learning ability, 191–93
    middle schools, 182–84
    rating of public schools, 200–201
    size, 186–86
    teachers, 186–89
    See also College, Educaton, *and* High
      school
Secular trend, 46–47
Self-concept
    career choice, 240
    changes during adolescence, 130–31
    defined, 130
    identity, 130
    multiple self-images, 131
    self-esteem, 131–36
Self-efficacy, 135
Self-esteem
    career choice, 250
    changes, 134–36
    defined, 131
    and delinquency, 299–300
    family communication, 116
    family structure, 132
    global, 131
    and learning, 135–36, 183
    self-efficacy, 135
    sex roles, 133–34
    stability, 134–36
    variables, 132–34
Sensation-seeking, 342, 343
Sex differences,
    body image, 51–53
    career choice, 247–50
    conformity, 165
    dating, 173
    delinquency, 292–94
    divorce effects, 100–101, 102–104
    drug use, 316, 319–20, 323, 325–26,
      331
    eating disorders, 354, 356
    friendship, 169–70
    household chores, 96–97
    identity, 148–50
    intimacy, 149–50, 169–70

moral development, 267
parent-adolescent conflicts, 119
popularity, 160–61
psychological disorders, 349, 354, 356, 358–60, 365, 368
sexual behavior, 211, 213, 216–17
suicide, 362
Sex role
  career choice, 247–50
  household chores, 96–97
  and identity, 149–50
  self-esteem, 133–34
Sexual attitudes, 208–14
  and behavior, 210–12
  changes, 208–10
  homosexual activity, 218
  honesty, 208
  love, 212–14
  and personal choice, 209
  premarital sex, 209
  sexual revolution, 209
Sexual behavior
  age of first intercourse, 211–12
  and attitudes, 210–12
  birth control, 219–22
  homosexuality, 217–19
  love, 212–14
  masturbation, 215–16
  oral-genital sex, 216
  physical intimacy, 216–17
  premarital pregnancy, 222–24
  problems, 219–26
  and religion, 283
  sexual intercourse, 216–17
  sexually transmitted diseases, 224–26
  and values, 214
  virgins, 211
Sexual conflicts, 27–29
Sex education, 226–30
  home, 228
  and premarital pregnancy, 230
  school, 228–30
  sources, 226–27
Sexual intercourse
  age of first intercourse, 211, 216
  birth control, 217, 219–22
  love, 212–14
  premarital pregnancy, 222–24
  and relationships, 216–17
Sexual maturation
  boys, 58–60
  girls, 53–57
Sexually transmitted diseases, 224–26
  AIDS, 209, 225–26
  genital herpes, 225
  gonorrhea, 224–25
  incidence, 224
  syphilis, 225

Single-parent families
  children and adolescents, 102
  effects on boys, 102–103, 105–106
  effects on girls, 103–105
  parents' self-assessment, 103
  single-fathers, 105–106
  single-mothers, 101–105
Social change
  and adolescents, 15–17, 39–40
  and concept of adolescents, 7–8
  identity, 147
  and moral values, 274–75
  women's movement, 17
  youth movements, 37–38
Social learning theory, 32–33
Socialization
  and cognitive development, 77–81
  defined, 32
  role-taking, 79
Socioeconomic differences
  career choice, 246–47
  delinquency, 296–99
  families, 102, 104, 123
  identity status, 147
  school, 196–199
Stepparents, 106–107
Suicide
  contributing factors, 362–63
  depression, 360, 362
  incidence, 360–61
  methods, 362
  prevention, 363–64
  sex differences, 362
  warning signs, 364

Teachers, 186–89
Teenage employment, 234–35
  hours worked, 237
  and school, 235–37
  types of jobs, 234–35
  value of work experience, 237–39
Television
  and aggression, 33
  influence, 35
  stereotypes of adolescents, 1
Tests, intelligence, 191–93
Therapy
  family, 308, 369–70
  psychotherapy, 369
Thinking. See Cognitive development
Tobacco, 324–26
  health hazard, 326
  use, 324–25
  why youth smoke, 325
Tranquilizers, 335
Transition to adulthood
  autonomous decision making, 381–83
  change, 389–90

Transition to adulthood (*cont.*)
   close relationships, 385–88
   continuity, 388–89
   developmental tasks, 375
   economic independence, 383–85
   individual differences, 390–92
   leaving home, 375–79
   living arrangements, 379
   returning home, 379–80
   youth, 12

Unethical behavior, 268–69
Unwed mothers, 104–105

Values
   career values, 244–45
   contemporary youth, 15–16
   moral values, 274–76
   sexual values, 214
   transmission to youth, 117–18
   value conflicts, 275
   value dichotomy, 160

Women's movement, 17
Working mothers, 93–95
   latchkey teens, 97–98
Workplace
   adolescents, 234–39
   working mothers, 93–95

Young adulthood, 374–75, 390
Youth, 12
Youth movements, 36–38

Name Index

Achenbach, T., 349, 350
Adams, G. R., 146, 147, 150
Adelson, J., 38, 39
Ageton, S. S., 298
Alter, J., 229
Altman, S. L., 245
Altrocchi, J., 339
Ames, L. B., 23
Amstey, M., 224
Anderson, C. S., 185, 186, 189
Anderson, S., 378
Angelev, J., 71
Angell, M., 361
Apter, D., 57
Archer, S., 145
Arenberg, D., 39
Arieti, S., 367
Asp, 197
Astin, A., 16
Ault-Riche, M., 308, 370
Ausubel, D. P., 122

Bachman,, J. G., 172, 174, 196, 246,
   274, 313, 314, 315, 317, 318, 320,
   331, 332, 335, 336, 341, 388
Bahr, H., 283
Bandura, A., 32, 135
Baron, R. A., 268, 269, 272
Barrett, S. A., 349, 361, 369
Batson, C. D., 278, 283
Bauer, K., 59, 213, 214
Baumaster, R., 129
Beck, A. T., 358
Bell, A. P., 219
Bernard, H., 143, 144, 145
Berndt, T., 154, 164, 167
Billy, J. O. G., 216
Blau, M., 292
Bleuler, M., 367
Block, J. H., 98, 296, 388
Bloom, B. S., 84
Blos, P., 30
Blumstein, P., 94, 95, 213, 385, 387
Blyth, D. A., 51, 61, 134, 169, 183
Boxer, A. M., 132
Braithwaite, J., 298

Braungart, R., 37
Brim, O. G., 390
Brinkerhoff, B., 99
Brody, C. J., 97
Brooks-Gunn, J., 56, 57
Brown, G. H., 15
Bruch, H., 355
Bryson, Y., 225
Bulcroft, R., 52
Bullough, V. L., 46
Burke, R. J., 122
Burton, N. W., 197
Byrne, D., 268, 269, 272

Calot, G., 223
Carli, L. L., 165
Carlsen, W. S., 249
Carlton-Ford, S., 134, 183
Carter, M. M., 292
Cash, T. F., 132
Chafetz, M., 303
Chassin, L., 131
Cherlin, A., 102, 106
Clifford, R., 215
Cohen, S., 303, 304, 312, 320,
   327
Colby, A., 264, 265
Cole, S., 237
Coleman, J. C., 165, 170
Coleman, J. S., 190
Collins, J. K., 173
Collins, J. K., 213, 215
Cooke, C., 197
Coons, P. M., 354
Corder, J., 95
Corey, L., 225
Costa, P., 389
Craighead, L. W., 250
Crittenden, A., 383
Cron, W. L., 241
Crook, R. H., 241
Crooks, R., 59, 213, 214
Cross, H. J., 336
Crouter, A. C., 246, 250
Csikszentmihalyi, M., 83, 153, 160
Cutrona, C., 168

Danner, F. W., 76
Davies, M., 196
Davis, K. E., 386
Davison, G. C., 353
Day, H. D., 148
Day, M. C., 76
DeBlassie, R., 341
DeLamater, J., 216
de Turck, M. A., 114
DeVaus, D. A., 274
Dean, A., 71
Dean, R. A., 280
Dearman, N. B., 181, 249, 253
Dienstbier, R. A., 271
Doane, J., 367
Doehrman, M., 38
Dornbush, S. M., 171, 188
Dostal, J. W., 106
Douglas, J. D., 71
Douglas, K., 39
Doyle, L., 334
Dreyer, P., 216, 217, 218
Dudley, M. G., 274
Dudley, R. L., 274
Dunne, F., 249
Dunphy, D. C., 161, 163, 164
Durrett, M., 250

Eagly, A., 165
Eckland, B. K., 186
Edelbrock, C. S., 350
Edwards, J. N., 175
Egeland, J. A., 358
Eisen, M., 131
Eisenberger, R., 270
Elder, G. H., 7
Elifson, K. W., 283
Elkind, D., 13, 78
Elliot, R., 249, 298
Enright, R., 80
Erikson, E., 9, 137, 138, 139, 140, 141, 149, 385
Eveland, L. K., 302
Everly, K., 229
Eversoll, D. A., 91

Falk, W. W., 250
Falkowski, C., 250
Farber, B. A., 149
Faust, M. S., 50, 64
Feather, N., 117
Feeney, S., 187
Feldman, R. S., 165
Fetters, W. B., 15
Fischer, J. L., 116, 150
Fitch, S. A., 146
Fleming, W., 378
Ford, K., 223

Forisha, B., 83
Foster, A. D., 223
Fox, M. F., 245, 250
Freeman, D., 35
Freeman, H. E., 117
Freeman, R., 238
Freeman, R., 383
Freund, A., 27, 28, 29
Freud, S., 25
Frisch, R. E., 46, 57
Froese, A., 363
Furstenburg, F., 221, 224

Gallup, A. M., 180, 187, 190, 194, 200, 201, 202, 203, 229, 250, 251, 253, 254
Gallup, G. H., 173, 229, 270, 276, 313
Garbarino, J., 107, 197
Garfinkel, B. C., 363
Gaylin, J., 173
Gelfand, D. M., 272
Gershoni, R., 168, 169, 170
Gesell, A. L., 22, 23
Getzels, J. W., 83
Giller, H., 294, 297, 299, 300, 306, 307
Gilliam, D., 13
Gilligan, C., 267
Ginsberg, A., 197
Ginzberg, E., 239
Gispert, M., 362
Gjerde, P., 98, 296
Glenn, N., 100
Gold, M., 107, 287, 288, 290, 292, 294, 296
Goleman, D., 362, 377
Goodwin, G., 114, 115
Gordon, T., 120
Gorman, M. E., 368
Gottesman, I. I., 366
Gould, R., 375, 377
Grabe, M., 185
Graham, J. W., 371
Grant, W. V., 199, 235, 236, 253, 254
Grattan, T. C., 6
Green, B. J., 250
Green, K., 165
Greenberg, J., 360, 362
Greenberger, E., 235
Gregg, C. H., 230
Greif, G., 105
Gressard, C. F., 340
Griffin, N., 131
Grossman, F. K., 245
Grotevant, H. D., 150, 247, 250
Gurman, A. S., 369

Haas, L., 95, 215, 218, 228
Hadaway, C. K., 283

Hadden, J. K., 280
Haley, J., 376
Hall, D. T., 242
Hall, G. S., 7, 21
Hammersmith, S. K., 219
Haney, B., 107
Hanks, M., 186
Hansen, S. L., 171
Hansen, W. B., 371
Harris, L., 323
Harris, M. W., 111, 186, 270
Hartman, D. P., 272
Hartup, W. W., 159, 165
Haskell, M. R., 296
Hassett, J., 268, 269, 275, 282
Havighurst, R., 31
Healy, C. C., 241
Helmreich, R., 83
Henshaw, S. K., 223
Hesse-Biber, S., 245, 250
Hetherington, M., 103
Hill, C. T., 214
Hill, J., 169, 183
Hodgson, J. W., 150
Hoffer, T., 190
Hoffman, M. L., 267, 271
Hoffman, M., 112
Hofman, J., 169, 170
Holmes, K., 225
Holmes, L., 35
Hood, J., 363
Hostetter, A. M., 358
Howard, K., 40, 111, 208, 369
Hunter, F., 166, 168
Huston, T. L., 274

Ibrahim, F. A., 266
Ilg, F. L., 23
Inhelder, B., 72
Isralowitz, R., 307

Jacklin, C. N., 83
Janda, L. H., 132
Jedlicka, F. D., 209
Jersild, A. T., 51
Jessop, D., 117
Jessor, R., 323
Jessor, S. L., 323
Johnson, C. A., 371
Johnson, R., 292
Johnson, V. E., 59, 219, 224, 225, 229
Johnston, J., 246
Johnston, L. D., 172, 174, 302, 313, 314,
    315, 317, 318, 319, 320, 329, 331,
    332, 335, 336, 341, 388
Jones, E., 339
Jones, L. V., 197

Jones-Witters, P., 312, 321, 324, 325,
    327, 331, 334, 335, 337, 338, 340
Jurich, A., 323

Kacerguis, M. A., 150
Kagan, J., 390
Kahn, H., 266
Kahn, J. R., 220
Kahoe, R., 276, 277, 278
Kalk, D. F., 269
Kalter, N., 100, 107
Kandel, D. B., 196, 318
Kanner, A., 370
Kantner, J. F., 223
Kaplan, H. B., 223
Kaufman, I. P., 304
Kelly, C., 114, 115
Kelly, J. B., 99, 100
Kennedy, S., 324
Kephart, W. M., 171
Kessler, G. R., 266
Kett, J., 6
Kiell, N., 3
Kilgore, S., 190
Kim, Y. J., 230
Kingston, P. W., 94
Kirby, D., 229
Kirkevold, B. C., 269
Kisker, E. E., 228, 229, 230
Klein, J. R., 57
Kleinhesselink, R. R., 336
Knapp, J. R., 51
Kohlberg, 261, 262, 263, 264, 265
Kohn, M. L., 247
Kohut, H., 29
Kolodny, R. C., 59, 219, 224, 225, 229
Kozak, C., 298
Kramer, K., 100
Kronick, D., 194
Kuhn, D., 71
Kulka, R. A., 101
Kumar, V., 361
Kumka, D., 245

Lapsley, D., 80
Larson, G. L., 81, 153
Larson, L. E., 155, 157
Larson, R., 160
Law, R., 199
Leo, J., 209, 224, 275
Lerner, B., 135
Lerner, R. M., 51
Levine, E., 298
Levinson, D., 375, 382, 383
Lewin, T., 248, 384
Lewin-Epstein, N., 235
Lewine, R. R. J., 365
Lewis, C., 125

Lipsitz, J. S., 183
Litt, I. F., 57
Livson, N., 62, 63
Lloyd, D., 198
Lockhard, J. S., 269
Long, T. E., 280
Louden, D. M., 134
Lounsberry, E., 380
Lovett, M. A., 225
Luepnitz, D., 98, 99
Lukasiewicz, J. M., 256
Lunzer, E. A., 83

MacCorquodale, P., 216
Maccoby, E. E., 83
Mann, D. W., 300
Marcia, J., 141, 143, 148
Martin, S. S., 223
Martin, T., 283
Martorano, S. C., 71
Maslow, A., 81
Masters, W. H., 59, 219, 224, 225, 229
Masterson, F. A., 270
McAllister, E. W. C., 276
McCabe, M. P., 173, 213, 215
McCarthy, J., 102, 106
McCartin, R., 236
McCormick, M. C., 223
McCrae, R., 389
McCrary, E., 97, 98
McCraw, A., 134
McGee, E. A., 223
McKinnon, J. W., 71, 75, 76
McLaughlin, S. D., 175, 224
Mead, M., 33, 35
Meadow, M. J., 276, 277, 278
Menaghan, E. G., 376
Meyer, K., 236
Millen, L., 168
Miller, G., 114
Miller, P. Y., 215
Miller, W. B., 299
Minuchin, P. P., 188
Minuchin, S., 356
Money, J., 46, 60
Montemayor, R., 120, 122, 131
Moore, A., 299
Moos, R., 189
Morash, M. A., 147
Morgan, E., 149
Morrison, A., 98, 296
Mortimer, J. T., 245
Munro, G., 147
Muuss, R. E., 272, 356

Nack, W., 10
Naedele, W. F., 275
Natriello, G., 188

Neale, J. M., 353
Neimark, E. D., 71, 76
Newman, J., 2
Newton, D. E., 229
Nicholi, A., 333
Nicol, S. E., 366
Nissim-Sabat, D., 80
Nock, S. L., 94
Norman, J., 111, 186, 270
Northman, J. E., 103, 296
Norwood, R., 172

O'Brien, R., 303, 304
O'Brien, R., 320, 327
Offer, D., 39, 40, 111, 208
Offer, J., 39
Olmedo, E. L., 191
Olson, D. H., 132
O'Malley, P. M., 172, 174, 196, 246, 302, 313, 314, 315, 317, 318, 319, 320, 331, 332, 335, 336, 341, 388
Openshaw, D. H., 132
Oppenheimer, M., 225
Orlos, J. B., 51
Orr, M. T., 229
Osburn, S. G., 290
O'Shea, D. W., 241
Ostrov, E., 40, 111, 208
Ostrow, R. J., 327
Owings, J. A., 15

Pardeck, J. T., 370
Parish, J., 132
Parish, T., 106, 132
Parke, R., 123
Parlee, M. B., 385
Patterson, G., 295
Peplau, L. A., 168, 214, 385
Peskin, H., 62, 63
Peter, J. B., 225
Petersen, A. C., 11, 24, 25, 44, 132
Peterson, D. M., 283
Peterson, K. L., 12, 374
Petronio, R. J., 287, 288, 290, 292, 294, 295
Phillips, D. P., 362
Piaget, J., 68, 71, 72
Pilkonis, P., 172
Pleck, J., 91
Plisko, V. W., 180, 181, 195, 249, 253
Poling, D., 276
Polit-O'Hara, D., 220
Polovy, P., 165
Poole, M., 203
Postman,, N., 13
Prescott, P., 304
Putnam, F., 354

Renner, J. W., 71, 74
Rice, F. P., 107
Richmond, L. H., 369
Roazen, P., 140
Robbins, C., 223
Robbins, S. L., 361
Robins, L. N., 357, 358
Robinson, I. E., 209
Roll, S., 168
Rollins, B. C., 132, 133
Ronberg, G., 7
Roscoe, B., 12, 374
Rosen, W., 330, 337, 342, 344
Rosenberg, F. R., 300
Rosenberg, M., 300
Rosenthal, D., 308, 370
Rothman, K. M., 142, 145, 149
Rubenstein, C., 168, 214, 383, 385, 387
Rubin, Z., 214
Ruble, D. N., 56, 57
Ruby, T., 199
Rumberger, R. W., 196, 197
Russell, C. S., 132
Rust, J. O., 134
Rutter, M., 186, 187, 189, 195
Rutter, M., 294, 296, 299, 300, 306, 307

St. Clair, S., 148
Sanford, J. P., 195
Sanik, M. M., 96
Santrock, J. W., 105
Sarason, B. R., 349
Sarason, I. G., 349
Sawin, D., 123
Saxe, J., 20
Scales, P., 229
Schellenbach, C., 107
Schill, W. J., 236
Schmid, R. E., 97, 217
Schooler, C., 247
Schreiber, F., 354
Schulenberg, J. E., 246, 250
Schwartz, P., 94, 95, 213, 385, 387
Scott, E. M., 341
Scott, V., 1
Sebald, H., 155, 156, 157, 159, 160, 165, 244
Sebes, J., 107
Selman, R. L., 79
Shah, F. K., 216, 217, 220, 221, 222
Shapiro, E. K., 188
Shapiro, H. S., 380
Shapiro, S., 223
Sharabany, R., 168, 169, 170
Shaver, P., 168, 385
Shigetomi, D. C., 272
Sieg, A., 35
Sifford, D., 281

Silvestri, G. T., 256
Simmons, R. G., 52, 134, 183
Simon, W., 215
Singer, M., 307
Slocum, J. W., 241
Slukla, D., 80
Smilgis, M., 226
Smith, E. T., 113
Smith, G. M., 325
Smith, M., 197
Smith, T., 117
Smyth, C., 183
Snyder, T. D., 199, 235, 236, 253, 254
Sorenson, R. C., 209
Spence, J., 83
Sprenkle, D. H., 132
Stafford, D. G., 71, 74
Stafford, K., 96
Staples, R., 104, 223
Starfield, B., 223
Steelman, L. C., 97
Steinberg, L., 235, 238
Stephan, C. S., 95
Stern, J. D., 180, 195, 296
Stern, M., 103
Sternberg, R., 191, 192
Story, M. D., 215
Stott, D., 295
Sullivan, A., 127, 377
Sullivan, K., 127, 377
Super, D., 240, 242
Svajian, P., 122
Swope, G. W., 280

Tanner, J. M., 45, 46, 47, 48, 49, 50, 51, 54, 60, 62, 64
Taube, C. A., 349, 361, 369
Taylor, B., 11, 44
Tenzer, A., 245
Thiel, K., 169
Thirer, J., 160, 161
Thomas, D. L., 132, 133
Thornbecke, W. L., 150, 247
Thornburg, H., 226, 227, 229, 290, 293
Thorne, C., 341
Trebilco, G. R., 253
Trickett, E., 189
Turanski, J. J., 340

Udry, J. R., 216

Van Slyck, M. R., 103, 296
Veldman, D. J., 195
Ventis, W. L., 278, 283
Vihko, R., 57
Vondracek, F. W., 246, 250
Vonnegut, M., 368

Walker, J. I., 335, 360
Wallerstein, J. S., 99, 100
Warshak, R. A., 105
Waterman, A., 146, 148
Watt, N., 367
Weil, A., 330, 337, 342, 344
Weinberg, M. S., 219
Weiner, I. B., 350, 353, 357, 359, 362, 365, 368
Weingarten, H., 101
Weir, T., 122
Weishaar, M. E., 250
Welsh, G. S., 84
West, D. J., 290
Westendorp, F., 294, 302
Westoff, C. F., 223
White, L. K., 99
Winer, J. A., 216
Winstead, B. A., 132
Wise, D., 238
Witters, W. L, 312, 321, 324, 325, 327, 331, 334, 335, 337, 338, 340

Wolf, F. M., 81
Wong, A. C., 71
Woodall, 252
Woodroof, J. T., 278, 283
Wright, L. S., 171, 363
Wright, S., 160, 161, 281

Yablonsky, L., 296
Yankelovich, D., 209, 247, 275, 382
Young, J. E., 358
Young, R. D., 131
Youniss, J., 71, 166, 168
Yussen, S. R., 267

Zabin, L., 210, 211, 212, 213, 221, 222
Zelnik, M., 216, 217, 220, 221, 222, 223, 230
Zern, D. S., 276
Zimbardo, P., 172
Zuckerman, M., 344
Zurcher, L., 147

About the Author

Eastwood Atwater is professor of psychology at Montgomery County Community College and lecturer in psychology at Gwynedd-Mercy College, both in the greater Philadelphia area. He teaches a course on adolescence at both schools, and also conducts a private practice that includes adolescents. He received his Ph.D. from the University of Chicago. Dr. Atwater belongs to the American Psychological Association, the Pennsylvania Psychological Association, and The Philadelphia Society of Clinical Psychologists. He is the author of several books, including two other textbooks: *Psychology of Adjustment,* Third Edition, and *Human Relations.* He is married and has two grown daughters.